AF356551

Deep Learning Applications with Practical Measured Results in Electronics Industries

Deep Learning Applications with Practical Measured Results in Electronics Industries

Special Issue Editors

Mong-Fong Horng
Hsu-Yang Kung
Chi-Hua Chen
Feng-Jang Hwang

MDPI • Basel • Beijing • Wuhan • Barcelona • Belgrade

Special Issue Editors

Mong-Fong Horng
Ph. D Program in Biomedical
Engineering, Kaohsiung
Medical University
Taiwan

Hsu-Yang Kung
Department of Management
Information Systems, National
Pingtung University of Science
and Technology
Taiwan

Chi-Hua Chen
College of Mathematics and
Computer Science,
Fuzhou University
China

Feng-Jang Hwang
School of Mathematical and
Physical Sciences, University of
Technology Sydney
Australia

Editorial Office
MDPI
St. Alban-Anlage 66
4052 Basel, Switzerland

This is a reprint of articles from the Special Issue published online in the open access journal *Electronics* (ISSN 2079-9292) from 2019 to 2020 (available at: https://www.mdpi.com/journal/electronics/special_issues/deep_learning_electronics_industry).

For citation purposes, cite each article independently as indicated on the article page online and as indicated below:

LastName, A.A.; LastName, B.B.; LastName, C.C. Article Title. *Journal Name* **Year**, *Article Number*, Page Range.

ISBN 978-3-03928-863-2 (Pbk)
ISBN 978-3-03928-864-9 (PDF)

Contents

About the Special Issue Editors

Mong Fong Horng received his Ph.D. degrees from the Computer Sciences and Information Engineering Department, National Cheng Kung University, Taiwan in 2003. He now is a professor with the department of Electronic Engineering at the National Kaohsiung University of Science and Technology and Kaohsiung Medical University, Taiwan. He is also a technical director of the Institute of Information Industrial. As of 2018, Dr. Horng had published 132 academic papers and 6 textbooks about his research. Based on the researches, Prof. Horng has been granted 14 Taiwan patents and 2 U.S. patents. He serves as a member of IEEE SMC Technical Committee on Computational Collective Intelligence, the president of the Taiwanese Association of Consumer Electronics (TACE) and Tainan Chapter, Signal Processing Society, IEEE. Dr. Horng contributed to the editorial boards of *International Journal of Knowledge Engineering and Soft Data Paradigms* published by Elsevier. Dr. Horng has received awards from the Ministry of Science and Technology and Ministry of Education due to his outstanding performance in industrial cooperation and supervision of student competition. Dr. Horng has hosted 12 research projects funded by the Ministry of Science and Technology, and a Special Issue of the *Journal of Innovative Computing, Information and Control* in 2011. His research interests include computational intelligence, the Internet of Things, network function virtualization, and the Internet.

Hsu-Yang Kung received his Ph.D. degree in Computer Science and Information Engineering from National Cheng-Kung University, Taiwan, ROC. He is currently a distinguished professor in the Department of Management Information Systems, National Pingtung University of Science and Technology, Taiwan, ROC. Prof. Kung has published around 300 academic papers and received 14 te best paper awards. He received the special talent award 8 times and the excellent team award 8 times in the Open Software Development Project Plan from the Ministry of Science and Technology. He received the outstanding performance and special contributions award in the Innovative Talents Promotion Program for the Communication Software from the Ministry of Education. He has led more than 100 industrial and academic research projects and owns 26 patents. He also served as a Guest Editor for the journals of *Agronomy, Electronics, Mathematical Problems in Engineering, Journal of Electrical and Computer Engineering, Advances in Multimedia,* and IEEE *Technology and Engineering Education.* His research interests include IoT middleware, cloud computing, wireless and mobile communications, and intelligent IoT applications in precision agriculture.

Chi-Hua Chen is a distinguished professor at Fuzhou University and a chair professor at Dalian Maritime University. He received his Ph.D. degree from National Chiao Tung University (NCTU) in 2013. He has served as an assistant professor for the National Tsing Hua University, NCTU, National Taipei University, and National Kaohsiung University of Science and Technology. He has served as a research fellow for the Telecommunication Laboratories of Chunghwa Telecom Co. Ltd. He has published over 230 academic articles and owns 50 patents. Some of these academic articles were published in IEEE *Internet of Things Journal,* IEICE Transactions, etc. He has hosted several projects that were funded by the National Natural Science Foundation of China, Fujian Province, etc. He serves as an editor for several SCI-indexed journals (e.g., IEEE *Access, Biosensors,* EURASIP *Journal on Wireless Communications and Networking, International Journal of Distributed Sensor Networks, Discrete Dynamics in Nature and Society, Mathematical Problems in Engineering,* etc.).

His recent research interests include the Internet of Things, machine learning and deep learning, mobile communications, and intelligent transportation systems.

Feng-Jang Hwang is a Senior Lecturer (Level C, Associate Professorship equivalency in the North American academic system) and the Leading PI of the Industrial Optimisation Group at the School of Mathematical and Physical Sciences, University of Technology Sydney. He earned his Ph.D. in Information Management from the National Chiao Tung University in 2011. F.J. was elected to the honorary membership of the Phi-Tau-Phi Scholastic Society twice. He was the nominee for the 2016 Australian Society for Operations Research Rising Star Award and the winner of the 2017 Albert Nelson Marquis Achievement Award. His research interests centre around production modelling and scheduling, supply chain and logistics optimisation, data-driven optimization, and computational intelligence. F.J. has published in leading journals including *Journal of Scheduling, Annals of Operations Research, Journal of the Operational Research Society, Computers & Operations Research, Discrete Optimization, Engineering Optimization, IEEE Access, International Journal of Sustainable Transportation,* etc. He has served as the guest editor for several international journals, including *International Journal of Distributed Sensor Network, EURASIP Journal on Wireless Communications and Networking, Electronics,* and *Agronomy.* He has been invited to give more than 30 research talks and conference presentations. His professional experience includes a logistics officer (ensign) at Taiwan Navy headquarters, a research assistant at the Institute of Statistical Science in Academia Sinica, as well as a postdoctoral fellow at National Tsing Hua University, National Chiao Tung University, and the Warwick Business School.

Editorial

Deep Learning Applications with Practical Measured Results in Electronics Industries

Mong-Fong Horng [1], Hsu-Yang Kung [2], Chi-Hua Chen [3,*] and Feng-Jang Hwang [4]

[1] Department of Electronic Engineering, National Kaohsiung University of Science and Technology, Kaohsiung 80778, Taiwan; mfhorng@nkust.edu.tw

[2] Department of Management Information Systems, National Pingtung University of Science and Technology, Pingtung 91201, Taiwan; kung@mail.npust.edu.tw

[3] College of Mathematics and Computer Science, Fuzhou University, Fuzhou 350100, China

[4] School of Mathematical and Physical Sciences, University of Technology Sydney, Ultimo, NSW 2007, Australia; Feng-Jang.Hwang@uts.edu.au

* Correspondence: chihua0826@gmail.com; Tel.: +86-18359183858

Received: 11 March 2020; Accepted: 12 March 2020; Published: 19 March 2020

Abstract: This editorial introduces the Special Issue, entitled "Deep Learning Applications with Practical Measured Results in Electronics Industries", of *Electronics*. Topics covered in this issue include four main parts: (I) environmental information analyses and predictions, (II) unmanned aerial vehicle (UAV) and object tracking applications, (III) measurement and denoising techniques, and (IV) recommendation systems and education systems. Four papers on environmental information analyses and predictions are as follows: (1) "A Data-Driven Short-Term Forecasting Model for Offshore Wind Speed Prediction Based on Computational Intelligence" by Panapakidis et al.; (2) "Multivariate Temporal Convolutional Network: A Deep Neural Networks Approach for Multivariate Time Series Forecasting" by Wan et al.; (3) "Modeling and Analysis of Adaptive Temperature Compensation for Humidity Sensors" by Xu et al.; (4) "An Image Compression Method for Video Surveillance System in Underground Mines Based on Residual Networks and Discrete Wavelet Transform" by Zhang et al. Three papers on UAV and object tracking applications are as follows: (1) "Trajectory Planning Algorithm of UAV Based on System Positioning Accuracy Constraints" by Zhou et al.; (2) "OTL-Classifier: Towards Imaging Processing for Future Unmanned Overhead Transmission Line Maintenance" by Zhang et al.; (3) "Model Update Strategies about Object Tracking: A State of the Art Review" by Wang et al. Five papers on measurement and denoising techniques are as follows: (1) "Characterization and Correction of the Geometric Errors in Using Confocal Microscope for Extended Topography Measurement. Part I: Models, Algorithms Development and Validation" by Wang et al.; (2) "Characterization and Correction of the Geometric Errors Using a Confocal Microscope for Extended Topography Measurement, Part II: Experimental Study and Uncertainty Evaluation" by Wang et al.; (3) "Deep Transfer HSI Classification Method Based on Information Measure and Optimal Neighborhood Noise Reduction" by Lin et al.; (4) "Quality Assessment of Tire Shearography Images via Ensemble Hybrid Faster Region-Based ConvNets" by Chang et al.; (5) "High-Resolution Image Inpainting Based on Multi-Scale Neural Network" by Sun et al. Two papers on recommendation systems and education systems are as follows: (1) "Deep Learning-Enhanced Framework for Performance Evaluation of a Recommending Interface with Varied Recommendation Position and Intensity Based on Eye-Tracking Equipment Data Processing" by Sulikowski et al. and (2) "Generative Adversarial Network Based Neural Audio Caption Model for Oral Evaluation" by Zhang et al.

Keywords: deep learning; machine learning; supervised learning; unsupervised learning; reinforcement learning; optimization techniques

1. Introduction

Machine learning and deep learning techniques have been the crucial tools when it comes to the feature extracting and event estimating for developing applications in the electronics industries [1–8]. Some techniques have been implemented in the embedded systems and applied to industry 4.0 applications, industrial electronics applications, consumer electronics applications, and other electronics applications. For instance, supervised learning techniques, including neural networks (NN) [9–19], convolutional neural networks (CNN) [20–26], and recurrent neural networks (RNN) [27–32], can be adopted for prediction applications and classification applications in the electronics industries. Unsupervised learning techniques, including restricted Boltzmann machine (RBM) [33,34], deep belief networks (DBN) [35], deep Boltzmann machine (DBM) [36], auto-encoders (AE) [37,38], and denoising auto-encoders (DAE) [39], can be used for denoising and generalization. Furthermore, reinforcement learning techniques, including generative adversarial networks (GANs) [40,41] and deep Q-networks (DQNs) [42], can be used to obtain generative networks and discriminative networks for contesting and optimizing in a zero-sum game framework. These techniques can provide the precise prediction and classification for electronics applications. Therefore, the aim of this Special Issue is to introduce the readers the state-of-the-art research work on deep learning applications with practical measured results in electronics industries.

This Special Issue had received a total of 45 submitted papers with only 14 papers accepted. A high rejection rate of 68.89% of this issue from the review process is to ensure that high-quality papers with significant results are selected and published. The statistics of the Special Issue are presented as follows.

- Submissions (45);
- Publications (14);
- Rejections (31);
- Article types: research article (13); review article (1).

Topics covered in this issue include the following four main parts: (I) environmental information analyses and predictions, (II) unmanned aerial vehicle (UAV) and object tracking applications, (III) measurement and denoising techniques, and (IV) recommendation systems and education systems. Four topics with accepted papers are briefly described below.

2. Environmental Information Analyses and Predictions

Four papers on environmental information analyses and predictions are as follows: (1) "A Data-Driven Short-Term Forecasting Model for Offshore Wind Speed Prediction Based on Computational Intelligence" by Panapakidis et al. [43]; (2) "Multivariate Temporal Convolutional Network: A Deep Neural Networks Approach for Multivariate Time Series Forecasting" by Wan et al. [44]; (3) "Modeling and Analysis of Adaptive Temperature Compensation for Humidity Sensors" by Xu et al. [45]; (4) "An Image Compression Method for Video Surveillance System in Underground Mines Based on Residual Networks and Discrete Wavelet Transform" by Zhang et al. [46].

Panapakidis et al. from Greece and Cyprus in "A Data-Driven Short-Term Forecasting Model for Offshore Wind Speed Prediction Based on Computational Intelligence" considered that the time series data of wind speed has the characters of high nonlinearity and volatilities. Therefore, an adaptive neuro-fuzzy inference system (ANFIS) and a feed-forward neural network (FFNN) were constructed to analyze the nonlinearity and volatilities of wind speed for short-term wind speed prediction. In their experiments, five cases were selected to predict the wind speeds of the 1-min-ahead and 10-min-ahead prediction horizons for the evaluation of the proposed method. The results show that all of mean absolute range normalized errors (MARNEs) of each case by the proposed method were lower than the MARNEs of each case by other methods (e.g., regression neural network, regression trees, support vector regression, etc.) [43].

Wan et al. from China in "Multivariate Temporal Convolutional Network: A Deep Neural Networks Approach for Multivariate Time Series Forecasting" considered that the long-term multivariate dependencies of time series data are hard to be captured. Therefore, a multivariate temporal convolution network (M-TCN) was proposed to combine convolutional layers and residual block for extracting the spatio-temporal features of environmental data. In the experiments, two benchmark datasets including a Beijing PM2.5 dataset and an ISO-NE Dataset were used to compare the M-TCN with other methods for evaluating the proposed method. The results show that the root mean squared errors (RMSEs) of each case by the M-TCN were lower than the RMSEs of each case with other methods (i.e., long short term memory (LSTM), convolutional LSTM (ConvLSTM), Temporal Convolution Network (TCN) and Multivariate Attention LSTM-FCN (MALSTM-FCN)) [44].

Xu et al. from China in "Modeling and Analysis of Adaptive Temperature Compensation for Humidity Sensors" considered that the nonlinear compensation of sensing data is required because the humidity sensitive materials may be sensitive to temperature with nonlinear relationships. Therefore, a genetic simulated annealing algorithm (GSA) was proposed and adopted into a back propagation neural network (BPNN)-based nonlinear compensation model to compensate the sensing data of different temperature ranges. In their experiments, 150 practical datasets were collected by a humidity sensor and used to train the proposed nonlinear compensation model; furthermore, 15 practical datasets were collected and analyzed to test the trained nonlinear compensation model for the performance evaluation of the proposed method. The results show that the errors the proposed method were lower than the errors of other methods (i.e., genetic algorithm-BPNN (GA-BPNN) and artificial fish-swarm algorithm-BPNN (AFSA-BPNN)) [45].

Zhang et al. from China in "An Image Compression Method for Video Surveillance System in Underground Mines Based on Residual Networks and Discrete Wavelet Transform" considered that the image compression can be used to transfer a large number of digital images through lower bandwidth underground channels for the applications of underground mines. Therefore, a neural network containing an encoder module and a decoder module with residual units was constructed, and a metric termed discrete wavelet structural similarity (DW-SSIM) was proposed for the loss function of the neural network. In the experiments, this study collected the images from the COCO 2014 dataset and the images of underground mines for training and testing. The results show that the peak signal-to-noise ratio (PSNR) and the structural similarity (SSIM) of the proposed method were higher than the PSNR and the SSIM of other methods (e.g., denoising-based approximate message passing (D-AMP), ReconNet and total variation augmented Lagrangian alternating direction algorithm (TVAL3)) [46].

3. UAV and Object Tracking Applications

Three papers on UAV and object tracking applications are as follows: (1) "Trajectory Planning Algorithm of UAV Based on System Positioning Accuracy Constraints" by Zhou et al. [47]; (2) "OTL-Classifier: Towards Imaging Processing for Future Unmanned Overhead Transmission Line Maintenance" by Zhang et al. [48]; (3) "Model Update Strategies about Object Tracking: A State of the Art Review" by Wang et al. [49].

Zhou et al. from China in "Trajectory Planning Algorithm of UAV Based on System Positioning Accuracy Constraints" considered that the location information cannot be accurately determined by UAVs with the limitation of system structure. Therefore, this study considered multi-constraints (e.g., vertical errors, horizontal errors, and flight distance) and proposed an improved genetic algorithm and an improved sparse A* algorithm to find the shortest trajectory length. In their experiments, two practical case studies were selected to evaluate the improved genetic algorithm and the improved sparse A* algorithm. The results show that the trajectory length could be reduced by 57.79% by the proposed methods [47].

Zhang et al. from China in "OTL-Classifier: Towards Imaging Processing for Future Unmanned Overhead Transmission Line Maintenance" considered that the transmission line-based robots equipped with cameras can only travel a line to inspect for maintenance. Therefore, an overhead transmission line classifier based on ResNet (deep residual network) and Faster-RCNN (faster regions with convolutional neural network) was proposed to analyze the images from robots for classification and inspection. In the experiments, 1558 images, which include 406 positive samples and 1152 negative samples, were collected for evaluating the proposed classification method. The results show that the area under curve (AUC) of the proposed classification method was higher than support vector machine (SVM). Furthermore, the precision-recall (PR) curve of the proposed classification method (i.e., ResNet) was also higher than the PR curve of the combination of VGG and Faster-RCNN [48].

Wang et al. from China in "Model Update Strategies about Object Tracking: A State of the Art Review" considered that tracking model update strategies were important factors for the robustness of image recognition. Therefore, the study conducted the literature review of target model update occasions, target model update strategies, and background model updates. Four update strategy types, which include (1) update strategies based on correlation filters, (2) update strategies based on dictionary learning and sparse coding, (3) update strategies based on bag-of-words, and (4) update strategies based on neural network models, were summarized and presented. The experimental results of different update strategies from recent publications were discussed, and it was concluded that the local representation, target re-detection, and background models were important factors for the improvement of object tracking [49].

4. Measurement and Denoising Techniques

Five papers on measurement and denoising techniques are as follows: (1) "Characterization and Correction of the Geometric Errors in Using Confocal Microscope for Extended Topography Measurement. Part I: Models, Algorithms Development and Validation" by Wang et al. [50]; (2) "Characterization and Correction of the Geometric Errors Using a Confocal Microscope for Extended Topography Measurement, Part II: Experimental Study and Uncertainty Evaluation" by Wang et al. [51]; (3) "Deep Transfer HSI Classification Method Based on Information Measure and Optimal Neighborhood Noise Reduction" by Lin et al. [52]; (4) "Quality Assessment of Tire Shearography Images via Ensemble Hybrid Faster Region-Based ConvNets" by Chang et al. [53]; (5) "High-Resolution Image Inpainting Based on Multi-Scale Neural Network" by Sun et al. [54].

Wang et al. from Spain and China in "Characterization and Correction of the Geometric Errors in Using Confocal Microscope for Extended Topography Measurement. Part I: Models, Algorithms Development and Validation" and "Characterization and Correction of the Geometric Errors Using a Confocal Microscope for Extended Topography Measurement, Part II: Experimental Study and Uncertainty Evaluation" considered that the measurement accuracy and error compensation are important issues for measuring machines. Therefore, Wang et al. proposed a mathematical model based on system kinematics for building the scale calibration of the X-coordinate and Y-coordinate in Part I; two experiments were designed based on Monte Carlo method to evaluate the proposed mathematical model and measure different target areas in Part II. In their experiments, 35 cylinders of point cloud were established in a 5×7 area and generated for evaluating the proposed mathematical model. The results show that the mean residuals and squared residuals of the proposed method were higher than those of other methods [50,51].

Lin et al. from China in "Deep Transfer HSI Classification Method Based on Information Measure and Optimal Neighborhood Noise Reduction" considered that high redundant spectral information in the hyperspectral images (HSIs) may interfere with the accuracy of image classification. Therefore, a deep learning method based on a dimensionality reduction method and convolutional neural networks was proposed to improve the accuracy of HIS classification. In the experiments, the dataset of Indian Pines and Salinas which were obtained by Airborne Visible/Infrared Imaging Spectrometer (AVIRIS) sensors were collected for evaluating the proposed method. The results show that the accuracy of the proposed method was higher than that of other methods (e.g., principal component analysis (PCA)) [52].

Chang et al. from Taiwan and India in "Quality Assessment of Tire Shearography Images via Ensemble Hybrid Faster Region-Based ConvNets" considered that the bubble defect detection is an important issue to filter out defective tires for the improvement of driving safety. Therefore, the combination of ensemble convolutional neural network and Faster-RCNN was proposed to detect bubble defects in the shearography images of tires. In their experiments, for the evaluation of the proposed method, 3279 tire images were selected as training data; 797 tire images were selected as testing data. The results show that the accuracy, sensitivity and specificity of the proposed method were higher than those of other methods (e.g., SVM, random forest, Haar-like AdaBoost, etc.) [53].

Sun et al. from China in "High-Resolution Image Inpainting Based on Multi-Scale Neural Network" considered that the blurred textures and the unpleasant boundaries may be obtained by the image inpainting method based on GAN in the cases of high resolution images. Therefore, this study applied the super-resolution using a generative adversarial network (SRGAN) to inpaint image and extract the features of textures for the improvement of image recognition. In the experiments, COCO and VOC datasets which included 135,414 images as training data and 200 images as testing data were selected to evaluate the proposed method. The results show that the PSNR and SSIM of the proposed method were higher than the PSNR and SSIM of other methods [54].

5. Recommendation Systems and Education Systems

Two papers on recommendation systems and education systems are as follows: (1) "Deep Learning-Enhanced Framework for Performance Evaluation of a Recommending Interface with Varied Recommendation Position and Intensity Based on Eye-Tracking Equipment Data Processing" by Sulikowski et al. [55] and (2) "Generative Adversarial Network Based Neural Audio Caption Model for Oral Evaluation" by Zhang et al. [56].

Sulikowski et al. from Poland in "Deep Learning-Enhanced Framework for Performance Evaluation of a Recommending Interface with Varied Recommendation Position and Intensity Based on Eye-Tracking Equipment Data Processing" considered that high correlations may exist between users' gaze data and interests in human-computer interaction for recommendation inferences. Therefore, this study collected eye-tracking data to train a deep learning neural network model for building an e-commerce recommendation system. In the experiments, 15,922 fixation records were generated by eye-tracking devices from 52 participants. The results show that the accuracies of training dataset and testing dataset were 98.4% and 98.2%, respectively [55].

Zhang et al. from China in "Generative Adversarial Network Based Neural Audio Caption Model for Oral Evaluation" considered that the massive human work is required by oral evaluation for testing children's language learning. Therefore, an automated expert comment generation method based on gated recurrent units (GRUs), LSTM networks and GANs was proposed to extract the features of orals and generate expert comments. In their experiments, the proposed neural audio caption model (NACM) and the proposed GAN-based NACM (GNACM) were implemented and compared; several oral audios from the children of 5-6 years old were collected for evaluating the proposed models. The results show that scores of GNACM were higher than the scores of NACM; furthermore, the average response time of GNACM was lower than that of NACM [56].

6. Conclusions and Future Work

Four main parts, including (I) environmental information analyses and predictions, (II) UAV and object tracking applications, (III) measurement and denoising techniques, and (IV) recommendation systems and education systems, are collected and discussed in this Special Issue. These articles utilized and improved the deep learning techniques (e.g., ResNet, Fast-RCNN, LSTM, ConvLSTM, GAN, etc.) to analyze and denoise measured data in a variety of applications and services (e.g., wind speed prediction, air quality prediction, underground mine applications, neural audio caption, etc.). Several practical experiments were given in these articles, and the results indicated that the performance of the improved deep learning methods could be higher than the performance of conventional machine learning methods [43–56].

In the future, the federated learning techniques can be considered to train deep learning and machine learning models across multiple decentralized servers for data privacy and data security in electronics industries. Furthermore, the optimization techniques (e.g., gradient descent algorithm, Adam optimization algorithm, particle swarm optimization algorithm [57,58], etc.) can be improved for finding the global optimal solution.

Author Contributions: M.-F.H., H.-Y.K., C.-H.C. and F.-J.H. edited the Special Issue, entitled "Deep Learning Applications with Practical Measured Results in Electronics Industries", of *Electronics*. M.-F.H., H.-Y.K., C.-H.C. and F.-J.H. wrote this editorial equally for the introduction of the Special Issue. All authors have read and agreed to the published version of the manuscript.

Funding: This work was partially supported by the National Natural Science Foundation of China (Nos. 61906043, 61877010, 11501114 and 11901100), Fujian Natural Science Funds (No. 2019J01243), and Fuzhou University (Nos. 510730/XRC-18075, 510809/GXRC-19037, 510649/XRC-18049 and 510650/XRC-18050). The work was also supported by the Ministry of Science and Technology of Taiwan under grant number MOST 108-2637-E-020-00 and MOST 107-2221-E-992 -059 -MY2.

Acknowledgments: We thank all authors who submitted their valuable papers to the Special Issue, entitled "Deep Learning Applications with Practical Measured Results in Electronics Industries", of *Electronics*. Furthermore, we thank all reviewers and the editorial team of *Electronics* for their great efforts and support.

Conflicts of Interest: The authors declare no conflict of interest.

References

1. Ma, Z.; Liu, Y.; Liu, X.M.; Ma, J.F.; Li, F.F. Privacy-preserving outsourced speech recognition for smart IoT devices. *IEEE Internet Things J.* **2019**, *6*, 8406–8420. [CrossRef]
2. Huang, Z.Y.; Yu, Y.L.; Xu, M.X. Bidirectional tracking scheme for visual object tracking based on recursive orthogonal least squares. *IEEE Access* **2019**, *7*, 159199–159213. [CrossRef]
3. Chen, C.H. An arrival time prediction method for bus system. *IEEE Internet Things J.* **2018**, *5*, 4231–4232. [CrossRef]
4. Dong, C.; Zhang, F.; Liu, X.M.; Huang, X.; Guo, W.Z.; Yang, Y. A locating method for multi-purposes HTs based on the boundary network. *IEEE Access* **2019**, *7*, 110936–110950. [CrossRef]
5. Chen, C.H. An explainable deep neural network for extracting features. *Science* **2019**. Available online: https://science.sciencemag.org/content/365/6452/416/tab-e-letters (accessed on 2 March 2020).
6. Zhang, S.C.; Xia, Y.S. Solving nonlinear optimization problems of real functions in complex variables by complex-valued iterative methods. *IEEE Trans. Cybern.* **2018**, *48*, 277–287. [CrossRef]
7. Chen, J.H.; Dong, C.; He, G.R.; Zhang, X.Y. A method for indoor Wi-Fi location based on improved back propagation neural network. *Turk. J. Electr. Eng. Comput. Sci.* **2019**, *27*, 2511–2525. [CrossRef]
8. Xia, Y.S.; Zhang, S.C.; Stanimirovic, P.S. Neural network for computing pseudoinverses and outer inverses of complex-valued matrices. *Appl. Math. Comput.* **2016**, *273*, 1107–1121. [CrossRef]
9. Mehta, P.P.; Pang, G.F.; Song, F.Y.; Karniadakis, G.E. Discovering a universal variable-order fractional model for turbulent Couette flow using a physics-informed neural network. *Fract. Calc. Appl. Anal.* **2019**, *22*, 1675–1688. [CrossRef]
10. Chen, C.H.; Hwang, F.J.; Kung, H.Y. Travel time prediction system based on data clustering for waste collection vehicles. *IEICE Trans. Inf. Syst.* **2019**, 1374–1383. [CrossRef]
11. Yu, Y.L.; Ye, Z.F.; Zheng, X.G.; Rong, C.M. An efficient cascaded method for network intrusion detection based on extreme learning machines. *J. Supercomput.* **2018**, *74*, 5797–5812. [CrossRef]
12. Liu, X.M.; Deng, R.H.; Yang, Y.; Iran, H.N.; Zhong, S.P. Hybrid privacy-preserving clinical decision support system in fog-cloud computing. *Future Gener. Comput. Syst.* **2018**, *78*, 825–837. [CrossRef]
13. Liao, X.W.; Zhang, L.Y.; Wei, J.J.; Yang, D.D.; Chen, G.L. Recommending mobile microblog users via a tensor factorization based on user cluster approach. *Wirel. Commun. Mob. Comput.* **2018**, *2018*, 9434239. [CrossRef]
14. Zhang, S.C.; Xia, Y.S. Two Fast complex-valued algorithms for solving complex quadratic programming problems. *IEEE Trans. Cybern.* **2016**, *46*, 2837–2847. [CrossRef] [PubMed]

15. Xia, Y.S.; Leung, H.; Kamel, M.S. A discrete-time learning algorithm for image restoration using a novel L-2-norm noise constrained estimation. *Neurocomputing* **2016**, *198*, 155–170. [CrossRef]

16. Xia, Y.S.; Wang, J. A bi-projection neural network for solving constrained quadratic optimization problems. *IEEE Trans. Neural Netw. Learn. Syst.* **2016**, *27*, 214–224. [CrossRef] [PubMed]

17. Wang, J.; Zhang, X.M.; Han, Q.L. Event-triggered generalized dissipativity filtering for neural networks with time-varying delays. *IEEE Trans. Neural Netw. Learn. Syst.* **2016**, *27*, 77–88. [CrossRef]

18. Zhang, S.C.; Xia, Y.S.; Wang, J. A complex-valued projection neural network for constrained optimization of real functions in complex variables. *IEEE Trans. Neural Netw. Learn. Syst.* **2015**, *26*, 3227–3238. [CrossRef]

19. Xia, Y.S.; Leung, H. A fast learning algorithm for blind data fusion using a novel L-2-norm estimation. *IEEE Sens. J.* **2014**, *14*, 666–672. [CrossRef]

20. Luo, H.F.; Chen, C.C.; Fang, L.N.; Zhu, X.; Lu, L.J. High-resolution aerial images semantic segmentation using deep fully convolutional network with channel attention mechanism. *IEEE J. Sel. Top. Appl. Earth Obs. Remote Sens.* **2019**, *12*, 3492–3507. [CrossRef]

21. Wang, Y.L.; Tang, W.Z.; Yang, X.J.; Wu, Y.J.; Chen, F.J. End-to-end automatic image annotation based on deep CNN and multi-label data augmentation. *IEEE Trans. Multimed.* **2019**, *21*, 2093–2106. [CrossRef]

22. Ke, X.; Shi, L.F.; Guo, W.Z.; Chen, D.W. Multi-dimensional traffic congestion detection based on fusion of visual features and convolutional neural network. *IEEE Trans. Intell. Transp. Syst.* **2019**, *20*, 2157–2170. [CrossRef]

23. Cheng, H.; Xie, Z.; Shi, Y.S.; Xiong, N.X. Multi-step data prediction in wireless sensor networks based on one-dimensional CNN and bidirectional LSTM. *IEEE Access* **2019**, *7*, 117883–117896. [CrossRef]

24. Zhang, G.; Hsu, C.H.R.; Lai, H.D.; Zheng, X.H. Deep learning based feature representation for automated skin histopathological image annotation. *Multimed. Tools Appl.* **2018**, *77*, 9849–9869. [CrossRef]

25. Dai, Y.F.; Guo, W.Z.; Chen, X.; Zhang, Z.W. Relation classification via LSTMs based on sequence and tree structure. *IEEE Access* **2018**, *6*, 64927–64937. [CrossRef]

26. Weng, Q.; Mao, Z.Y.; Lin, J.W.; Guo, W.Z. Land-use classification via extreme learning classifier based on deep convolutional features. *IEEE Geosci. Remote Sens. Lett.* **2017**, *14*, 704–708. [CrossRef]

27. Cheng, H.J.; Xie, Z.; Wu, L.H.; Yu, Z.Y.; Li, R.X. Data prediction model in wireless sensor networks based on bidirectional LSTM. *EURASIP J. Wirel. Commun. Netw.* **2019**, *2019*, 203. [CrossRef]

28. Chen, C.H.; Kung, H.Y.; Hwang, F.J. Deep learning techniques for agronomy applications. *Agronomy* **2019**, *9*, 142. [CrossRef]

29. Wu, L.; Chen, C.H.; Zhang, Q. A Mobile Positioning Method Based on Deep Learning Techniques. *Electronics* **2019**, *8*, 59. [CrossRef]

30. Xia, Y.S.; Wang, J. Robust regression estimation based on low-dimensional recurrent neural networks. *IEEE Trans. Neural Netw. Learn. Syst.* **2018**, *29*, 5935–5946. [CrossRef]

31. Chen, G.Y.; Gan, M.; Chen, G.L. Generalized exponential autoregressive models for nonlinear time series: Stationarity, estimation and applications. *Inf. Sci.* **2018**, *438*, 46–57. [CrossRef]

32. Xia, Y.S.; Wang, J. Low-dimensional recurrent neural network-based Kalman filter for speech enhancement. *Neural Netw.* **2015**, *67*, 131–139. [CrossRef] [PubMed]

33. Zhang, C.Y.; Chen, C.L.P.; Chen, D.W.; Ng, K.T. MapReduce based distributed learning algorithm for Restricted Boltzmann Machine. *Neurocomputing* **2016**, *198*, 4–11. [CrossRef]

34. Chen, C.L.P.; Zhang, C.Y.; Chen, L.; Gan, M. Fuzzy restricted Boltzmann machine for the enhancement of deep learning. *IEEE Trans. Fuzzy Syst.* **2015**, *23*, 2163–2173. [CrossRef]

35. Zhang, Y.C.; Le, J.; Liao, X.B.; Zheng, F.; Li, Y.H. A novel combination forecasting model for wind power integrating least square support vector machine, deep belief network, singular spectrum analysis and locality-sensitive hashing. *Energy* **2019**, *168*, 558–572. [CrossRef]

36. Zeng, X.X.; Chen, F.; Wang, M.Q. Shape group Boltzmann machine for simultaneous object segmentation and action classification. *Pattern Recognit. Lett.* **2019**, *111*, 43–50. [CrossRef]

37. Wang, Y.L.; Tang, W.Z.; Yang, X.J.; Wu, Y.J.; Chen, F.J. An efficient method for autoencoder-based collaborative filtering. *Concurr. Comput. Pract. Exp.* **2019**, *31*, e4507. [CrossRef]

38. Chen, C.H. Reducing the dimensionality of time-series data with deep learning techniques. *Science* **2018**. Available online: http://science.sciencemag.org/content/313/5786/504/tab-e-letters (accessed on 2 March 2020).

39. Xie, H.S.; Zhang, Y.F.; Wu, Z.S. Fabric Defect Detection Method Combing Image Pyramid and Direction Template. *IEEE Access* **2019**, *7*, 182320–182334. [CrossRef]

40. Guo, W.Z.; Wang, J.W.; Wang, S.P. Deep Multimodal Representation Learning: A Survey. *IEEE Access* **2019**, *7*, 63373–63394. [CrossRef]

41. Guo, W.Z.; Cai, J.Y.; Wang, S.P. Unsupervised discriminative feature representation via adversarial auto-encoder. *Appl. Intell.* **2019**. [CrossRef]

42. Mnih, V.; Kavukcuoglu, K.; Silver, D.; Rusu, A.A.; Veness, J.; Bellemare, M.G.; Graves, A.; Riedmiller, M.; Fidjeland, A.K.; Ostrovski, G.; et al. Human-level control through deep reinforcement learning. *Nature* **2015**, *518*, 529–533. [CrossRef] [PubMed]

43. Panapakidis, I.P.; Michailides, C.; Angelides, D.C. A Data-Driven Short-Term Forecasting Model for Offshore Wind Speed Prediction Based on Computational Intelligence. *Electronics* **2019**, *8*, 420. [CrossRef]

44. Wan, R.; Mei, S.; Wang, J.; Liu, M.; Yang, F. Multivariate Temporal Convolutional Network: A Deep Neural Networks Approach for Multivariate Time Series Forecasting. *Electronics* **2019**, *8*, 876. [CrossRef]

45. Xu, W.; Feng, X.; Xing, H. Modeling and Analysis of Adaptive Temperature Compensation for Humidity Sensors. *Electronics* **2019**, *8*, 425. [CrossRef]

46. Zhang, F.; Xu, Z.; Chen, W.; Zhang, Z.; Zhong, H.; Luan, J.; Li, C. An Image Compression Method for Video Surveillance System in Underground Mines Based on Residual Networks and Discrete Wavelet Transform. *Electronics* **2019**, *8*, 1559. [CrossRef]

47. Zhou, H.; Xiong, H.-L.; Liu, Y.; Tan, N.-D.; Chen, L. Trajectory Planning Algorithm of UAV Based on System Positioning Accuracy Constraints. *Electronics* **2020**, *9*, 250. [CrossRef]

48. Zhang, F.; Fan, Y.; Cai, T.; Liu, W.; Hu, Z.; Wang, N.; Wu, M. OTL-Classifier: Towards Imaging Processing for Future Unmanned Overhead Transmission Line Maintenance. *Electronics* **2019**, *8*, 1270. [CrossRef]

49. Wang, D.; Fang, W.; Chen, W.; Sun, T.; Chen, T. Model Update Strategies about Object Tracking: A State of the Art Review. *Electronics* **2019**, *8*, 1207. [CrossRef]

50. Wang, C.; Gómez, E.; Yu, Y. Characterization and Correction of the Geometric Errors in Using Confocal Microscope for Extended Topography Measurement. Part I: Models, Algorithms Development and Validation. *Electronics* **2019**, *8*, 733. [CrossRef]

51. Wang, C.; Gómez, E.; Yu, Y. Characterization and Correction of the Geometric Errors Using a Confocal Microscope for Extended Topography Measurement, Part II: Experimental Study and Uncertainty Evaluation. *Electronics* **2019**, *8*, 1217. [CrossRef]

52. Lin, L.; Chen, C.; Yang, J.; Zhang, S. Deep Transfer HSI Classification Method Based on Information Measure and Optimal Neighborhood Noise Reduction. *Electronics* **2019**, *8*, 1112. [CrossRef]

53. Chang, C.-Y.; Srinivasan, K.; Wang, W.-C.; Ganapathy, G.P.; Vincent, D.R.; Deepa, N. Quality Assessment of Tire Shearography Images via Ensemble Hybrid Faster Region-Based ConvNets. *Electronics* **2020**, *9*, 45. [CrossRef]

54. Sun, T.; Fang, W.; Chen, W.; Yao, Y.; Bi, F.; Wu, B. High-Resolution Image Inpainting Based on Multi-Scale Neural Network. *Electronics* **2019**, *8*, 1370. [CrossRef]

55. Sulikowski, P.; Zdziebko, T. Deep Learning-Enhanced Framework for Performance Evaluation of a Recommending Interface with Varied Recommendation Position and Intensity Based on Eye-Tracking Equipment Data Processing. *Electronics* **2020**, *9*, 266. [CrossRef]

56. Zhang, L.; Shu, C.; Guo, J.; Zhang, H.; Xie, C.; Liu, Q. Generative Adversarial Network Based Neural Audio Caption Model for Oral Evaluation. *Electronics* **2020**, *9*, 424. [CrossRef]

57. Liu, N.; Pan, J.; Nguyen, T. A Bi-population QUasi-Affine TRansformation Evolution Algorithm for Global Optimization and Its Application to Dynamic Deployment in Wireless Sensor Networks. *EURASIP J. Wirel. Commun. Netw.* **2019**, *2019*, 175. [CrossRef]

58. Li, Y.F.; Li, J.B.; Pan, J.S. Hyperspectral image recognition using SVM combined deep learning. *J. Internet Technol.* **2019**, *20*, 851–859. [CrossRef]

 *electronics*

Article

A Data-Driven Short-Term Forecasting Model for Offshore Wind Speed Prediction Based on Computational Intelligence

Ioannis P. Panapakidis [1,*], **Constantine Michailides** [2] **and Demos C. Angelides** [3]

[1] Department of Electrical Engineering, Technological Educational Institute of Thessaly, 41110 Larisa, Greece
[2] Department of Civil Engineering and Geomatics, Cyprus University of Technology, 3036 Limassol, Cyprus; c.michailides@cut.ac.cy
[3] Department of Civil Engineering, Aristotle University of Thessaloniki, 54124 Thessaloniki, Greece; dangelid@civil.auth.gr
* Correspondence: panap@teilar.gr; Tel.: +30-2410-684325

Received: 27 March 2019; Accepted: 6 April 2019; Published: 10 April 2019

Abstract: Wind speed forecasting is an important element for the further development of offshore wind turbines. Due to its importance, many researchers have proposed different models for wind speed forecasting that differ in terms of the time-horizon of the forecast, types and number of inputs, complexity, structure, and others. Wind speed series present high nonlinearity and volatilities, and thus an effective model should successfully deal with those features. An approach to deal with the nonlinearities and volatilities is to utilize a time series processing technique such as the wavelet transform. In the present paper, an ensemble data-driven short-term wind speed forecasting model is developed, tested and applied. The term "ensemble" refers to the combination of two different predictors that run in parallel and the prediction is obtained by the predictor that leads to the lowest error. The proposed model utilizes the wavelet transform and is compared with other models that have been presented in the related literature and outperforms their accuracy. The proposed forecasting model can be used effectively for 1 min and 10 min ahead horizon wind speed predictions.

Keywords: computational intelligence; offshore wind; forecasting; machine learning; neural networks; neuro-fuzzy systems

---

1. Introduction

1.1. Motivation and State-of-the-Art

The rapid implementation of wind turbines across the globe corresponds to a set of challenges during power systems operation and planning. This is due to intermittent nature of wind potential. By the end of 2018, the total European Union-installed offshore wind capacity reached 19 GW. Total investments in offshore wind in 2018 were more than 10.3 billion € (WindEurope [1], 2018). This includes investments in construction of projects, transmission assets, and refinancing. More than 91% of all offshore wind installations were located in shallow or intermediate water depths with a mean water depth equal to 27.1 m. The capacities span from few to hundred megawatts and the installations differ in terms of hub site, distance to shore, water depth and others (Snyder et al. [2], 2009). The reduction of installation and maintenance costs but also the reliable assessment of energy production of offshore wind parks will signify the next phase of their deployment.

Wind energy leads to disturbances of the balance between generation and demand sides. A wind speed prediction system is a potential solution to the aforementioned situation. Apart from the balance, accurate predictions can lead to lower costs during the installation in the offshore field of wind turbines and can strengthen their reliability during their operation (Soman et al. [3], 2010). A large number of papers exist in the literature aiming at developing robust forecasting systems of wind speed. The literature can be divided to research that focuses on (a) wind speed predictions, (b) wind farm power predictions, and (c) both wind speed and farm power predictions.

According to (Jung and Broadwater [4], 2014) the forecasting horizon can be classified into the following categories; (a) very short-term: few minutes up to 30 min ahead; (b) short-term: 30 min up to 6 h ahead; (c) medium-term: 6 h to 1 day ahead; and (d) long-term: 1 day up to 1 week (or more) ahead. Short-term forecasts are exploitable in day-ahead power system operations such as scheduling and commitment of power units.

The forecasting models are distinguished into physical and statistical. Physical models take into consideration parameters like the topology of the ground (for on-site parks) and topology of the wind park and temperature. They use the outputs of a Numerical Weather Prediction (NWP) model and provide final forecasts. NWP models are used by meteorologists and usually they provide predictions for the next 48 up to 172 h ahead. For the case of wind power predictions, the wind speed prediction is obtained directly by the NWP and the wind power prediction is obtained by using the power curve of a wind turbine. Physical models are appropriate for long-term predictions. However, it is difficult to scale the forecasts per wind turbine or per wind farm. Also, physical models are complex in terms of inputs requirements and execution time. On the other hand, statistical models are favored in short-term prediction horizons. The wind is treated as a regression of its past values. A relatively large number of historical values are needed to train the models and define their optimal composition. Statistical models refer to time series models, Artificial Neural Networks, Bayesian Networks, Support Vector Machines, and others. Time series models refer to autoregressive models, autoregressive models with moving average, and others. The main advantages are their potential for removing the trend of time series and their availability in software packages. However, there is a difficulty in extracting the optimal structure of the model. Also, time series models require a large number of historical values and are not definitely suitable for highly nonlinear series. Neural networks are suitable for nonlinear series and are a favorable scheme in many forecasting problems. Support vector machines are also a well-known forecasting engine but they demand large durations for their training and their parameters are optimized by a relatively complex process. Bayesian networks are more appropriate for small data sets. Finally, various statistical models can be integrated to form ensemble forecasting models [4–11].

In (Sfetsos [12], 2002), a comparison takes place between a persistent model, autoregressive integrated moving average (ARIMA) and neural network. The latter outperforms the rest. No external variables are considered. The comparison takes place in two different sets that refer to one month each. In (Başaran and Filik [13], 2017) the authors consider three cases of inputs for the neural network, i.e., using only past wind speed values, using past wind speed values and temperature and, finally, using wind speed, temperature, and pressure. The test refers to five days and two intervals for predictions are taken into account, namely 30 and 90 s ahead. The case with wind speed, temperature, and pressure leads to the lower errors. No comparison with other models is presented. In (More and Deo [14], 2003) the authors test a feed-forward neural network, a recurrent neural network, and an ARIMA model to forecast daily, weekly, and monthly wind speeds at two coastal locations in India using only past wind speed values. The feed-forward neural network wins the competition. In (Li and Shi [15], 2010), the authors examine three types of neural networks, namely, the adaptive linear element, backpropagation, and radial basis function. The wind data used are the hourly mean wind speed collected at two observation sites in a United States of America (USA) location. The results show that even for the same wind dataset, no single neural network model outperforms others universally in terms of all evaluation metrics. Moreover, the selection of the type of neural networks for best performance is also depends upon the data sources. In (Zeng and Qiao [16], 2011) a support vector machine model is presented for wind power forecasting. Instead of predicting wind power directly, the model first predicts the wind speed, which is then used to predict the wind power by using the power–wind speed characteristics of the wind turbine generators. Simulation studies are carried out to validate the proposed model for very short-term and short-term predictions by using the data obtained from the National Renewable Energy Laboratory of USA. The model is compared with feed-forward neural network and radial basis network. The prediction is held using only past wind speed values. In (Zhou et al. [17], 2011), the authors present a least-squares support vector machine for one-step ahead wind speed forecasting. Three kernels, namely linear, Gaussian, and polynomial kernels, are implemented. The support vector machine's parameters considered include the training sample size, order, regularization parameter, and kernel parameters. The support vector machine's version are compared with a persistence model and provide better forecasts. The Adaptive Neuro-Fuzzy Inference System (ANFIS) is utilized in (Fazelpour et al. [18], 2016), and is compared with a feed-forward neural network and radial basis network in hour-ahead forecasting in a location in Tehran, Iran. No exogenous parameters are used. ANFIS results in better forecasts. In (Fortuna et al. [19], 2016), the clustering tool is used to form wind speed classes. Then, two models, namely the Hidden Markov Model and the Nonlinear Autoregressive are compared for predicting the class of each new wind speed data entry. In general, wind speed series present volatilities and stochasticity. Depending on the data set, an analysis on the wind speed characteristics can take place. For instance, in (Fortuna et al. [20], 2014), the authors provide a fractal analysis on wind speed observations. Exploitable information can be derived for such analysis for further modeling.

1.2. Contribution of the Present Paper

A variety of forecasting techniques have been proposed so far from different research groups. In the present paper a relatively simple yet efficient model for short-term wind speed forecasting based on real measured wind speed data is developed, applied, and proposed. The used data set involves inconsistencies of the time sequence of the wind speed series due to missing data. Various experiments take place that refer to different input combinations. Also, the Discrete Wavelet Transform (DWT) is utilized in order to decompose the initial series into a set of wavelet components for strengthening the forecasting credibility [21,22]. The model is composed by an ANFIS and a Feed-Forward Neural Network (FFNN) [23,24]. In the majority of the studies of the literature, the prediction is accomplished using only past values. In order to fully examine the level of influence of external variables such as temperature and speed directions on the prediction accuracy, in the present paper various input combinations of wind speed, wind direction, and air temperature are examined. Overall, the purpose of the paper is to test the performance of a proposed hybrid computational intelligence model in the wind speed forecasting problem under the limitation of using incomplete data for the training and validation of the model.

2. Short-Term Wind Speed Forecasting Hybrid Model

2.1. Description

In this section, an efficient forecasting model is developed and proposed. The model consists of an FFNN trained by the Levenberg–Marquardt algorithm and an ANFIS [25]. Neural network-based forecasting systems are a favorable scheme in recent years in predictions over traditional time series models. Numerous applications in load and price forecasting studies have brought forth the advantages of neural networks. Recently, neural networks have been used in wind power predictions. For full mathematical description of the FFNN the reader is referred to (Graupe [24], 2007). A general illustration of an FFNN is shown in Figure 1.

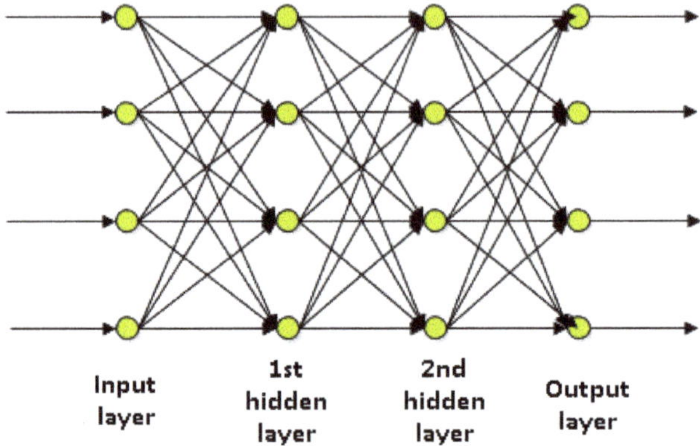

Figure 1. A Feed-Forward Neural Network (FFNN) with 4 inputs, 4 outputs, and 2 hidden layers.

Another common forecasting system is ANFIS [23]. ANFIS is based on a fuzzy rule-based inference mechanism. It is composed of five layers and each layer contains several nodes. The nodes are described by a node function. Let O_i^j be the output of the *i*-th node in layer *j*. In the 1st layer, every node I is an adaptive node with node function:

$$O_i^1 = \mu A_i(x), i = 1, 2 \tag{1}$$

or

$$O_i^1 = \mu B_{i-2}(y), i = 3, 4 \tag{2}$$

where x or y is the input of the *i*th node and A_i or B_{i-2} is a linguistic label associated with the node. Hence, O_i^j is the membership grade of a fuzzy set A_1, A_2, B_1 or B_2 and it specifies the degree to which the input x or y satisfies the quantifier A or B. Any continuous and piecewise differential function can be used as node function in the 1st layer. In the 2nd layer, each node Π multiplies the inputs and sends the product in output:

$$O_i^2 = w_i = \mu A_i(x) \mu B_i(y), i = 1, 2 \tag{3}$$

In the 3rd layer, each node N computes the ratio

$$O_i^3 = \overline{w}_i = \frac{w_i}{w_1 + w_2}, i = 1, 2 \tag{4}$$

In the 4th layer, each node computes the contribution of the *i*th rule to the overall output:

$$O_i^4 = \overline{w}_i z_i = \overline{w}_i(a_i x + b_i y + c), i = 1, 2 \tag{5}$$

where $\overline{w}_i$ is the output of the 3rd layer and a_i, b_i, c are a set of parameters.

Finally, in the 5th layer, the node Σ computes the final output as the summation of all inputs:

$$O_i^5 = \sum_i \overline{w}_i z_i = \frac{\sum_i w_i z_i}{\sum_i w_i} \tag{6}$$

ANFIS topology is displayed in Figure 2.

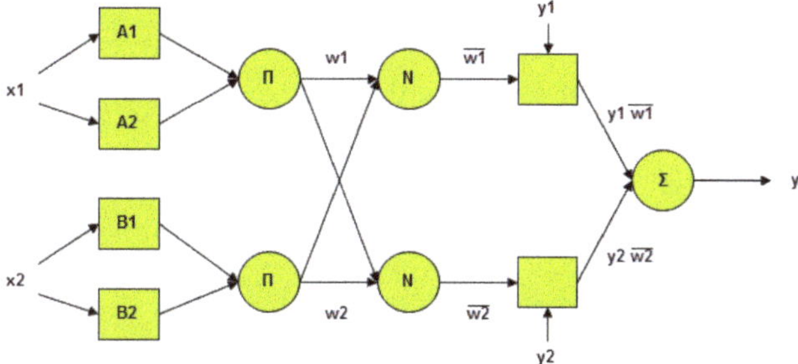

Figure 2. A general illustration of Adaptive Neuro-Fuzzy Inference System (ANFIS).

The proposed forecasting model combines the independent forecasts of FFNN and ANFIS. A schematic representation of the hybrid model is shown in Figure 3. The models are trained separately. The training set is used to define the optimal model parameters. For instance, for the case of the FFNN the parameters that need to be defined are the number of hidden layers, the number of neurons in the hidden layer, and the type of activation function in the hidden and output layers. While for the case of ANFIS the required parameters that need to be defined are the type of inference mechanism, the training epochs, the number of fuzzy rules, the type of membership function, and the values of a_i, b_i, c. Real monitored environmental data measured with a monitoring system are used in the present paper for developing the forecasting model. The monitoring system is placed in the coastal area of Neos Marmaras, Greece. Details about the monitoring system (e.g., sensors used and verification) can be found in (Michailides et al. [26], 2013). The training and test sets cover the periods 01/04/2013–10/08/2013 and 01/09/2013–24/12/2013, respectively. The test set is used for the comparison of the models. No filling of incomplete or missing data took place. Also, no other preprocessing of the data took place. The aim is to build a model applied to raw data obtained from a real measurement system.

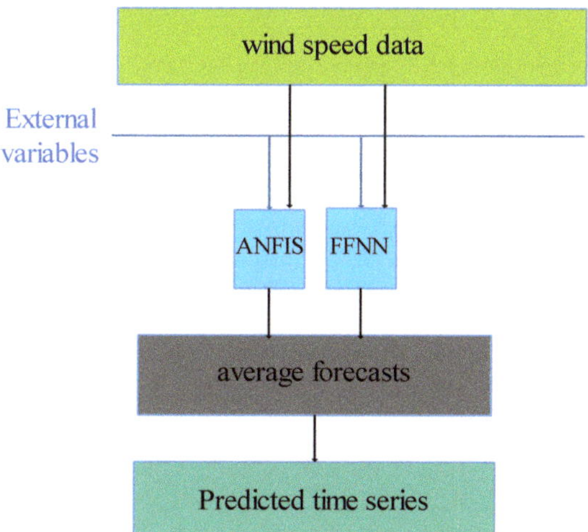

Figure 3. Structure of the proposed forecasting model.

The proposed model is applied to two different predictions test examples, i.e., 1 min-ahead and 10 min ahead. Five cases with different types of inputs are examined, namely Case#1, Case#2, Case#3, Case#4, and Case#5. The inputs considered in each case are described in the following.

- Case#1: wind speed
- Case#2: wind speed and wind direction
- Case#3: wind speed and temperature
- Case#4: wind speed, wind direction and temperature
- Case#5: wavelet components of wind speed.

The topology of the proposed model using the wavelet components is shown in Figure 4.

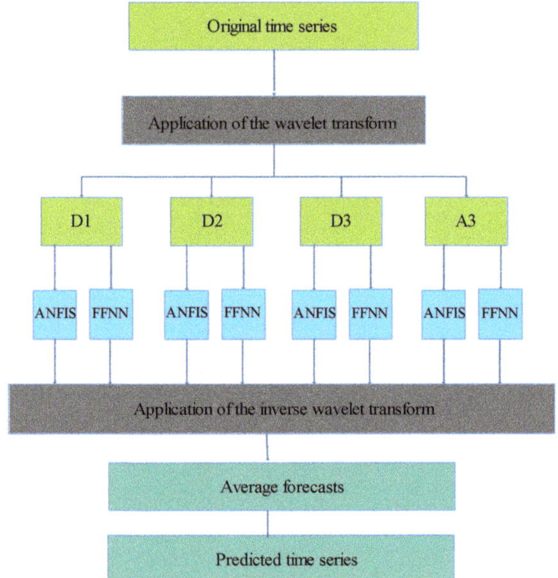

Figure 4. Structure of the proposed forecasting model using the wavelet components.

Therefore, we examined ten different cases referring to the two prediction horizons.

2.2. Performance Assessment

The performance assessment includes a set of mathematical criteria that measure the prediction errors. To fully examine the proposed model performance, we used a set of different mathematical criteria. Let p_m^a and p_m^f be the actual and predicted wind speed values of the m-th day of the test set, $m = 1, 2, \ldots, M$, respectively. The indicator considered for the assessment are the Absolute Error (AE), the Mean Absolute Error (MAE), the Root Mean Squared Error (RMSE), and the Mean Absolute Range Normalized Error (MARNE) as defined in Equation (10). The AE is defined as

$$\text{AE} = \sum_{m=1}^{M} \left| p_m^a - p_m^f \right| \tag{7}$$

The MAE corresponds to the sum of all AEs:

$$\text{MAE} = \frac{1}{M} \sum_{m=1}^{M} \left| p_m^a - p_m^f \right| \tag{8}$$

The RMSE is expressed as

$$\text{RMSE} = \frac{1}{M}\sqrt{\sum_{m=1}^{M}\left(p_m^{\text{a}} - p_m^{\text{f}}\right)^2}\tag{9}$$

The MARNE is the absolute difference between the actual and forecast wind speed, normalized to the maximum wind speed:

$$\text{MARNE} = \frac{1}{M}\sum_{m=1}^{M}\frac{\left|p_m^{\text{a}} - p_m^{\text{f}}\right|}{\max(p_m^{\text{a}})}\times 100\tag{10}$$

As benchmarks for the proposed model test, the individual applications of FFNN and ANFIS are used.

3. Simulation Results

3.1. Wind Speed Forecasting

Computational intelligence-based systems are a favourable scheme in recent years in various variable predictions, such as electric load, over traditional time series models. However, a careful selection of inputs and a proper training phase are essential for the model's successful implementation and utilization. The selection of the types of inputs is crucial to the forecasting success. In the present study, various input combinations are examined. The objective is to test computational intelligence-based models for the case of incomplete data. Three models are compared that refer to an FFNN, an ANFIS, and a proposed FFNN-ANFIS. After the decision of the types of inputs, i.e., Case#1–Case#5, the next test is to define the number of inputs. This number refers to the historical values of the used parameters: wind speed, temperature and wind direction. With the application of the Sample Autocorrelation Function (SAF), the historical values are evaluated based on the correlation of the present value. Figure 5 displays the SAF that resembles the minute-ahead wind speed set. Only the first 20 values are displayed. It is shown that the correlation is decreasing progressively when the lagged value becomes more time distant. The same conclusions are drawn from the 10 min ahead set. The first five values are selected as inputs for the models. Also, the corresponding values of temperature and wind direction are proportionally selected.

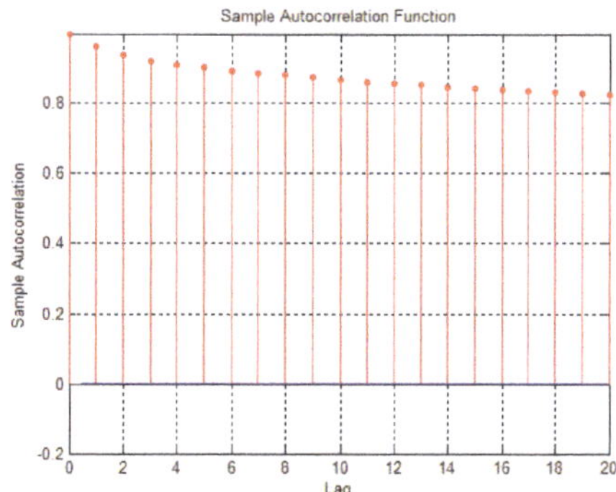

Figure 5. Sample autocorrelation function values of the wind speed data set.

Employing the minute-ahead set and the data set of Case#1, a series of experiments were conducted for the purpose of defining the optimal FFNN and ANFIS structures. The optimal FFNN structure has one hidden layer. The tangent sigmoid function is used both for the hidden and output layers. The number of training epochs is set equal to 100. The optimal number of neurons in the hidden layer is defined also by series of simulations. It differs among the various cases. Thus, a series of FFNN executions took place to track the number of hidden layers that minimize the RMSE indicator.

Concerning the optimal ANFIS topology, the Sugeno inference method is selected together with Gaussian membership functions.

The scores of the forecasting models on the assessment indicators are presented at Tables 1 and 2. Table 1 refers to the 1-min-ahead horizon and Table 2 to the 10-min-ahead horizon. The rows of the Tables correspond to the different cases. According to the results of Table 1, the proposed model outperforms the FFNN and ANFIS in all test cases, highlighting the significance of using combined forecasts. The prediction accuracy improvement that is obtained with the proposed model is more evident in the data sets of Case#3, Case#4, and Case#5. FFNN leads to better results compared to ANFIS in Case#2, Case#3, and Case#5 when using the MAE indicator. Also, the FFNN leads to better results in Case#1 according to the RMSE and MARNE measures. However, it scores in MARNE = 4.2353% in Case#2, a value that is higher than the respective of ANFIS. While in most experiments the FFNN appears more robust, it can be suggested over the ANFIS in the minute-ahead wind speed prediction problem.

Table 1. Comparison of the forecasting models considering the minute-ahead horizon.

	MAE			RMSE			MARNE (%)		
	FFNN	**ANFIS**	**FFNN-ANFIS**	**FFNN**	**ANFIS**	**FFNN-ANFIS**	**FFNN**	**ANFIS**	**FFNN-ANFIS**
Case#1	0.3683	0.3675	0.3673	0.5576	0.5578	0.5563	2.3365	2.3311	2.3298
Case#2	0.3676	0.3682	0.3664	0.7510	0.5592	0.5551	2.5952	2.3356	2.3242
Case#3	0.4123	0.4443	0.3887	0.7237	0.9163	0.6435	2.6153	2.8185	2.4658
Case#4	0.4406	0.4210	0.4091	0.9206	0.8044	0.7511	2.7947	2.6703	2.5952
Case#5	0.1021	0.1324	0.0812	0.1528	0.2052	0.1301	0.6477	0.8403	0.5601

Table 2. Comparison of the forecasting models considering the 10-min-ahead horizon.

	MAE			RMSE			MARNE (%)		
	FFNN	**ANFIS**	**FFNN-ANFIS**	**FFNN**	**ANFIS**	**FFNN-ANFIS**	**FFNN**	**ANFIS**	**FFNN-ANFIS**
Case#1	0.4316	0.4296	0.4287	0.6258	0.6265	0.6240	4.2280	4.2088	4.2008
Case#2	0.4323	0.4295	0.4292	0.6253	0.6263	0.6239	4.2353	4.2077	4.2054
Case#3	0.4706	0.4985	0.4605	0.7569	0.9373	0.6851	4.6104	4.8841	4.5115
Case#4	0.4654	0.4936	0.4528	0.7155	0.8719	0.6689	4.5594	4.8363	4.4367
Case#5	0.1227	0.1572	0.1101	0.1693	0.2226	0.1556	1.077	1.5409	0.9967

Among the types of data inputs, Case#5 leads to considerably lower errors indicating the benefit of transforming the volatile wind data into the wavelet domain. This is evident in all measures and especially in MARNE. Considering the MAE indicator, Case#2 leads to better predictions if the latter is held with FFNN or the hybrid model. On the contrary, the data of Case#1, i.e., using only wind speed values is more suitable for ANFIS. Using wind speed, temperature, and wind direction as inputs, the prediction is less credible. This implies again for FFNN and the proposed model. ANFIS scores MAE = 0.4443 with the data of Case#3. The aforementioned conclusions are identical when the evaluation is held with the RMSE or MARNE. Therefore, by combining wind speed and direction data the forecasting procedure is strengthened. The use of temperature is not recommended for the test set under study. According to the above analysis, the combination of the FFNN and ANFIS that is fed with the wind speed data transformed in the wavelet domain is the recommended model for minute-ahead forecasts under the limitation of many incomplete data entries.

According to the findings presented in Table 2, it is evident that the 10-min-ahead wind speed prediction problem is a more difficult task. A possible reason for this is the decrease that is accomplished between current and past values of the minute time frame measurements. This means that the 10-min data are less correlated since the minute correlation lowers due to the averaging of the one-minute data for the purpose of transforming them in the 10-min intervals. The proposed FFNN-ANFIS model is more accurate than the rest in all types of data sets. Again the 10-min-ahead problem benefits from the implementation of the wavelet transform. The data of Case#5 provide more robust predictions independently of the model used.

A further comparison of the models is held via the AE distribution over time. The MAE indicator receives one value for a specific prediction, for example for a given number of neurons in the hidden layer. It is essential to examine the error distribution over the focusing period. Using the AE indicator, the analysis can be scaled to minutes. This concept strengthens the conclusions drawn from the models comparison. Figure 6 presents the AE distribution per Case. The figure refers to the 1-min-ahead prediction horizon while the forecasting is achieved with the proposed FFNN-ANFIS model. The discrete peaks correspond to large error values, which can be considered as indicators of the model's poor performance for the specific minute. The data of Case#1–Case#4 lead to some high peaks of the AE shape. These are mainly met in December days. The lowest errors are mostly gathered in September days. As the time horizon progresses, the peaks become more frequent. Hence, extreme weather conditions worse the credibility of the predictions. Some late autumn and winter wind speeds are difficult to effectively be predicted in the coastal site under study. The mean values of AE are 0.3673, 0.3664, 0.4123, 0.4210, and 0.1021 for Case#1, Case#2, Case#3, Case#4, and Case#5, respectively. Some parallel conclusions with the above results can be made for the 10-min-ahead horizon. The corresponding results are graphically presented in Figure 7. In the 10-min-ahead problem, the implementation of the wavelet transform is more advantageous compared to the minute-ahead case. Case#5 increases the accuracy by a large portion. The mean values of AE are 0.4288, 0.4293, 0.4706, 0.4654, and 0.1227 for Case#1, Case#2, Case#3, Case#4, and Case#5, respectively.

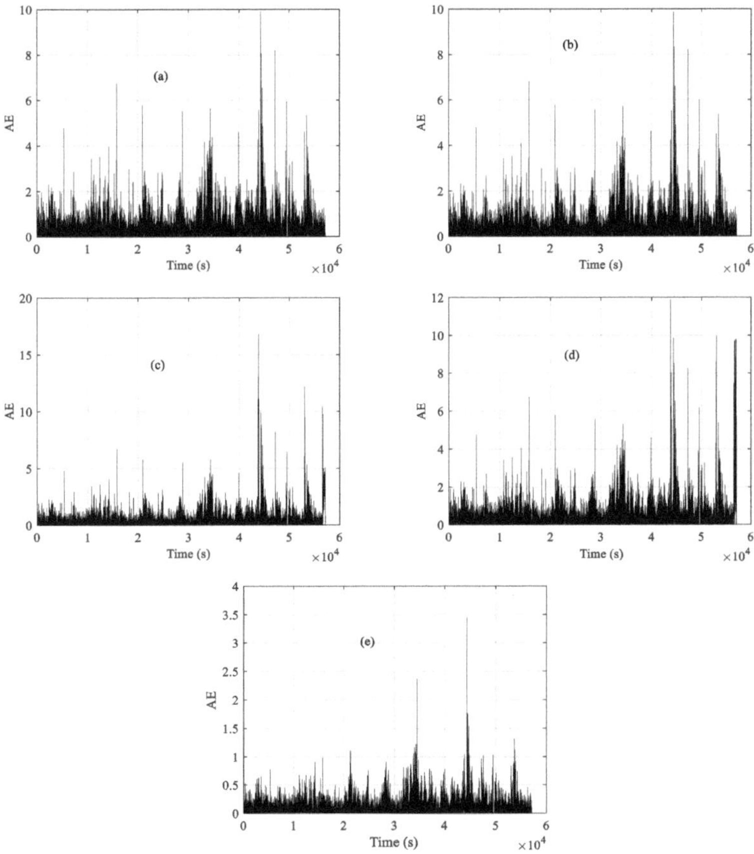

Figure 6. Absolute error of the proposed model corresponding at (**a**) Case#1, (**b**) Case#2, (**c**) Case#3, (**d**) Case#4, and (**e**) Case#5 of the 1-min-ahead prediction horizon.

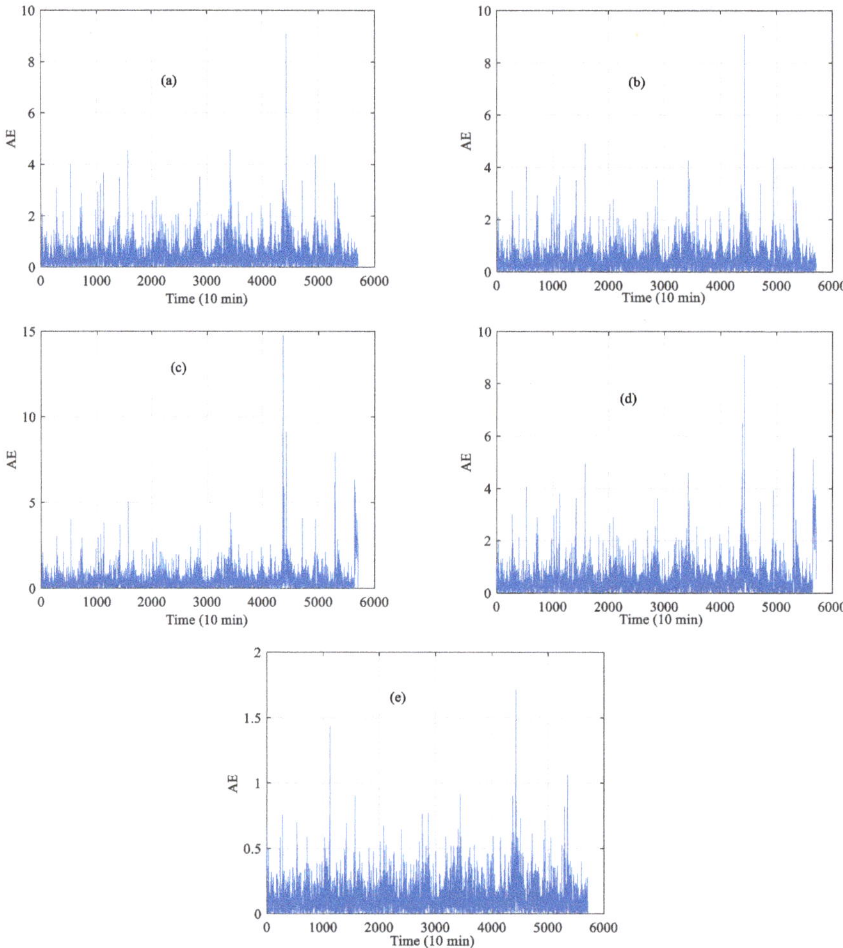

Figure 7. Absolute Error per time frame of the proposed model corresponding at (**a**) Case#1, (**b**) Case#2, (**c**) Case#3, (**d**) Case#4, (**e**) Case#5 of the 10-min-ahead prediction horizon.

In order to examine the relationship between the accuracy and the direction of the wind, we measured the AE per direction degree. Figure 8 shows the comparison among Case#2 and Case#5 for the 1-min-ahead forecasts. Case#2 refers to the combination of wind speed and direction. The predictions refer to the proposed model. Case#5 involves only to the transformed wind speed data. It is plotted here for the sake of comparison. The lowest AE of Case#2 is 5.97×10^{-6} and occurred for $312.48°$. The next lowest AE degrees are $200.73°$, $104.56°$, $151.90°$, and $281.14°$. The larger values of AE are presented for $162.47°$, $42.17°$, $55.08°$, and $222.70°$. According to these findings, a preliminary conclusion is that there is no strong correlation between the direction and the forecasting error. For example, it can be strongly supported that normal to the monitoring system winds (e.g., $90°$) are less predictable compared to other directions. This statement is also supported from the data of Case#5. The lowest errors refer to $134.67°$, $122.06°$, $73.07°$, $317.51°$, and $46.48°$ directions, while, the highest ones are occurred for $27.01°$, $35.56°$, $41.31°$, $57.65°$, and $31.57°$. In this data set, the less accurate prediction is presented for wind direction degrees below $60°$.

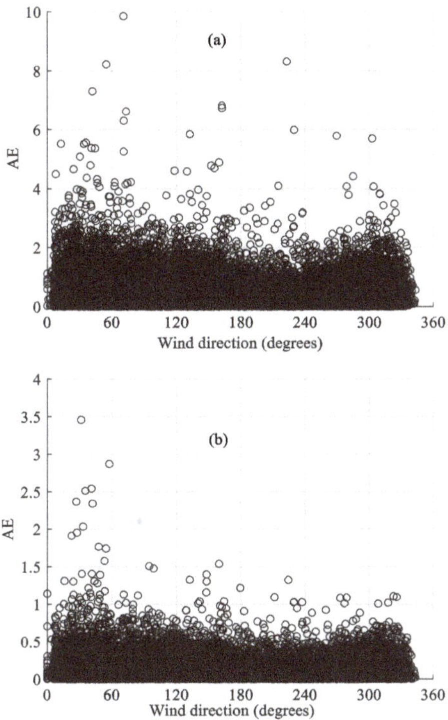

Figure 8. Absolute error per wind direction of the proposed model corresponding at (**a**) Case#2 and (**b**) Case#5 of the 1-min-ahead prediction horizon.

As illustrative examples of the proposed model's behavior, Figures 9 and 10 present the actual and forecasted wind speed curve of the test set for 1-min- and 10-min-ahead horizons, respectively. The forecasted wind speed sequences of the two figures succeed by a large portion to accurately simulate the actual data, which is another one indicator of the robustness of the model.

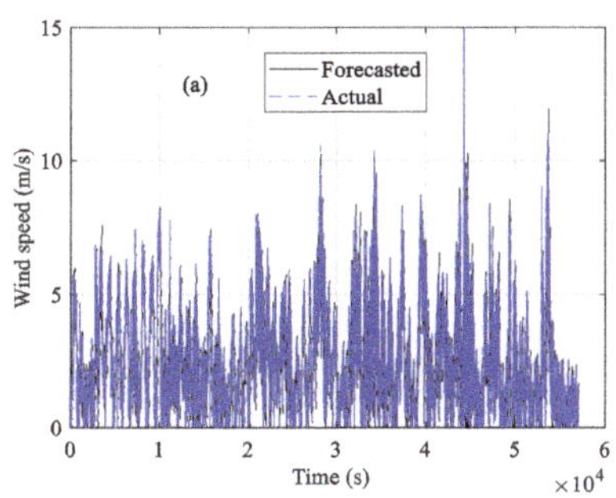
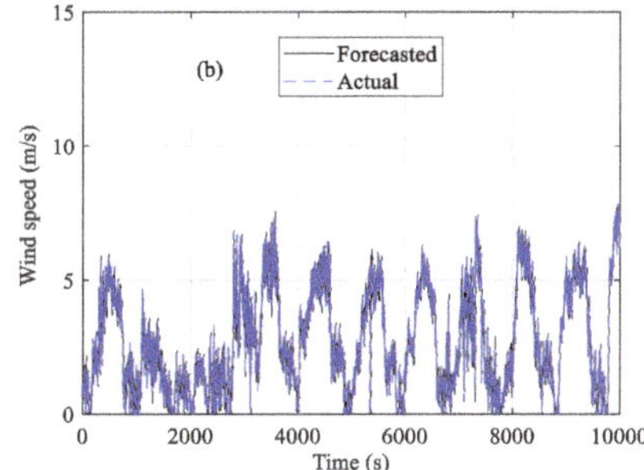

Figure 9. *Cont.*

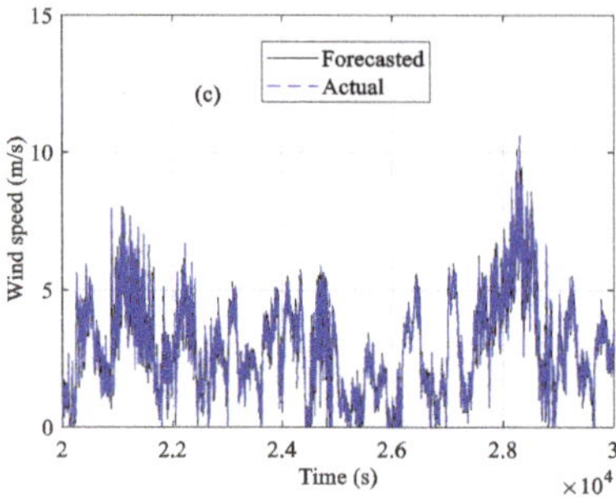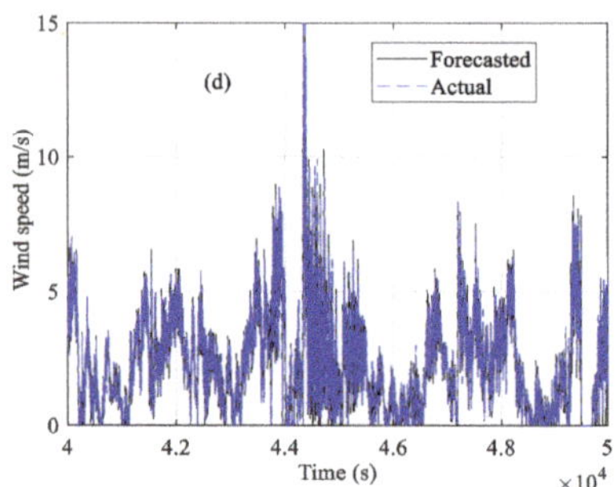

Figure 9. Forecasted and actual series of the test set referring to the 1-min-ahead prediction horizon for all the examined time period (**a**), between 0 and 10,000 s (**b**), between 20,000 and 30,000 s (**c**), and between 40,000 and 50,000 s (**d**).

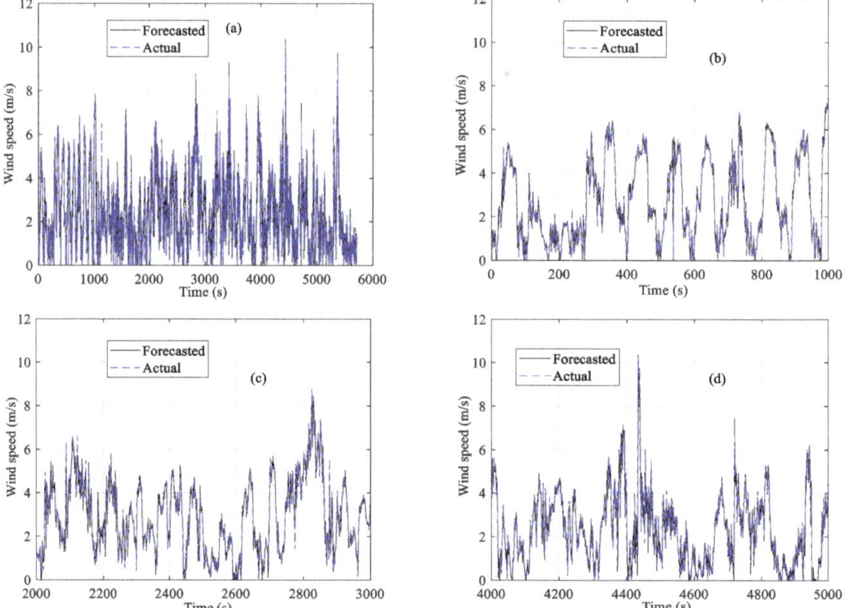

Figure 10. Forecasted and actual series of the test set referring to the 10 min ahead prediction horizon for all the examined time period (**a**), between 0 and 1000 s (**b**), between 2000 and 3000 s (**c**), and between 4000 and 5000 s (**d**).

3.2. Comparison with Other Forecasting Models

To fully evaluate the proposed model, a comparison is made with the following models; Group Method of Data Handling Neural Network (GMDHNN) [27], Regression Neural Network (GRNN) [28], Regression Trees (RTs) [29], Relevance Vector Machine (RVM) [30], and Support Vector Regression (SVR) [31]. Tables 3 and 4 present the scores of GMDHNN, GRNN and RTs, RVM and SVR, respectively, on the error metrics, for the 1-min-ahead predictions. Correspondingly, Tables 5 and 6 present the scores of GMDHNN, GRNN and RTs, RVM and SVR, respectively, on the error metrics MAE, RMSE, and MARNE for 10-min-ahead predictions. Among these models, SVR and GMDHNN display comparative results with the hybrid model. The latter outperforms all the other models. It can be noticed that GRNN and RTs result in high errors and thus, for the problem under study are not recommended.

Table 3. Evaluation of Group Method of Data Handling (GMDH), Regression Neural Network (GRNN), and Regression Trees (RTs) considering the 1-min-ahead horizon.

	MAE			RMSE			MARNE (%)		
	GMDHNN	GRNN	RTs	GMDHNN	GRNN	RTs	GMDHNN	GRNN	RTs
Case#1	0.3688	0.3846	0.4432	0.5689	0.5721	0.6331	2.3411	2.3869	3.1256
Case#2	0.3701	0.3910	0.4509	0.7051	0.7644	0.8109	2.4578	2.4771	3.2672
Case#3	0.3698	0.4256	0.4781	0.7189	0.7367	0.7992	2.5871	2.6225	3.4155
Case#4	0.3944	0.4201	0.4541	0.8902	0.9012	1.1091	2.7112	2.9904	3.6203
Case#5	0.1094	0.1388	0.1692	0.1481	0.1556	0.2012	0.6289	0.6552	0.7154

Table 4. Evaluation of Relevance Vector Machine (RVM) and Support Vector Regression (SVR) considering the 1-min-ahead horizon.

	MAE		RMSE		MARNE (%)	
	RVM	SVR	RVM	SVR	RVM	SVR
Case#1	0.3721	0.3704	0.5614	0.5377	2.3482	2.3345
Case#2	0.3933	0.3865	0.7597	0.7029	2.4754	2.3419
Case#3	0.3885	0.3893	0.7408	0.6911	2.6172	2.5678
Case#4	0.3821	0.3783	0.8928	0.8709	2.9232	2.6213
Case#5	0.1277	0.1178	0.1542	0.1421	0.6524	0.5772

Table 5. Evaluation of GMDH, GRNN, and RTs considering the 10-min-ahead horizon.

	MAE			RMSE			MARNE (%)		
	GMDHNN	GRNN	RTs	GMDHNN	GRNN	RTs	GMDHNN	GRNN	RTs
Case#1	0.4292	0.4525	0.5898	0.6258	0.6553	0.8093	4.2051	4.4326	5.7812
Case#2	0.4302	0.5597	0.5901	0.6272	0.6715	0.8095	4.2149	4.5031	5.7812
Case#3	0.4704	0.6927	0.6157	0.6862	0.7342	0.8461	4.6334	4.7862	6.0319
Case#4	0.4601	0.6056	0.6108	0.6715	0.6989	0.8012	4.5130	4.6902	5.7568
Case#5	0.1238	0.1624	0.1799	0.1601	0.1833	0.2109	1.2671	1.4884	1.6117

Table 6. Evaluation of RVM and SVR considering the 10-min-ahead horizon.

	MAE		RMSE		MARNE (%)	
	RVM	SVR	RVM	SVR	RVM	SVR
Case#1	0.4342	0.4298	0.6359	0.6268	4.2541	4.2206
Case#2	0.4351	0.4307	0.6440	0.6268	4.2630	4.2198
Case#3	0.5370	0.4681	0.8649	0.7529	5.2605	4.5365
Case#4	0.4988	0.4644	0.7898	0.7459	5.1133	4.5301
Case#5	0.1424	0.1308	0.2037	0.1654	1.5893	1.1276

4. Discussion and Concluding Remarks

Offshore wind turbine installations are continually gathering the research interest since they are considered an efficient mechanism for covering the electrical needs of various isolated loads. The present study emphasizes on the development of an effective method for very short-term wind speed forecasting under the limitation of wind speed series that do not present consistency in time, i.e., there are interruptions in the date sequence. Real measured data are used for the training of the developed method. An ensemble data-driven short-term wind speed forecasting model is developed, tested, and applied. The term "ensemble" refers to the combination of two different predictors that run in parallel and the prediction is obtained by the predictor that leads to the lowest error. The proposed model utilizes the wavelet transform and is compared with other models that have been presented in the related literature. The main conclusions of the present study:

- The proposed forecasting model can be used effectively for 1 min and 10 min ahead horizon wind speed predictions.
- The exogenous variables (i.e., wind speed direction and air temperature) decrease the prediction accuracy. The best results are obtained using the DWT.
- The highest errors are met on winter days and especially in instances with high wind speed.
- There is no correlation among the forecasting error and the wind direction.
- The hybrid model (combination of FFNN and ANFIS) leads to better forecasts in all examined data set cases.
- The proposed model outperforms the accuracy of other forecasting models that have been presented in the related literature.

The research of the present paper will be further expanding by checking the implementation of the forecasting problem in the Wind Farm Layout Optimization (WLFO) problem incorporating wake effects with the use of specific mathematical or numerical models. Forecasted wind speed time series can serve as inputs to the problem. By estimating future wind speed values, the WLFO can be modified to a scenario-based problem where different wind speed forecasts can lead to various WLFO solutions and thus, assessing the level of influence of the future wind speed variations in the outputs of the optimization problem.

Author Contributions: I.P.P. performed the research and wrote the first draft of the paper. C.M. and D.C.A. set the objectives of the research, provided guidance for the research, and revised the paper.

Funding: This research received no external funding.

Conflicts of Interest: The authors declare no conflicts of interest.

References

1. WindEurope. *Offshore Wind in Europe Key Trends and Statistics 2018*. Available online: https://windeurope.org/wp-content/uploads/files/about-wind/statistics/WindEurope-Annual-Offshore-Statistics-2018.pdf (accessed on 27 March 2019).
2. Snyder, B.; Kaiser, M.J. A comparison of offshore wind power development in Europe and the US: Patterns and drivers of development. *Appl. Energy* **2009**, *86*, 1845–1856. [CrossRef]
3. Soman, S.S.; Zareipour, H.; Malik, O.; Mandal, P. A review of wind power and wind speed forecasting methods with different time horizons. In Proceedings of the 2010 North American Power Symposium, Arlington, TX, USA, 26–28 September 2010; pp. 1–8.
4. Jung, J.; Broadwater, R.P. Current status and future advances for wind speed and power forecasting. *Renew. Sustain. Energy Rev.* **2014**, *31*, 762–777. [CrossRef]
5. Okumus, I.; Dinler, A. Current status of wind energy forecasting and a hybrid method for hourly predictions. *Energy Conves. Manag.* **2016**, *123*, 362–371. [CrossRef]
6. Lei, M.; Shiyan, L.; Chuanwen, J.; Hongling, L.; Ya, Z. A review on the forecasting of wind speed and generated power. *Renew. Sustain. Energy Rev.* **2009**, *13*, 915–920. [CrossRef]
7. Chang, W.Y. A literature review of wind forecasting methods. *J. Power Energy Eng.* **2014**, *2*, 161–168. [CrossRef]
8. Foley, A.M.; Leahy, P.G.; Marvuglia, A.; McKeogh, E.J. Current methods and advances in forecasting of wind power generation. *Renew. Energy* **2012**, *37*, 1–8. [CrossRef]
9. Bhaskar, M.; Jain, A.; Srinath, N.V. Wind speed forecasting: Present status. In Proceedings of the 2010 International Conference on Power System Technology, Hangzhou, China, 24–28 October 2010; pp. 1–7.
10. Murthal, D. Wind power forecasting: A survey. *Int. J. Eng. Res. Gen. Sci.* **2016**, *4*, 802–806.
11. Tascikaraoglu, A.; Uzunoglu, M. A review of combined approaches for prediction of short-term wind speed and power. *Renew. Sustain. Energy Rev.* **2014**, *34*, 234–254. [CrossRef]

12. Sfetsos, A. A novel approach for the forecasting of mean hourly wind speed time series. *Renew. Energy* **2002**, *27*, 163–174. [CrossRef]

13. Başaran, Ü.; Filik, T. Wind speed prediction using artificial neural networks based on multiple local measurements in Eskisehir. *Energy Procedia* **2017**, *107*, 264–269.

14. More, A.; Deo, M.C. Forecasting Wind with Neural Networks. *Mar. Struct.* **2003**, *16*, 35–49. [CrossRef]

15. Li, G.; Shi, J. On comparing three artificial neural networks for wind speed forecasting. *Appl. Energy* **2010**, *87*, 2313–2320. [CrossRef]

16. Zeng, J.W.; Qiao, W. Support vector machine-based short-term wind power forecasting. In Proceedings of the 2011 IEEE/PES Power Systems Conference and Exposition, Phoenix, AZ, USA, 20–23 March 2011; pp. 1–8.

17. Zhou, J.Y.; Shi, J.; Li, G. Fine tuning support vector machines for short-term wind speed forecasting. *Energy Convers. Manag.* **2011**, *52*, 1990–1998. [CrossRef]

18. Fazelpour, F.; Tarashkar, N.; Rosen, M.A. Short-term wind speed forecasting using artificial neural networks for Tehran. *Iran. Int. J. Energy Environ. Eng.* **2016**, *7*, 377–390. [CrossRef]

19. Fortuna, L.; Guariso, G.; Nunnari, S. One day ahead prediction of wind speed class by statistical models. *Int. J. Renew. Energy Res.* **2016**, *6*, 1137–1145.

20. Fortuna, L.; Nunnari, S.; Guariso, G. Fractal order evidences in wind speed time series. In Proceedings of the 2014 International Conference on Fractional Differentiation and Its Applications, Catania, Italy, 23–25 June 2014; pp. 1–6.

21. Mallat, S. A theory for multiresolution signal decomposition-the wavelet representation. *IEEE Trans. Pattern Anal. Mach. Intell.* **1989**, *11*, 674–693. [CrossRef]

22. Amjady, N.; Keynia, F. Short-term load forecasting of power systems by combination of wavelet transform and neuro-evolutionary algorithm. *Energy* **2009**, *34*, 46–57. [CrossRef]

23. Jang, J.S.R. ANFIS: Adaptive-network-based fuzzy inference system. *IEEE Trans. Syst. Man Cybern.* **1993**, *23*, 665–685. [CrossRef]

24. Graupe, P. *Principles of Artificial Neural Networks*, 2nd ed.; World Scientific: Singapore, 2007.

25. Levenberg, K. A method for the solution of certain non-linear problems in least squares. *Q. J. Appl. Math.* **1944**, *2*, 164–168. [CrossRef]

26. Michailides, C.; Loukogeorgaki, E.; Angelides, D.C. Monitoring the response of connected moored floating modules. In Proceedings of the Twenty-third International Offshore and Polar Engineering, Anchorage, AK, USA, 30 June–5 July 2013; pp. 869–876, ISBN 978-1-880653-99-9.

27. Abdel-Aal, R.E.; Elhadidy, M.A.; Shaahid, S.M. Modeling and forecasting the mean hourly wind speed time series using GMDH-based abductive networks. *Renew. Energy* **2009**, *34*, 1686–1699. [CrossRef]

28. Kumar, G.; Malik, H. Generalized regression neural network based wind speed prediction model for western region of India. *Procedia Comput. Sci.* **2016**, *93*, 26–32. [CrossRef]

29. Troncoso, A.; Salcedo-Sanz, S.; Casanova-Mateo, C.; Riquelme, J.C.; Prieto, L. Local models-based regression trees for very short-term wind speed prediction. *Renew. Energy* **2015**, *81*, 589–598. [CrossRef]

30. Sun, G.; Chen, Y.; Wei, Z.; Li, X.; Cheung, K.W. Day-ahead wind speed forecasting using relevance vector machine. *J. Appl. Math.* **2014**, *2014*, 1–6. [CrossRef]

31. Botha, N.; van der Walt, C.M. Forecasting wind speed using support vector regression and feature selection. In Proceedings of the 2017 Pattern Recognition Association of South Africa and Robotics and Mechatronics, Bloemfontein, South Africa, 29 November–1 December 2017; pp. 181–186.

 *electronics*

Article

Multivariate Temporal Convolutional Network: A Deep Neural Networks Approach for Multivariate Time Series Forecasting

Renzhuo Wan [1], Shuping Mei [1], Jun Wang [1], Min Liu [2] and Fan Yang [1,*]

[1] Nano-Optical Material and Storage Device Research Center, School of Electronic and Electrical Engineering, Wuhan Textile University, Wuhan 430200, China

[2] State Key Laboratory of Powder Metallurgy, School of Physics and Electronics, Central South University, Changsha 410083, China

* Correspondence: yangfan@wtu.edu.cn

Received: 7 July 2019; Accepted: 5 August 2019; Published: 7 August 2019

Abstract: Multivariable time series prediction has been widely studied in power energy, aerology, meteorology, finance, transportation, etc. Traditional modeling methods have complex patterns and are inefficient to capture long-term multivariate dependencies of data for desired forecasting accuracy. To address such concerns, various deep learning models based on Recurrent Neural Network (RNN) and Convolutional Neural Network (CNN) methods are proposed. To improve the prediction accuracy and minimize the multivariate time series data dependence for aperiodic data, in this article, Beijing PM2.5 and ISO-NE Dataset are analyzed by a novel Multivariate Temporal Convolution Network (M-TCN) model. In this model, multi-variable time series prediction is constructed as a sequence-to-sequence scenario for non-periodic datasets. The multichannel residual blocks in parallel with asymmetric structure based on deep convolution neural network is proposed. The results are compared with rich competitive algorithms of long short term memory (LSTM), convolutional LSTM (ConvLSTM), Temporal Convolution Network (TCN) and Multivariate Attention LSTM-FCN (MALSTM-FCN), which indicate significant improvement of prediction accuracy, robust and generalization of our model.

Keywords: deep learning; multivariate time series forecasting; multivariate temporal convolutional network

1. Introduction

With the explosive growth of Internet of Things (IoT) applications and big data, multivariate time series is becoming ubiquitous in many fields, e.g., aerology [1], meteorology [2], environment [3], multimedia [4], power energy [5], finance [6], and transportation [7]. The precise trend forecasting, as well as for potential hazardous events, based on historical dynamical data are a major challenge, especially for aperiodic multivariate time series. One of the crucial reasons is aperiodic and nonlinearity among variables, which is incapable by models to capture and have self-adaption of the complex data features. Traditional methods such as Autoregressive (AR) [8] models and Gaussian Process (GP) [9] may fail. As an important part of the field of artificial intelligence, deep neural networks (DNNs) provide state-of-the-art accuracy on many tasks [10] and has been developed intensively in natural language processing (NLP), computer vision (CV), time series classifications and time series forecasting.

Enlightened by algorithms used in NLP (i.e., Sequence to Sequence [11,12] and Attention mechanism) and CV (i.e., Dilated convolution network [13] and residual structure [14]), in this paper, the M-TCN model is proposed for aperiodic multivariate time-series prediction, which constructs the aperiodic data as sequence-to-sequence and a novel multichannel and asymmetric residual blocks network. The model is cross validated by a rich set of existing competitive models with an aperiodic time series dataset. The reminder of the article is organized as follows: Section 2 reviews the background work. Section 3 presents the methodology of the proposed model. In Section 4, the experiment is analyzed and discussed. Finally, conclusions and outlook are drawn in Section 5.

2. Background

One of the major challenges of multivariate time series forecasting is nonlinearity and aperiodic of data originated by short-term and long-term dynamical behavior. Various models have been established based on classical statistic methods or machine learning algorithms.

The prominent classical univariate time series model is Autoregressive (AR) with classical statistic algorithms, as well as its progeny. The AR method is well used to stationary time series. The improved models, such as autoregressive integrated moving average (ARIMA) [15], autoregressive moving average (ARMA) [16], and vector auto-regression (VAR) [17], were developed by including flexible exponential smoothing techniques. However, for long-term temporal patterns, these models are inevitably prone to overfitting and high computational cost, especially for high-dimensional inputs.

Alternative methods by treating the time series forecasting problems as general regression with time-varying parameters were applied by machine learning models, e.g., linear support vector regression (SVR) [18], random forest [19], ridge regression [20] and LASSO [21] models. Those models are practically more efficient due to high quality off-the-shelf solutions in machine learning community. Still, machine learning based models may be incapable of including complex nonlinearity dependences of multivariate large datasets.

Meanwhile, the well-built deep neural networks of Convolutional Neural Networks (CNNs) and Recurrent Neural Networks (RNNs) have been widely applied in time series forecasting, which are attributed to the open source deep learning frameworks, such as Keras (Keras, available online: https://keras.io), TensorFlow (TensorFlow, available online: https://tensorflow.org) and PyTorch (PyTorch, available online: https://pytorch.org), including flexible and sophisticated mathematical libraries. Some representative models are long short-term memory (LSTM) [22] and its inheritors, convolutional LSTM (ConvLSTM) [23] and Multivariate Attention LSTM-FCN (MALSTM-FCN) [24], which overcome the challenges involved in training a recurrent neural network for a mixture of long and short-term horizons. However, these models are time consuming and non-robust for aperiodic data forecasting.

Another novel method for time-series forecasting is a hybrid multiscale approach, such as empirical mode decomposition (EMD) [25], ensemble EMD (EEMD) [26], multi-level wavelet decomposition network (mWDN) [27] and variational mode decomposition (VMD) [28]. These methods are used to decompose data into different frequencies' components to facilitate forecasting. However, the pre-design decomposition K value is an essential prerequisite as an input of training models, which is not versatile for complicated multivariate time series prediction.

Recently, a general architecture for a predictive sequences model by convolutional and recurrent architecture on sequence modeling tasks, the Temporal Convolution Network (TCN) [29], is proposed. The prominent characteristics of TCNs are casualness in convolution architecture design and sequence length. In addition, it is also convenient to build a very deep and wide network by a combination of residual network and extended convolution. Under this background, our model is designed based on TCN and tested for PM2.5 and electric power forecasting.

For comparison, Table 1 contrasts the advantages and challenges of some common methods for multivariate time series prediction.

Table 1. Summary of advantages and challenges of time series prediction methods.

Method	Advantages	Challenges
AUTOREGRESSIVE [8]	Simple and efficient for lower order models	Nonlinear, multivariable and non-stationary
SVR [18]	Nonlinear and high-dimensional	Selection of free parameters, NOT suitable for big data
Hybrid VMD and ANN [30]	Strong explanatory power of mathematics	Pre-processing is complex, poor generalization ability
LSTM [22]	mixture of long- and short-term memory	Huge computing resource
TCN [29]	Large scale parallel computing mitigating the gradient of explosion and greater flexibility in model structure	Long-term memory

3. Methodology

In this section, the time series forecasting problem is formulated first. In addition, then the baseline models, ConvLSTM and Multivariate LSTM FCN are presented to be used as the methods in our comparative evaluation. Finally, M-TCN model is introduced.

3.1. Sequence Problem Statement

From the nature of machine learning, to minimize the expected error, it requires obtaining an ideal nonlinear mapping from a historical dataset to a current state, especially for hazard events forecasting. The prerequisite is to employ enough characteristic parameters to feature the various phenomena, which makes the current state strictly dependent on the historical dataset. The problem of multivariable time series prediction is defined as the problem of sequence to sequence in this paper. Before defining the network structure, more formally, given an input sequence time series signal $X = (x_1, x_2, \cdots, x_T)$ with $x_t \in \mathbb{R}^n$, where n is the variable dimension, we aim at predicting corresponding outputs $Y = (y_1, y_2, \cdots, y_h)$ at each time. The target of sequence modeling network is to obtain a nonlinear mapping to the prediction sequence from the current state as:

$$(y_1, y_2, \cdots, y_h) = f(x_1, x_2, \cdots, x_T). \tag{1}$$

3.2. Baseline Test

To build a baseline test benchmark, the traditional models, naive forecast, average approach forecast and seasonal persistent forecast models are included for a cross evaluation.

Naive forecast model: It takes the value from the last hour prior to the forecast period (e.g., 24 h) and uses it as the value of a dataset for each hour in the forecast period (e.g., 1 to 24 h). Using the naive approach, forecasts are produced that are equal to the last observed value. This model is defined as:

$$\hat{y}_{T+1} = y_T, \tag{2}$$

where y_T is the past data, and $\hat{y}_{T+1}$ is the next time value.

Average approach forecast model: In this model, the predictions of all future values are equal to the mean of the past data. This method can be used for any type of data available in the past and defined as:

$$\hat{y}_{T+1} = \bar{y} = (y_1 + \ldots + y_T)/T, \tag{3}$$

where $(y_1, y_2, \cdots, y_T)$ is the past data, and $\hat{y}_{T+1}$ is the next time predicted value.

Seasonal persistent forecast model: It defines the same time period a year ago as the predicted value. This method accounts for seasonality by setting each prediction to be equal to the last observed value of the same season. This model is defined as:

$$\hat{y}_{T+1} = y_{T-Y}, \tag{4}$$

where y_{T-Y} is the past data, and $\hat{y}_{T+1}$ is the next time predicted value.

3.3. ConvLSTM Encoder–Decoder Model

A convolutional LSTM (ConvLSTM) encoder–decoder network is built in this work, which reconstructs the input sequence and predicts the future sequence simultaneously. The ConvLSTM input layer is designed to be a 4D tensor [*timestep, row, column, channel*], where *timestep* is the number of subsequences, *row* is the one-dimensional shape of each subsequence, *column* is the hours in each subsequence and *channel* is the features that we are working with as input. The encoding ConvLSTM compresses the whole input sequence into a hidden state tensor and the decoding LSTM unfolds this hidden state to give the final prediction. An overview of the ConvLSTM is shown in Figure 1.

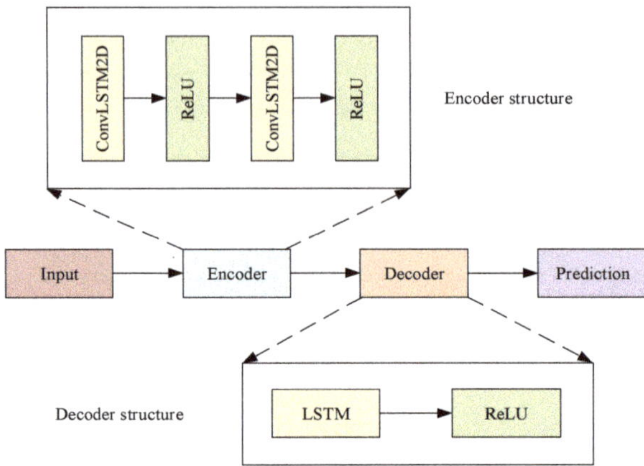

Figure 1. An overview of the ConvLSTM Encoder–Decoder network (ConvLSTM).

Multivariate ALSTM Fully Convolutional Networks models are comprised of temporal convolutional blocks and an LSTM block, as depicted in Figure 2. The feature extractor consists of three stacked temporal convolutional blocks. In addition, the first two convolutional blocks conclude with a squeeze and excite block.

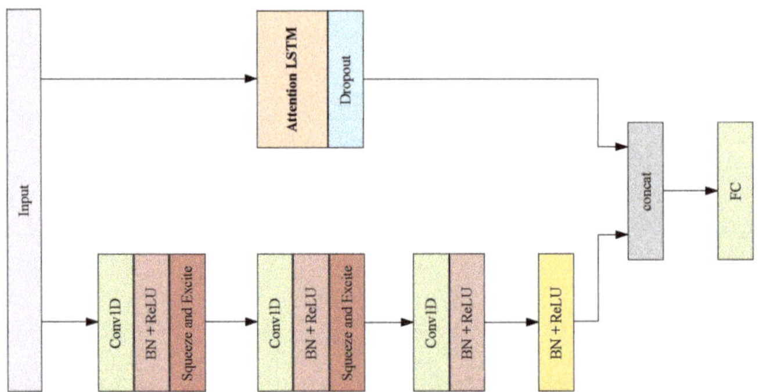

Figure 2. Modified multivariate attention LSTM-FCN (MALSTM-FCN) network structure for time series forecasting.

We consider this model structure as a parallel structure of CNN (temporal convolutional blocks) and RNN (LSTM block). In order to study the regression problem, the final softmax layer used for classification is changed to a fully connected layer with 24 nodes.

3.4. M-TCN Model

The main characteristic of CNN is a local feature by convolving filters. For time series forecasting, the local correlation is reflected in the continuous change over a period of time within a small time slot. In addition, RNN models, such as LSTM, have always been considered as the best standard method to solve sequence problems; however, RNNs cannot be parallel, resulting in huge time-consumption compared to that of CNN. From those considerations, the overall framework of the model is designed based on CNN. Our aim is to distill the best practices in designing convolutional networks to be flexible and stable frameworks with a simple architecture and high efficiency for multivariate time series forecasting. The distinguishing characteristics of M-TCN are: (1) the input and output lengths of our network could be determined to be flexible for various scenarios; (2) M-TCN uses the 1D convolution instead of causal convolutions; (3) M-TCN augmented with two different asymmetric residual blocks; (4) M-TCN constructs a sub-model for each feature of input data, and the prediction is accomplished by a combination of all sub-models. We call this typical structure a multihead model. In this work, what we emphasize is the methodology on how to build effective networks (i.e., Multihead model) using a combination of network (augmented with two different residual blocks) and dilated convolutions. The following are details of the network structure.

3.4.1. 1D Convolutions

TCN uses causal convolutions, where an output at time t is convolved only with elements from time t and earlier in the previous layer. In Figure 3, causal convolution is used to assume that all data must have a one-to-one causal relationship in chronological order. Given an input sequence time series signal $X = (x_1, x_2, x_3, x_4, x_5)$ with $x_t \in \mathbb{R}^n$ where n is the variable dimension, x_t does not strict causality in chronological order. While x_1 and x_5 may have a direct logical connection, causal convolution will make the relationship between x_1 and x_5 affected by x_2, x_3, x_4. This design was limited by the absolute order of time-series and inefficient for accurate characteristics learning at a relative time. Thus, in our model, only a 1D convolutional network is adopted to avoid this situation.

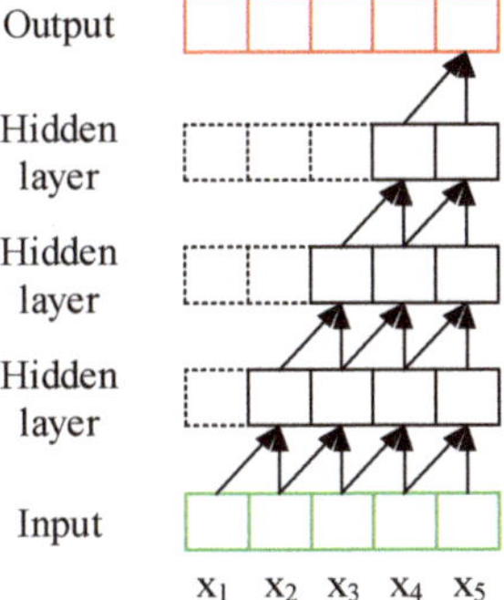

Figure 3. Visualization of a stack of causal convolutional layers.

3.4.2. Dilated Convolutions

The dilated convolutions algorithm [13] is used in our model. Since the traditional convolution operation process is to convolute the sequence once and then pool, which reduces the size of the sequence and enlarges the receptive field at the same time. One of the main faults is that some sequential information will be lost during the pooling process, while the advantage of dilated convolutions is that they don't need the pooling process and gradually increase the field of perception through a series of dilated convolutions, thus leading to the output of each convolution encompasses rich information for long-term tracking. Thus, the dilated convolutions could be well applied in the problem of long information dependence of sequence, such as voice and signal processing, environment forecasting, etc. Dilated convolution is defined as

$$F(s) = (\mathbf{x} *_d f)(s) = \sum_{i=0}^{k-1} f(i) \cdot \mathbf{x}_{s-d \cdot i}, \tag{5}$$

where d is the dilation factor, k is the filter size, and $s - d \cdot i$ accounts for the direction of the past. A filter $f : \{0, \ldots, k-1\} \rightarrow \mathbb{N}$. Figure 4 depicts dilated 1D convolutions for dilations 1, 2 and 4.

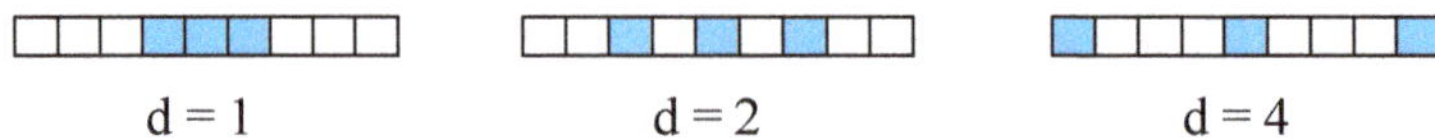

Figure 4. Visualization of 1D convolutions with different dilation factors.

3.4.3. Residual Block

A novel structure is designed by a multilayer and sequential residual network and parallel residual blocks. The core of ResNet [14] is to create a shortcut for information dissemination in front and back layers. A basic Residual block is used in the TCN network; however, the jump connection in ResNet, resulting in only a small number of residual blocks' learning useful information, and thus the basic residual block structure is not adapted for time series prediction. An alternative way is to increase the convolution kernel size for a better prediction; however, the computational load increases sharply. In [31], an asymmetric block structures were introduced both for MobileNetV3-Large and MobileNetV3-Small. By this way, asymmetric factors will be generated in the whole network structure and may make a positive impact on the in-depth learning models. The optimal asymmetric structure needs Neural Architecture Search(NAS) [32,33]; however, it is computationally expensive. In a more direct way, two asymmetric residual blocks in parallel are constructed. The architectural elements in our model are shown in Figure 5.

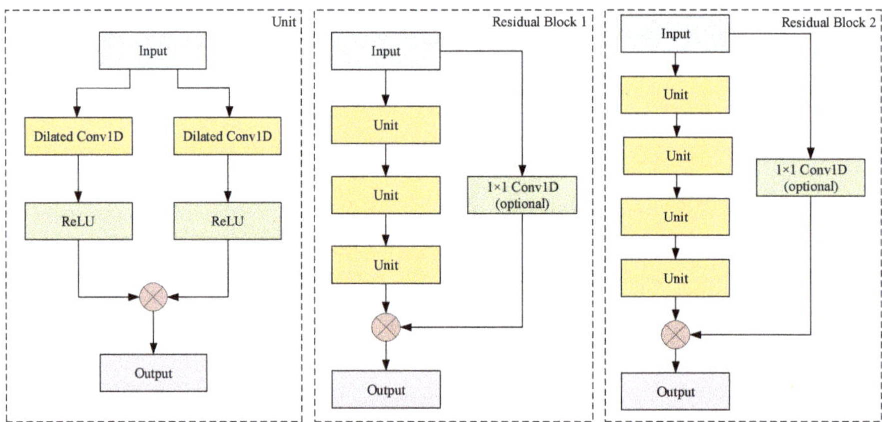

Figure 5. Residual Block in our network. (**left**) details of the Unit architecture. (**middle**) Residual Block 1; (**right**) Residual Block 2.

The Unit for our model is shown in Figure 5 (left). The Unit has two channels. Each channel has dilated convolution and nonlinearity, for which we used the rectified linear unit (ReLU) [34]. The residual block 1 is shown in Figure 5 (middle). Within a residual block, the model has three units. The output is the sum of the results of two channel operations. The residual block 2 is shown in Figure 5 (right), which has the same basic structure as residual block 1, but one more unit layer is implemented. To be more precise, a dilated convolution with different dilation factors and filter size k = 3 are constructed both for residual blocks. In addition, an optional 1 × 1 convolution is introduced to adjust the dimensions of different feature maps (see Figure 5 (middle, right)) for summation.

The Unit takes the same input with two different convolutions, and then adds up the results. The convolutional layer consists of multiple kernels with different sizes. The k-th filter sweeps through the input data X, which can be formulated as:

$$\mathrm{ReLU}(x) = \max(0, x), \tag{6}$$

$$h_{1k} = \text{ReLU}\left(W_k * X + b_k\right), \tag{7}$$

$$h_{2k} = \text{ReLU}\left(W_k * X + b_k\right), \tag{8}$$

$$h_k = h_{1k} + h_{2k}, \tag{9}$$

where h_{1k} is the result of channel 1, h_{2k} is the result of channel 2, and h_k is result of unit. * stands for a convolutional operation.

A residual block contains a channel, which passes through a series of conversion functions $\mathcal{F}$, and the final output is added to the input X of the block:

$$o = (\mathbf{x} + \mathcal{F}(\mathbf{x})). \tag{10}$$

3.4.4. Fully Connected Layers

Fully connected layers can be replaced by global average pooling (GAP) for better efficiency and accuracy in image recognition tasks. However, fully connected layers are essential in prediction tasks and can easily change the length of the output sequence. Formally, a statistic $z \in R^C$ is generated by shrinking X through its spatial dimensions $H \times W$, such that the output z is calculated by:

$$z = GAP\left(x_c\right) = \frac{1}{H \times W} \sum_{i=1}^{H} \sum_{j=1}^{W} x_c(i, j). \tag{11}$$

The whole spatial feature on a channel is averaged as a global feature. Each feature map is averaged into one value, thus the local information of the whole feature value is lost, which has a negative impact on the prediction problem.

The full connection layer is shown in Figure 6, which not only establishes the position relationship between feature maps, but also retains the internal feature information of the same feature map. This will have a beneficial impact on the prediction problem. The disadvantage is that the parameters are greatly increased.

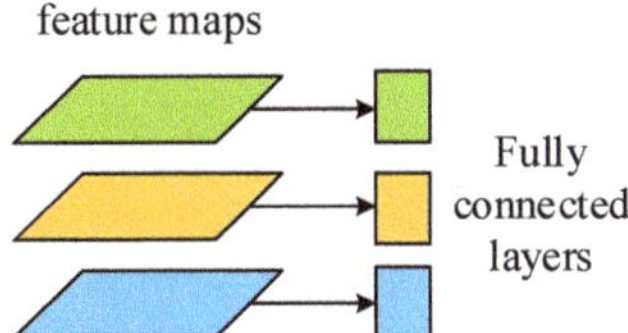

Figure 6. Relation between full connection layers and feature maps.

3.4.5. Multi-Head Model

The model is further extended so that each input variable has a separate sub-model, named after a multi-headed model. This sub-model for each input variable has to be defined first. Each sub-model learns the information with different features in the sequence separately. In addition, the outputs of those models are then combined in series to form a very long vector, which is interpreted by some fully connected layers before the prediction is made. An overview of multi-head temporal convolutional network (M-TCN) architecture is shown in Figure 7. To provide more detail, an overview of convolutions is shown in Figure 8.

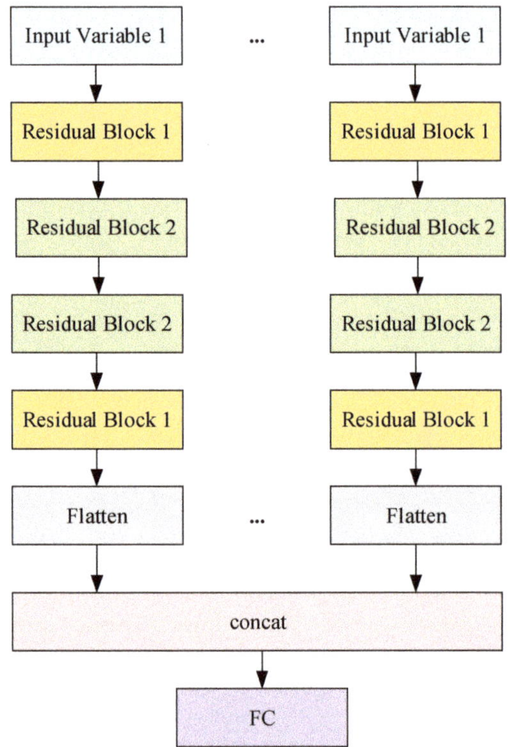

Figure 7. An overview of the M-TCN network.

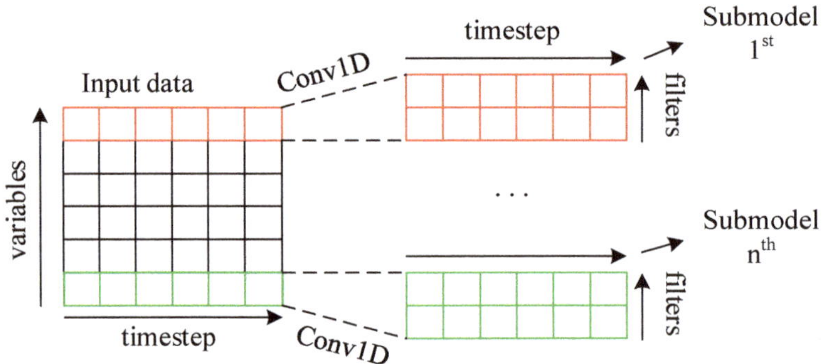

Figure 8. An overview of convolutions.

3.5. Training Procedure

The training procedure can be described as Algorithm 1.

Meaning represented by each parameter. min_lr: minimum learning rate; $initial\_lr$: initial learning rate; factor: factor by which the learning rate will be reduced; wait: number of epochs with no improvement after which learning rate will be reduced; new_lr: new learning rate; epoch: number of epochs to train the model; $best\_score$: minimum RMSE.

Algorithm 1: Training procedure.

1: min_lr = 1e-4; epoch = 200; $initial\_lr$ = $initial\_lr$
2: factor
3: **for** n $<$ epoch **do**
4:    wait += 1
5:    if $best\_score$ > RMSE
6:        $best\_score$ = RMSE
7:        save model
8:        if wait >= 10
9:            if $initial\_lr$ > min_lr
10:               min_lr = $initial\_lr$ × factor
11:               new_lr = max(new_lr, min_lr)
12:               wait = 0

---

4. Experiments

In this section, we first describe two datasets for empirical studies. All of the data are available online. Then, the parameter settings of model and evaluation metrics are introduced in our studies. Finally, the proposed M-TCN model against different baseline models is compared.

4.1. Datasets

Two benchmark datasets are used which are publicly available. Table 2 summarizes the corpus statistics.

Beijing PM2.5 Dataset (available online: https://archive.ics.uci.edu/ml/datasets/Beijing+PM2.5+Data): It contains hourly PM2.5 data and the associated meteorological data in Beijing, China. The exogenous time series include dew point, temperature, and atmospheric pressure, combined wind direction, cumulated wind speed, hours of snow, and hours of rain. In total, we have 43,824 multivariable sequences. For this dataset, the hourly PM2.5 data are used as a predictive value.

ISO-NE Dataset (available online: https://www.iso-ne.com/isoexpress/web/reports/load-and-demand): The time range of the dataset is between March 2003 and December 2014. The ISO-NE Dataset includes hourly demand, prices, weather data and system load. The dataset contains two variables, which are hourly electricity demand in MW and dry-bulb temperature in °F. For this dataset, the hourly electricity demand is used as a predictive value.

Table 2. Dataset statistics.

Datasets	Length of Time Series	Total Number of Variables	Sample Rate
ISO-NE	103,776	2	1 h
Beijing PM2.5	43,824	8	1 h

In our experiments, ISO-NE datasets have been split into training set (from 1 March 2003 to 31 December 2012), valid set (the whole year of 2013) and test set (the whole year of 2014) in a chronological order. In addition, the Beijing PM2.5 Dataset has been split into a training set (from January 2, 2010 to December 31, 2012), valid set (the whole year of 2013) and test set (the whole year of 2014) in a chronological order.

4.2. Data Processing

According to the characteristics of each dataset, it is necessary to preprocess the data. Each of the datasets is normalized with a mean of 0 and a standard deviation of 1.

For the Beijing PM2.5 Dataset, PM2.5 is NA in the first 24 h. We will, therefore, need to remove the first row of data. There are also a few scattered "NA" values later in the dataset, and we use zero to fill in missing values. The wind speed feature is label encoded (integer encoded). We apply the new dataset to every algorithm in later experiments.

4.3. Evaluation Criteria

Three evaluation metrics, root mean squared error (RMSE), root relative squared error (RRSE) and empirical correlation coefficient (CORR) for multivariate forecasting, are used and defined as:

$$\text{RMSE} = \sqrt{\frac{1}{N} \sum_{i=1}^{N} \left(y_t^i - \hat{y}_t^i \right)^2}, \tag{12}$$

$$\text{RRSE} = \frac{\sqrt{\sum_{(i,t)\in\Omega_{Test}} (Y_{it} - \hat{Y}_{it})^2}}{\sqrt{\sum_{(i,t)\in\Omega_{Test}} (Y_{it} - \text{mean}(Y))^2}}, \tag{13}$$

$$\text{CORR} = \frac{1}{n} \sum_{i=1}^{n} \frac{\sum_t (Y_{it} - \text{mean}(Y_i))(\hat{Y}_{it} - \text{mean}(\hat{Y}_i))}{\sqrt{\sum_t (Y_{it} - \text{mean}(Y_i))^2 (\hat{Y}_{it} - \text{mean}(\hat{Y}_i))^2}}, \tag{14}$$

where $Y, \hat{Y} \in \mathbb{R}^{n \times T}$ are ground value and system prediction value, respectively, and Ω_{Test} is the set of time stamps used for testing. For RMSE and RRSE, the lower value is better, while, for CORR, the higher value is better for evaluation.

4.4. Walk-Forward Validation

In the test set, the Walk-Forward Validation method is adopted, but the model is not updated. In this case, a model is needed to predict a period of time, and then the actual data of the current period is provided to the model, so that it can be used as the basis for the prediction of subsequent periods. This is not only applicable to the way the model is used in practice, but also conducive to the model using the best available data.

In the experiment, the output length is set to 24. For multi-step prediction problems, we evaluate each prediction time step separately. Table 3 summarizes the actual value and predicted value. Models can be trained and evaluated as follows.

Step 1: Starting at the beginning of the test set, the last set of observations in the training set is used as input of the model to predict the next set of data (the first set of true values in the validation set).

Step 2: The model makes a prediction for the next time step.
Step 3: Get real observation and add to history for predicting the next time.
Step 4: The prediction is stored and evaluated against the real observation.
Step 5: Go to step 1.

Table 3. Dataset Statistics, where h is hour, d is day.

Input (Actual Value)	Output (Predicted Value)
Current 24 h	Next, 24 h
1d 1 h–1 d 24 h	2 d 1 h–2 d 24 h
2d 1 h–1 d 24 h	3 d 1 h–2 d 24 h
...	...

4.5. Experimental Details

To be more specific, most models chose input length from $\{24, 72, 168\}$, and the batch size is set to 100. The mean squared error is the default loss function for forecasting tasks. Adam [35] is adopted as optimization strategy, with an initial learning rate set to 0.001. In addition, the learning rate is reduced by a factor of every 10 epochs of no improvement in the validation score, until the final learning rate was reached.

For the LSTM model, a single hidden layer with $\{50, 100, 200\}$ units is defined. The number of units in the hidden layer is unrelated to the number of time steps in the input sequences. Finally, an output layer will directly predict a vector with 24 elements, one for each hour in the output sequence. SGD [36] is adopted as an optimizer. The learning rate is set to 0.05 with a reduction rate by a factor of 0.3.

In the ConvLSTM Encoder–Decoder model, input data have the shape of [*timestep, row, column, channel*]. *Timestep* is chosen from $\{1, 3, 7\}$. *Row* is set to 1. *Column* is chosen from $\{24, 72, 168\}$. *Channel* is chosen from $\{2, 8\}$. SGD is adopted as the optimization algorithm. The learning rate is set as the same in LSTM. For this network, the 1-layer network contains one ConvLSTM layer with 64 hidden states, the 2-layer network contains one ConvLSTM layer with 128 hidden states, and the 3-layer network has 200 hidden states in the LSTM layers. All the input-to-state and state-to-state kernels are of size 1×3.

For the MALSTM-FCN network, the optimal number of LSTM hidden states for each dataset was found via grid search over $\{8, 50, 100, 200\}$. The FCN block is comprised of three blocks of 128-256-128 filters. The models are trained using a batch size of 128. The convolution kernels are initialized following the work of [24].

For the TCN network, the optimal number of hidden units per layer for each dataset was found via grid search over $\{30, 50, 100\}$. The convolution kernels are of size 1×3.

In our M-TCN model, Adam is adopted as an optimization strategy with an initial learning rate set to 0.001(ISO-NE Dataset), while, for Beijing PM2.5, SGD is adopted as an optimization strategy with an initial learning rate set to 0.05.

The implementations of M-TCN are built based on Keras library with the Tensorflow backend. We run all the experiments on a computer with a single NVIDIA 1080 GPU (Santa Clara, CA , USA).

4.6. Experimental Results

Table 4 summarizes the results on multivariate testing sets in the metrics RMSE, RRSE and CORR across all forecast hours. The output sequence length is set to 24, which means that the horizons were set from the 1st hour to the 24th hour for forecasting over the Beijing PM2.5 and ISO-NE Electricity data. In the time series forecasting, larger horizons shall make the prediction harder. Thus, our experiments give a detailed analysis of the results in this large horizon. The best results for each data and metric pair are highlighted in bold. To demonstrate the effectiveness of the models, the results are compared with three baseline methods by the Naive, Average and Seasonal persistent model, as well as four competitive algorithms of LSTM, ConvLSTM, TCN and MALSTM-FCN. For RMSE and RRSE, the lower value is better, while the higher value is better for CORR. Overall performance of neural network based models is better than traditional methods. The performance of M-TCN is comparable with LSTM and MALSTM-FCN and outperforms both of them by about 10%~20% for both datasets. Furthermore, the ConvLSTM model has weak generalization ability, and its prediction ability varies greatly on different datasets.

Figure 9 presents the results on RMSE for both datasets at a larger horizon from the 1st hour to the 24th hour. It is obvious that M-TCN is better than others and RRSE maintains a steady increase without obvious fluctuation in the long-term forecasting period.

Table 4. Results summary (in RMSE, RSE and CORR) of all methods with two datasets.

Methods	Metrics	Beijing PM2.5 Dataset	ISO-NE Dataset
		Length = 24	Length = 24
Naive	RMSE	80.55	2823.35
	RRSE	0.8608	1.0526
	CORR	0.6736	0.5330
Average	RMSE	87.89	2363.07
	RRSE	0.9393	0.8810
	CORR	0.4972	0.4885
Seasonal Persistent	RMSE	123.45	1654.38
	RRSE	1.3193	0.6168
	CORR	0.1722	0.8314
LSTM	RMSE	68.07	783.90
	RRSE	0.7275	0.2923
	CORR	0.6877	0.9573
ConvLSTM	RMSE	82.32	687.17
	RRSE	0.8798	0.2562
	CORR	0.4873	0.9670
TCN	RMSE	112.35	720.12
	RRSE	1.1453	0.2685
	CORR	0.0075	0.9636
MALSTM-FCN	RMSE	71.54	680.95
	RRSE	0.7646	0.2539
	CORR	0.6463	0.9677
M-TCN	RMSE	**65.35**	**648.48**
	RRSE	**0.6984**	**0.2418**
	CORR	**0.7163**	**0.9707**

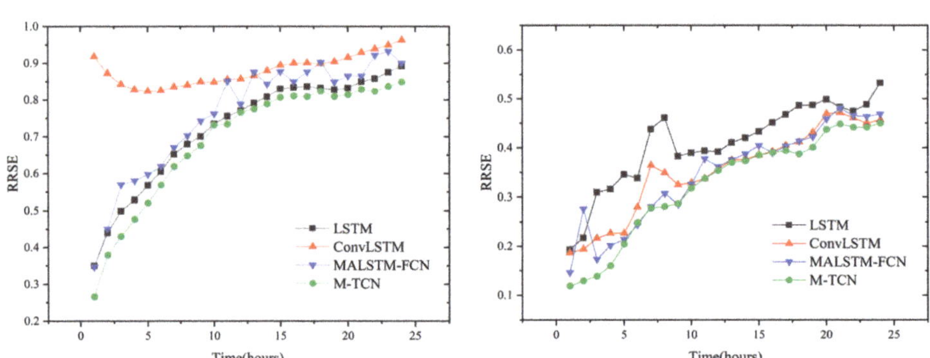

Figure 9. The RMSE for each lead time from hour 1 to hour 24 vs. different algorithms over Beijing PM2.5 (**left**) and ISO-NE Dataset Dataset (**right**).

4.7. Spectrum Analysis

In order to further study the performance of the model, we analyzed the spectrum of the test set and the prediction data. Spectrum refers to the representation of a time domain signal in frequency domain, which can be used for discrete Fourier transform of sequence data. Discrete Fourier Transform (DFT) of k points are computed as:

$$X(k) = DFT[X(n)] = \sum_{n=0}^{N-1} X(n)W^{nk} \quad (0 \leq k \leq N-1), \tag{15}$$

$$W = e^{-j\left(\frac{2\pi}{N}\right)}, \tag{16}$$

where $X(k)$ is the time series.

More detailed calculations include:

$$X(kf_1) = DFT[x(nT_s)] = \sum_{n=0}^{N-1} X(nT_s)\, e^{-j\left(\frac{2\pi}{N}\right)nk},$$ (17)

$$f_1 = \frac{1}{T_1},$$ (18)

$$T_s = \frac{T_1}{N},$$ (19)

where T_1 is signal time, f_1 is the frequency interval, N is the number of signal sampling, and T_s is the signal sampling interval time.

The amplitude spectrum analysis of these datasets is performed, so as to check the existence of repetitive patterns in the datasets. The hourly PM2.5 and ISO-NE data of test set and predictions are plotted in the frequency domain as shown in Figures 10 and 11 separately, where **Freq** is the frequency with a unit of 1/Hour and Am is the amplitude in dB. Sampling frequency is set to 8760 (the same as test set time variable length). Sampling frequency is set to 8760 (the same as the time variable length set by the test), which ensures that the frequency and time correspond to each other numerically. Both figures show that frequency domain is irregular continuous waveform indicating a non-periodic of PM2.5 and ISO-NE datasets. As can be clearly seen, PM2.5 data have no periodicity, which brings great errors to accurate prediction. Since the ISO-NE data change regularly from 1 to 1000 h, the prediction effect is the best.

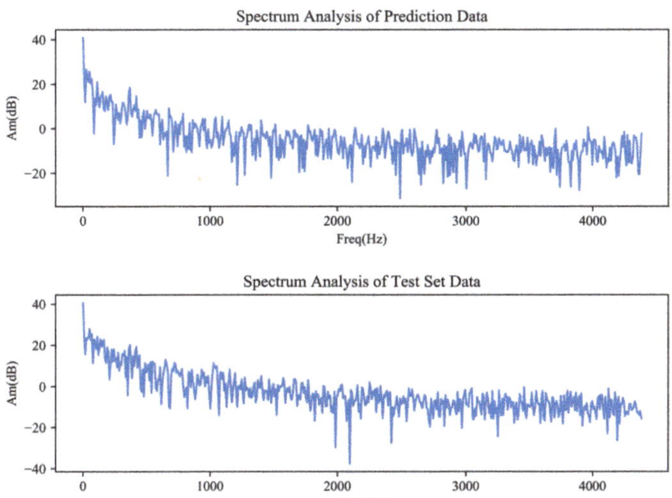

Figure 10. Amplitude Spectrum of Beijing PM2.5 Dataset. Freq: the hourly data in frequency domain (1/Hour); Am: the amplitude of data in both datasets.

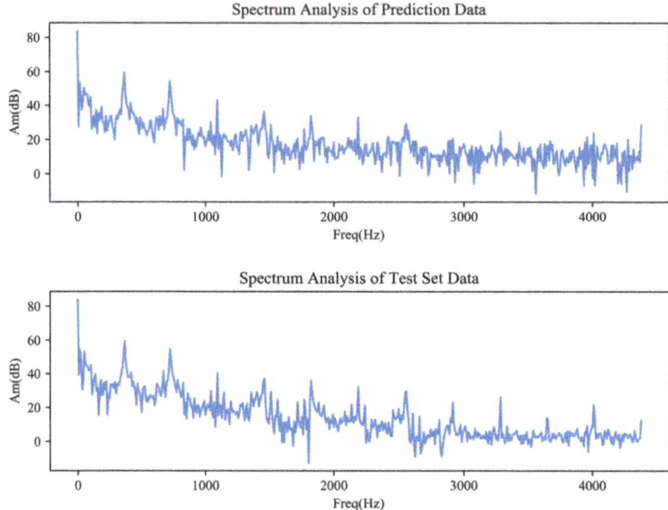

Figure 11. Amplitude Spectrum of ISO-NE Dataset. Freq: the hourly data in frequency domain (1/Hour); Am: the amplitude of data in both datasets.

4.8. Ablation Tests

Furthermore, to demonstrate the efficiency of our model structure, a careful further study is performed. Specifically, we add each component one at a time in our framework. M-TCN with different components are defined as follows:

Model/w/BN: The model adds a Batch Normalization (BN) [37] component. In this test, Batch Normalization was applied to the input of each nonlinearity, in a convolutional way, while keeping the rest of the architecture constant. Figure 12 (left) describes this model in detail.

Model/r/GAP: In the model, the full connection layer is replaced by the global average pooling. Figure 12 (right) describes this model in detail.

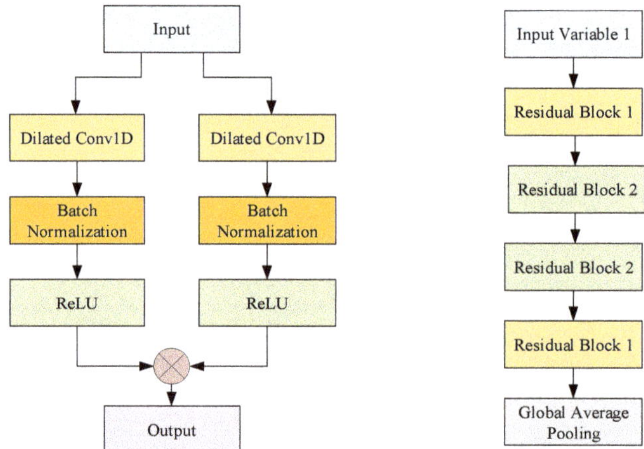

Figure 12. (**left**) Model/w/BN: detail architecture of the Unit. (**right**) Model/r/GAP: the full connection layer is replaced by the global average pooling.

The test results measured using RRSE are shown in Figure 13. Comparing the results, we see that, in both datasets, BN cannot help the network achieve higher accuracy. Adding the BN components in (Model/w/BN) caused big performance drops on both datasets. All of the components of the M-TCN model together lead to the robust performance of our approach on the Beijing PM2.5 dataset.

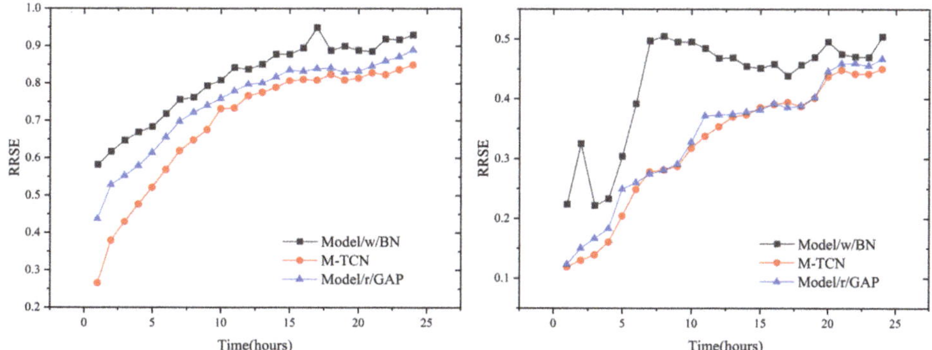

Figure 13. (**left**) RRSE of models over the Beijing PM2.5 dataset. (**right**) RRSE of models over the ISO-NE dataset.

4.9. Model Efficiency

s/epoch denotes the time required for each epoch (in seconds). Boldface indicates the best result. In Table 5, M-TCN proves to be quite competitive.

Table 5. Model training efficiency.

Methods	Beijing PM2.5 Dataset	ISO-NE Dataset
	s/epoch	s/epoch
M-TCN	**29**	**39**
LSTM	95	270
ConvLSTM	33	99

5. Conclusions

The multivariate time series forecasting is investigated by introducing a novel M-TCN model, in order to compare with traditional models and especially deep learning (generic recurrent architectures such as LSTM; generic convolutional architecture such as TCN; hybrid architectures such as ConvLSTM and MALSTM-FCN.). In M-TCN, the dilated network is employed as a meta-network and asymmetric residual blocks are constructed. The proposed approach significantly improved the results in time series forecasting on benchmark datasets of Beijing PM2.5 and ISO-NE. Our research focuses on the trade-off between implementation complexity and prediction accuracy. With in-depth analysis and empirical evidence, the results indicate a prominent efficiency of M-TCN.

For future research, we will focus on the extraction technology based on higher-order statistical features instead of fully connected layers, which can reduce the parameters of the model and training time.

Author Contributions: Conceptualization, R.W. and S.M.; methodology, R.W.; software, S.M.; validation, S.M. and J.W.; formal analysis, S.M.; investigation, R.W. and F.Y.; resources, M.L.; data curation, R.W. and F.Y.; writing—original draft preparation, S.M.; writing—review and editing, R.W. and F.Y.; visualization, S.M.; supervision, R.W.; project administration, F.Y.; funding acquisition, R.W. and M.L.

Funding: This work was supported by the National Natural Science Foundation of China (Grant No. 11505130 and 21872174), the Project of Innovation-Driven Plan in Central South University (2017CX003), State Key Laboratory of Powder Metallurgy, Shenzhen Science and Technology Innovation Project (JCYJ20180307151313532), Thousand Youth Talents Plan of China and Hundred Youth Talents Program of Hunan.

Conflicts of Interest: The authors declare no conflict of interest.

References

1. Seyed, B.L.; Behrouz, M. Comparison between ANN and Decision Tree in Aerology Event Prediction. In Proceedings of the International Conference on Advanced Computer Theory & Engineering, Phuket, Thailand, 20–22 December 2008; pp. 533–537. [CrossRef]
2. Simmonds, J.; Gómez, J.A.; Ledezma, A. Data Preprocessing to Enhance Flow Forecasting in a Tropical River Basin. In *Engineering Applications of Neural Networks*; Springer: Cham, Switzerland, 2017; pp. 429–440.
3. Mohamad, S. Artificial intelligence for the prediction of water quality index in groundwater systems. *Model. Earth Syst. Environ.* **2016**, *2*, 8.
4. Amato, F.; Castiglione, A.; Moscato, V.; Picariello, A.; Sperlì, G. Multimedia summarization using social media content. *Multimed. Tools Appl.* **2018**, *77*, 17803–17827. [CrossRef]
5. Kadir, K.; Halim, C.; Harun, K.O.; Olcay, E.C. Modeling and prediction of Turkey's electricity consumption using Artificial Neural Networks. *Energy Convers. Manag.* **2009**, *50*, 2719–2727.
6. Wu, Y.; José, M.H.; Ghahramani, Z. Dynamic Covariance Models for Multivariate Financial Time Series. *arXiv* **2013**, arXiv:1305.4268.
7. Yu, R.; Li, Y.; Shahabi, C.; Demiryurek, U.; Liu, Y. Deep learning: A generic approach for extreme condition traffic forecasting.In Proceedings of the 2017 SIAM International Conference on Data Mining, Houston, TX, USA; 27–29 April 2017; pp. 777–785.
8. Akaike, H. Fitting autoregressive models for prediction. *Ann. Inst. Stat. Math.* **1969**, *21*, 243–247. [CrossRef]
9. Frigola, R.; Rasmussen, C.E. Integrated pre-processing for bayesian nonlinear system identification with gaussian processes. In Proceedings of the IEEE Conference on Decision and Control, Florence, Italy, 10–13 December 2013; pp. 552–560.
10. Alom, M.; Tha, T.; Yakopcic, C.; Westberg, S.; Sidike, P.; Nasrin, M.; Hasan, M.; Essen, B.; Awwal, A.; Asari, V. A State-of-the-Art Survey on Deep Learning Theory and Architectures. *Electronics* **2019**, *8*, 292. [CrossRef]
11. Liu, L.; Finch, A.M.; Utiyama, M.; Sumita, E. Agreement on Target-Bidirectional LSTMs for Sequence-to-Sequence Learning. In Proceedings of the Thirtieth Aaai Conference on Artificial Intelligence, Phoenix, AZ, USA, 12–17 February 2016; pp. 2630–2637.
12. Gehring, J.; Auli, M.; Grangier, D.; Yarats, D.; Dauphin, Y.N. Convolutional Sequence to Sequence Learning. *arXiv* **2017**, arXiv:1705.03122.
13. Yu, F.; Koltun, V.; Funkhouser, T. Dilated Residual Networks. In Proceedings of the IEEE Conference on Computer Vision and Pattern Recognition (CVPR), Honolulu, HI, USA, 21–26 July 2017; pp. 472–480. [CrossRef]
14. He, K.; Zhang, X.; Ren, S.; Sun, J. Deep Residual Learning for Image Recognition. In Proceedings of the IEEE Conference on Computer Vision and Pattern Recognition (CVPR), Las Vegas, NV, USA, 26 June–1 July 2016; pp. 770–778. [CrossRef]
15. Ediger, V.; Akar, S. ARIMA forecasting of primary energy demand by fuel in Turkey. *Energy Policy* **2007**, *35*, 1701–1708. [CrossRef]
16. Rojas, I.; Valenzuela, O.; Rojas, F.; Guillen, A.; Herreraet, L.; Pomares, H.; Marquez, L.; Pasadas, M. Soft-computing techniques and ARMA model for time series prediction. *Neurocomputing* **2008**, *71*, 519–537. [CrossRef]
17. Kilian, L. New introduction to multiple time series analysis. *Econ. Rec.* **2006**, *83*, 109–110.
18. Sapankevych, N.; Sankar, R. Time Series Prediction Using Support Vector Machines: A Survey. *IEEE Comput. Intell. Mag.* **2009**, *4*, 24–38. [CrossRef]
19. Hamidi, O.; Tapak, L.; Abbasi, H.; Abbasi, H.; Maryanaji, Z. Application of random forest time series, support vector regression and multivariate adaptive regression splines models in prediction of snowfall (a case study of Alvand in the middle Zagros, Iran). *Theor. Appl. Climatol.* **2018**, *134*, 769–776. [CrossRef]
20. Lima, C.; Lall, U. Climate informed monthly streamflow forecasts for the Brazilian hydropower network using a periodic ridge regression model. *J. Hydrol.* **2010**, *380*, 438–449. [CrossRef]

21. Li, J.; Chen, W. Forecasting macroeconomic time series: LASSO-based approaches and their forecast combinations with dynamic factor models. *Int. J. Forecast.* **2014**, *30*, 996–1015. [CrossRef]

22. Hochreiter, S.; Schmidhuber, J. Long Short-Term Memory. *Neural Comput.* **1997**, *9*, 1735–1780. [CrossRef]

23. Shi, X.; Chen, Z.; Wang, H.; Yeung, DY.;Wong, W.K.; Woo, W.C. Convolutional LSTM Network: A Machine Learning Approach for Precipitation Nowcasting. In Proceedings of the Neural Information Processing Systems Conference, Montreal, QC, Canada, 7–12 December 2015; pp. 802–810.

24. Karim, F.; Majumdar, S.; Darabi, H.; Harforda, S. Multivariate LSTM-FCNs for Time Series Classification. *Neural Netw.* **2019**, *116*, 237–245. [CrossRef]

25. Huang, N.E.; Zheng, S.; Long, S.R.; Wu, M.C.; Shih, H.H.; Zheng, Q.; Yen, N.-C.; Tung, C.C.; Liu, H.H. The empirical mode decomposition and the Hilbert spectrum for nonlinear and non-stationary time series analysis. *Proc. Math. Phys. Eng. Sci.* **1998**, *454*, 903–995. [CrossRef]

26. Wu, Z.; Huang, N.E. Ensemble empirical mode decomposition: A noise-assisted data analysis method. *Adv. Adapt. Data Anal.* **2009** 1, 1–41. [CrossRef]

27. Wang, J.; Wang, Z.; Li, J.; Wu, J. Multilevel Wavelet Decomposition Network for Interpretable Time Series Analysis. In Proceedings of the 24th ACM SIGKDD International Conference, London, UK, 19–23 August 2018; pp. 2437–2446. [CrossRef]

28. Dragomiretskiy, K.; Zosso, D. Variational Mode Decomposition. *IEEE Trans. Signal Process.* **2014**, *62*, 531–544. [CrossRef]

29. Bai, S.; Kolter, J.Z.; Koltun, V. An Empirical Evaluation of Generic Convolutional and Recurrent Networks for Sequence Modeling. *arXiv* **2018**, arXiv:1803.01271.

30. Dou, C.; Zheng, Y.; Yue, D.; Zhang, Z.; Ma, K. Hybrid model for renewable energy and loads prediction based on data mining and variational mode decomposition. *IET Gener. Transm. Distrib.* **2018**, *12*, 2642–2649. [CrossRef]

31. Howard, A.; Sandler, M.; Chu, G.; Chen, L.-C.; Chen, B.; Tan, M.; Wang, W.; Zhu, Y.; Pang, R.; Vasudevan, V.; et al. Searching for MobileNetV3. *arXiv* **2018**, arXiv:1905.02244.

32. Cai, H.; Zhu, L.; Han, S. ProxylessNAS: Direct Neural Architecture Search on Target Task and Hardware. *arXiv* **2018**, arXiv:1812.00332.

33. Elsken, T.; Metzen, J.H.; Hutter, F. Neural Architecture Search: A Survey. *arXiv* **2018**, arXiv:1808.05377.

34. Nair, V.; Hinton, G. Rectified linear units improve restricted Boltzmann machines. In Proceedings of the 27th International Conference on International Conference on Machine Learning, Haifa, Israel, 21–24 June 2010; pp. 807–814.

35. Kingma, D.; Ba, J. Adam:A method for stochastic optimization. *arXiv* **2014**, arXiv:1412.6980.

36. Sutskever, I.; Martens, J.; Dahl, G.E.; Hinton, G. On the importance of initialization and momentum in deep learning. In Proceedings of the 30th International Conference on International Conference on Machine Learning, Atlanta, GA, USA, 16–21 June 2013; pp. 1139–1147.

37. Ioffe, S.; Szegedy, C. Batch normalization: Accelerating deep network training by reducing internal covariate. In Proceedings of the 32nd International Conference on Machine Learning, Lille, France, 6–11 July 2015; pp. 448–456.

Article

Modeling and Analysis of Adaptive Temperature Compensation for Humidity Sensors

Wei Xu, Xiaoyu Feng and Hongyan Xing *

Jiangsu Key Laboratory of Meteorological Observation and Information Processing, Nanjing University of Information Science & Technology, Nanjing 210044, China; xw@nuist.edu.cn (W.X.); 20162281511@nuist.edu.cn (X.F.)
* Correspondence: xinghy@nuist.edu.cn

Received: 2 March 2019; Accepted: 8 April 2019; Published: 11 April 2019

Abstract: In addition to being sensitive to humidity, humidity sensors with moisture sensitive elements are also sensitive to ambient temperature. The fusion of temperature and humidity data is an effective way to improve the accuracy of humidity sensors. In view of the problem of insufficient adaptive ability and poor universality in the current compensation algorithm, a piecewise processing of measured error at different temperatures by using multiple linear regression is proposed in this paper. The least squares method and back propagation (BP) neural network improved by a genetic simulated annealing algorithm (GSA-BP) were used to compensate the measured humidity data of different temperature ranges. The efficiency of the GSA-BP algorithm was tested, and the compensation function model was established. The compensation accuracy was also compared with the accuracies obtained by other methods. The experimental results show that the adaptive segmentation compensation method can significantly improve the measured error of the humidity sensor over a wide temperature range.

Keywords: humidity sensor; data fusion; nonlinear optimization; multiple linear regression; GSA-BP

1. Introduction

Automatic weather stations monitor changes in the climate environment in real time. The meteorological sensors are susceptible to ambient influences and their measurement errors exist objectively [1]. Usually, the humidity sensor used in automatic weather stations is a voltage output type polymer film humidity sensitive capacitance sensor, which senses the humidity through the humidity sensitive capacitor and then converts it into a voltage amount by the conversion circuit [2]. The humidity sensitive capacitor is mainly composed of an upper electrode, a humidity sensitive material, a lower electrode and a glass substrate. The humidity sensitive material is a high-molecular-weight polymer with a dielectric constant that changes with the relative humidity of the external environment. In addition to being sensitive to ambient humidity, humidity sensitive materials are also sensitive to temperature. The temperature coefficient is not a constant but a variable. Nonlinear compensation for measured data of the sensor is often required [3,4].

The humidity sensor manufacturer and meteorological calibrator will compensate for the influence of temperature on the measurement results, but the compensation effect is not ideal under low temperature (−20 °C) or high temperature (+50 °C) conditions, and the compensation algorithm is not universal over a wide temperature range. It is important to study an efficient and adaptive compensation method for improving the calibration efficiency.

In recent years, many scholars have compensated sensors using both hardware and software and have achieved some notable results. References 5 and 6 proposed using a conditioning chip and concentric wheatstone bridge circuit to compensate [5,6], but this hardware compensation circuit is subject to electronic components' temperature drift and process technology constraints, which results in high cost and poor compensation. Software compensation has become a research hotspot because of its low cost, strong applicability and high compensation accuracy. In 2014, reference 7 proposed a combination of hardware and software. The circuit was first designed and compensated by the extreme learning machine (ELM) [7]. The hardware compensation circuit itself would be affected by the ambient temperature, the ELM algorithm easily produced the over fitting problem, and the optimal effect could not be obtained. Reference 8 proposed using principal component analysis (PCA) to improve the back propagation (BP) neural network for nonlinear compensation [8]. PCA was the most widely used method of reducing the dimension and error correction. In practical applications, when gross corruptions existed, PCA could not grasp the real subspace structure of the data well, and the algorithm had no universality. In 2015, reference 9 used a particle swarm optimization (PSO) algorithm to optimize the nonlinear compensation method of the BP neural network. PSO had no crossover and mutation operations [9], and the search speed was fast, but it lacked dynamic speed adjustment and would easily fall into a local extremum. The ability to adapt to ambient temperature was not strong. In 2016, reference 10 proposed using the least squares support vector machine (LS-SVM) to compensate [10]. Compared with the artificial neural network, the LS-SVM could overcome the shortage of long training time and was faster than SVM in solving equations. The solution satisfied the extreme condition, but it could not guarantee that it was a global optimal solution, and there was still the problem of easily falling into a local extremum. All of these compensation methods simply applied an algorithm to the sensor compensation and did not account for the influence of temperature, making the compensation method less adaptive, and making it difficult to guarantee the superiority of the compensation algorithm over a wide temperature range.

In recent years, we have conducted considerable research on the nonlinear compensation of humidity sensors. From 2012 to 2017, we proposed an improved BP neural network nonlinear compensation method, which used a genetic algorithm (GA) to optimize the weight and threshold [11,12]. The method avoided the BP neural network plunging into a local extremum, but the compensation speed was slow when the amount of humidity data was large. Furthermore, combined with the influence of temperature on the humidity sensor, a method of segmentation compensation was proposed, and the compensation speed was fast [13], but the segmentation node was artificially selected. The intelligence and adaptive ability were not high. On the basis of our previous research, and inspired by the idea of a multi-information approach, this paper proposes an adaptive nonlinear compensation method for humidity sensors. According to the influence regularity of temperature on humidity sensors, the sensor was compensated by adaptive segmentation, and the effects of various compensation methods were compared and studied.

2. Principle of Compensation

Through a large number of experimental tests, the measured error of the humidity sensor has been shown to be linear near room temperature and nonlinear at high and low temperatures. Some sensors are also nonlinear near room temperature. In the experiment, the humidity sensor was put into the temperature and humidity test chamber, which can adjust the temperature and humidity simultaneously. The standard humidity value was calculated from the measured values of the precise dew point instrument, temperature sensor and pressure gauge. The temperature at which the water vapor in the air becomes dewdrops is called the dew point temperature. The dew point temperature is a means of expressing air humidity. The standard humidity value can be calculated by the experimentally measured dew point value, temperature value and pressure value. In this experiment, the data acquisition range of the data collector was 0.1 uV–100 V, the sensitivity was 100 nV, and the error was ±0.002%. The temperature was measured by a second-class standard PT100 temperature sensor with an allowable error value of ±0.15 °C. During the experiment, all equipment and instruments were verified with high measurement accuracy and could obtain accurate, scientific and reasonable data. The measured value of the humidity sensor was read by the high precision data acquisition unit. The temperature and humidity test chamber was adjusted, and the points near the preset value were observed and recorded. The experimental principle is shown in Figure 1.

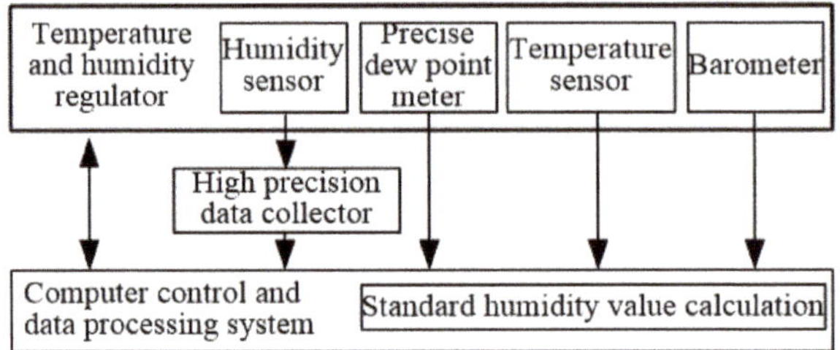

Figure 1. Error experiment of humidity sensor at different temperatures.

The humidity-sensitive capacitive sensor was placed in the temperature and humidity regulating chamber. The humidity was set to 10% RH, and the temperature was changed. After the temperature was stabilized, the measured value of the humidity sensor was read. The humidity was set to 30% RH, 50% RH, 70% RH and 90% RH, and the above steps were repeated. The measuring error curve of the humidity sensor at different temperatures was obtained as shown in Figure 2.

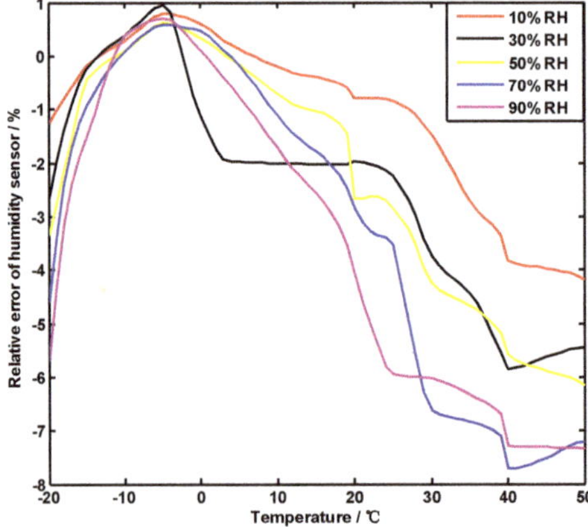

Figure 2. Measured error curve of the humidity sensor at different temperatures.

From the error curve of Figure 2, it can be seen that the error of the humidity sensor obviously increases in low and high temperature regions and has nonlinear characteristics. The original measurement value of the sensor was compensated according to different ambient temperatures, such that the compensated value was close to the standard value. Setting the ambient temperature T, the relationship between the original measured value of the humidity sensor H_I and the compensated humidity value H_C is Equation (1).

$$H_C = f(H_I, T) \tag{1}$$

H_C and T are both single-valued functions of H_I, then the inverse function $H_I = f^{-1}(H_C, T)$ exists. So the introduced influence parameter T and the measured value H_I were used as data to be compensated, and the regression analysis was performed by the binary linear regression function. The regression effect and the value of the segmentation point T_1, T_2 were determined based on the value of the coefficient of determination R^2. The least squares method and GSA-BP neural network were used to compensate the different temperature intervals. The function of segmentation compensation is similar to Equation (2).

$$H_C = f(H_I, T) = \begin{cases} f_1(H_I, T), T \in [T_0, T_1] \\ f_2(H_I, T), T \in [T_1, T_2] \\ f_3(H_I, T), T \in [T_2, T_3] \end{cases} \tag{2}$$

Among these values, $T_0 \leq T_1 \leq T_2 \leq T_3$, $f_2(H_I, T)$ is the compensation function of the temperature range with good linear regression effect. A simple and efficient least squares method was used for line fitting. $f_1(H_I, T)$ and $f_3(H_I, T)$ are compensated functions at low temperature and high temperature, respectively, and the GSA-BP neural network was used. The influence of temperature T was effectively reduced, and the measured data were approximated to the stand value of humidity. The overall idea of optimal compensation is shown in Figure 3.

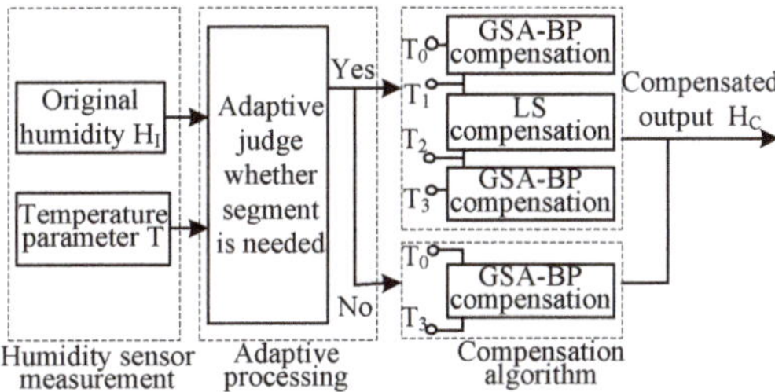

Figure 3. Compensation principle of humidity sensor.

3. Adaptive Segmentation Based on Multiple Linear Regression

For the humidity measurement errors at different temperatures, the key to improve the accuracy of compensation is to judge whether the segmentation compensation is necessary and find the best segmentation point. The linear regression method mainly determines how to obtain the best fitting line through the sample. The process is a mathematical optimization method, and it searches for the best function of data by minimizing the square of error [14]. In the regression analysis, two or more independent variables were included, and these independent variables can be approximated by a straight line. As seen from Figure 2, in the temperature experiment of the humidity sensor, there is a certain linear relationship between the measured error of humidity sensor and the temperature value in some intervals. The multivariate linear regression method can be used to analyze and compensate errors in linear intervals [15]. The regression model between the expected ideal humidity value, the measured value and the temperature was established. The coefficient of determination was calculated and whether it needed segmentation compensation according to its value was determined.

A multiple linear regression model for temperature compensation was established, such as Equation (3).

$$H_{Oi} = \beta_0 + \beta_1 H_{Ii} + \beta_2 T_i \tag{3}$$

where H_{Oi} is the expected humidity value of the regression, H_{Ii} is the measured value of the humidity sensor, T_i is the value of the ambient temperature, i represents the i-th group data, $i = 1, 2, 3, \cdots, n$, there are n sets of data, and $\beta_0, \beta_1, \beta_2$ are the regression coefficients. The actual regression model can be expressed as Equation (4).

$$\hat{H}_{Oi} = \hat{\beta}_0 + \hat{\beta}_1 H_{Ii} + \hat{\beta}_2 T_i \tag{4}$$

where $\hat{\beta}_0, \hat{\beta}_1$ and $\hat{\beta}_2$ are the estimated values of β_0, β_1 and β_2, respectively. $\hat{H}_{Oi}$ is the estimated value of H_{Oi}, and the residual of them is ε_i. It can be known from the least squares method that $\hat{\beta}_0, \hat{\beta}_1$ and $\hat{\beta}_2$ should minimize the sum of squares of the residuals ε_i.

$$Q = \sum \varepsilon_i^2 = \sum \left(H_{Oi} - \hat{H}_{Oi}\right)^2 = \sum \left(H_{Oi} - \hat{\beta}_0 - \hat{\beta}_1 H_{Ii} - \hat{\beta}_2 T_i\right)^2 \tag{5}$$

According to the extremum principle of multivariate function, when Q gets the minimum value, the partial derivatives of Q to $\hat{\beta}_0$, $\hat{\beta}_1$ and $\hat{\beta}_2$ are all equal to zero. Then there is Equation (6).

$$\begin{cases} \dfrac{\partial Q}{\partial \hat{\beta}_0} = 2\sum\left(H_{Oi} - \hat{\beta}_0 - \hat{\beta}_1 H_{Ii} - \hat{\beta}_2 T_i\right)(-1) = 0 \\[2ex] \dfrac{\partial Q}{\partial \hat{\beta}_1} = 2\sum\left(H_{Oi} - \hat{\beta}_0 - \hat{\beta}_1 H_{Ii} - \hat{\beta}_2 T_i\right)(-H_{Ii}) = 0 \\[2ex] \dfrac{\partial Q}{\partial \hat{\beta}_2} = 2\sum\left(H_{Oi} - \hat{\beta}_0 - \hat{\beta}_1 H_{Ii} - \hat{\beta}_2 T_i\right)(-T_i) = 0 \end{cases} \tag{6}$$

Then Equation (7) is derived.

$$\begin{cases} n\hat{\beta}_0 + \hat{\beta}_1 \sum H_{Ii} + \hat{\beta}_2 \sum T_i = \sum H_{Oi} \\[2ex] \hat{\beta}_0 \sum H_{Ii} + \hat{\beta}_1 \sum H_{Ii}^2 + \hat{\beta}_2 \sum T_i H_{Ii} = \sum H_{Ii} H_{Oi} \\[2ex] \hat{\beta}_0 \sum T_i + \hat{\beta}_1 \sum T_i H_{Ii} + \hat{\beta}_2 \sum T_i^2 = \sum T_i H_{Oi} \end{cases} \tag{7}$$

Its matrix form is

$$\begin{bmatrix} n & \sum H_{Ii} & \sum T_i \\ \sum H_{Ii} & \sum H_{Ii}^2 & \sum T_i H_{Ii} \\ \sum T_i & \sum T_i H_{Ii} & \sum T_i^2 \end{bmatrix} \begin{bmatrix} \hat{\beta}_0 \\ \hat{\beta}_1 \\ \hat{\beta}_2 \end{bmatrix} = \begin{bmatrix} \sum H_{Oi} \\ \sum H_{Ii} H_{Oi} \\ \sum T_i H_{Oi} \end{bmatrix} \tag{8}$$

In Equation (8), two of the matrices can be written as

$$\begin{bmatrix} n & \sum H_{Ii} & \sum T_i \\ \sum H_{Ii} & \sum H_{Ii}^2 & \sum T_i H_{Ii} \\ \sum T_i & \sum T_i H_{Ii} & \sum T_i^2 \end{bmatrix} = \begin{bmatrix} 1 & 1 & \cdots & 1 \\ H_{I1} & H_{I2} & \cdots & H_{In} \\ T_1 & T_2 & \cdots & T_n \end{bmatrix} \begin{bmatrix} 1 & H_{I1} & T_1 \\ 1 & H_{I2} & T_2 \\ \vdots & \vdots & \vdots \\ 1 & H_{In} & T_n \end{bmatrix} \tag{9}$$

$$\begin{bmatrix} \sum H_{Oi} \\ \sum H_{Ii} H_{Oi} \\ \sum T_i H_{Oi} \end{bmatrix} = \begin{bmatrix} 1 & 1 & \cdots & 1 \\ H_{I1} & H_{I2} & \cdots & H_{In} \\ T_1 & T_2 & \cdots & T_n \end{bmatrix} \begin{bmatrix} H_{O1} \\ H_{O2} \\ \vdots \\ H_{On} \end{bmatrix} \tag{10}$$

Assume

$$\begin{bmatrix} 1 & H_{I1} & T_1 \\ 1 & H_{I2} & T_2 \\ \vdots & \vdots & \vdots \\ 1 & H_{In} & T_n \end{bmatrix} = X \tag{11}$$

Then, Equations (9) and (10) can be rewritten as

$$\begin{bmatrix} n & \sum H_{Ii} & \sum T_i \\ \sum H_{Ii} & \sum H_{Ii}^2 & \sum T_i H_{Ii} \\ \sum T_i & \sum T_i H_{Ii} & \sum T_i^2 \end{bmatrix} = X'X \tag{12}$$

$$\begin{bmatrix} \sum H_{Oi} \\ \sum H_{Ii} H_{Oi} \\ \sum T_i H_{Oi} \end{bmatrix} = X'H_O \tag{13}$$

Substitute Equation (12) and Equation (13) into Equation (8)

$$X'X \begin{bmatrix} \hat{\beta}_0 \\ \hat{\beta}_1 \\ \hat{\beta}_2 \end{bmatrix} = X'H_O \tag{14}$$

Thus, the regression coefficients are obtained by Equation (15)

$$\begin{bmatrix} \hat{\beta}_0 \\ \hat{\beta}_1 \\ \hat{\beta}_2 \end{bmatrix} = (X'X)^{-1}X'H_O \tag{15}$$

By taking the regression coefficients into the multiple linear regression model, a compensation function can be obtained. To verify the rationality of the model, the coefficient of determination R^2 is used to estimate the fit of the model to the measured data. In multiple regression analysis, the coefficient of determination is the square of the path coefficient, that is

$$R^2 = \frac{SSR}{SST} = \frac{\sum_i (\hat{H}_{Oi} - \overline{H}_{Oi})^2}{\sum_i (H_{Oi} - \overline{H}_{Oi})^2} \tag{16}$$

In Equation (16), $\overline{H}_{Oi}$ is the average expected humidity value of H_{Oi}. The total dispersion square sum SST reflects the discrete state of all expected humidity values $\overline{H}_{Oi}$. The regression square sum SSR reflects the difference after regression. The sum of squared residuals is $SSE = \sum_i \varepsilon_i^2$, so $SST = SSR + SSE$. Then, Equation (16) can be rewritten to Equation (17)

$$R^2 = \frac{SSR}{SSR + SSE} = 1 - \frac{SSE}{SST} = 1 - \frac{\sum_i \varepsilon_i^2}{\sum_i (H_{Oi} - \overline{H}_{Oi})^2} \tag{17}$$

The coefficient of determination R^2 represents the interpretation degree of the estimated value $\hat{H}_{Oi}$ to the ideal value H_{Oi}. The larger the R^2, the closer the regression curve is to the ideal humidity value [16].

Therefore, the regression coefficients $\hat{\beta}_0$, $\hat{\beta}_1$, $\hat{\beta}_2$ and coefficient of determination R^2 are calculated by using the actual measured humidity value H_I, temperature value T_i and the expected ideal humidity value H_{Oi}.The coefficient of determination R^2 is used to judge whether to segment and to determine the segmentation point. The specific implementation steps are as follows.

1. The error curve of humidity sensor is linear at room temperature. So the initial temperature value T_x is determined to be about 25 °C. The measured error of humidity sensor at the ambient temperature 25 °C is −6%−−1% when the humidity is 10%–90% RH.
2. Read the previous temperature value T_{x-1} and the next temperature value T_{x+1} and their respective humidity values.
3. Each time a set of temperature and corresponding humidity data are read, a regression is performed to obtain a coefficient of determination R^2.
4. If $R^2 \geq 0.911$, return to step 2, if $R^2 < 0.911$, get the segmentation points T_1 and T_2.
5. When the temperature interval of the linear interval is greater than 5 °C, that is, $T_1 - T_2 > 5$, the segmentation compensation will be performed according to Figure 3. If $T_1 - T_2 \leq 5$, the whole process will be compensated by GSA-BP. An adaptive segmentation flowchart based on multiple linear regression is shown in Figure 4.

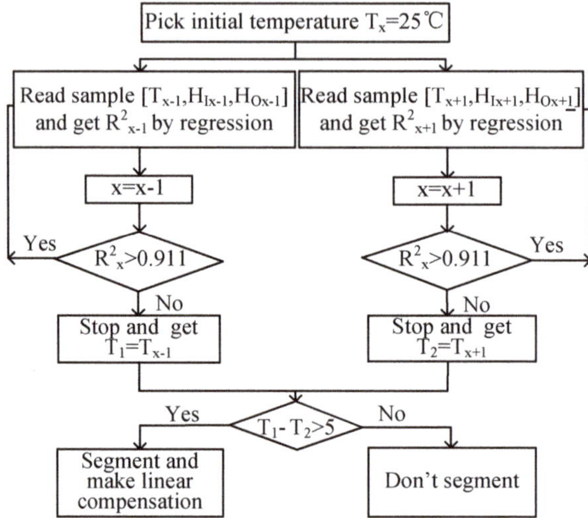

Figure 4. Flow chart of searching segmentation point.

The multiple linear regression method is used to analyze the humidity measurement errors at different temperatures. For the intervals with better linearity, the multiple regression model established by Equation (4) is used to compensate the errors.

4. Nonlinear Compensation Model

The nonlinear interval adopts the BP neural network with strong nonlinear mapping ability to compensate. The input of the BP neural network is temperature T and the measured value H_I, and the output is the compensated humidity value H_C. The humidity sensor temperature compensation model was established, and the BP neural network was trained by multiple sets of data. To improve the local minimum of the BP neural network, a genetic simulated annealing algorithm was used.

The genetic simulated annealing algorithm is an optimization algorithm that combines a genetic algorithm and a simulated annealing algorithm. The local search ability of the genetic algorithm is limited, but the ability to grasp the overall search process is strong. The simulated annealing algorithm has strong local search ability and can prevent the search process from falling into the local optimal solution. However, little is known regarding the state of the entire search space. It is inconvenient to make the search process enter the optimal search area, which makes the simulated annealing algorithm less efficient. However, if the genetic algorithm is combined with the simulated annealing algorithm, a global search algorithm with excellent performance can be developed [17,18].

The BP neural network based on the genetic simulated annealing algorithm (GSA-BP) is mainly divided into the determination of BP network structure and the selection of weight and threshold. The specific compensation steps are the following:

1. Determine the topology structure of the BP neural network. The BP neural network is set to a three-layer network structure. The original measured value of temperature T and humidity sensor H_I are the network inputs, and the number of input nodes n_1 is 2. The number of hidden layer nodes n_2 is set to 7 according to the compensation effect. The output of the network is the humidity value H_C after compensation, and the number of output nodes n_3 is 1. The number of optimized parameters of the genetic simulated annealing algorithm is determined as follows: $(n_1 + 1)n_2 + (n_2 + 1)n_3 = 29$.
2. Initialize the genetic simulated annealing algorithm. The population size with weights and thresholds M is 60; the maximum number of iterations MAXGEN is 2000; the crossover probability

P_c is 0.6; the mutation probability P_m is 0.1; the initial temperature T_0 is 100; the end temperature T_e is 0.99; the temperature cooling coefficient ∂ is 0.99.

3. Initialize the weights and thresholds of the BP neural network and calculate fitness. The initial population of the genetic simulated annealing algorithm is generated by combining the initial weights and thresholds of the BP neural network initialization with the original measured values H_I and temperature T. Each individual in the genetic simulated annealing algorithm represents all the weights and threshold of a network, and the algorithm then calculates the fitness of each individual through a fitness function. The fitness function of this paper adopts the fitness stretching method, and the fitness of the i-th individual after improvement is calculated by Equation (18).

$$fit(i) = \frac{e^{f_i/T}}{\sum\limits_{i=1}^{M} e^{f_i/T}} \tag{18}$$

Among these values, f_i is the i-th individual fitness before improvement, $f_i = \frac{1}{H_{Oi}-H_{Ci}}$. T_0 and T are the initial temperature and the current temperature in the simulated annealing algorithm respectively, $T = T_0(0.99^{gen-1})$. *gen* is the current genetic evolution algebra, and M is the population size. H_{Oi} and H_{Ci} are the standard humidity values expected and the humidity values actually obtained by the network of the i-th individual.

4. The genetic simulated annealing algorithm finds individuals with optimal fitness based on a series of operations such as selection, crossover, mutation and annealing. Compare the current fitness and historical best fitness of each individual in the population. If the current value is better, the current value is the best value of the history, and save the individual as the best value of history, otherwise the best value will not change.

5. The genetic simulated annealing algorithm obtains the optimal individual as the initial weight and threshold of the BP neural network. The optimized BP neural network is used to train the humidity data at different temperatures.

Figure 5 shows the process of the compensation method of the humidity sensor. After inputting the original measured values of the humidity sensors and ambient temperature, the GSA optimizes the weights and thresholds of the BP neural network. Finally, the compensated humidity value is obtained.

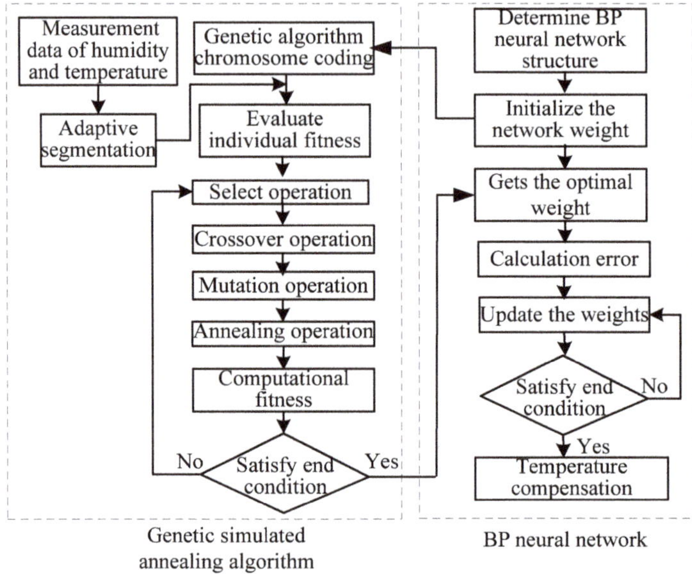

Figure 5. Flow chart of compensation algorithm based on genetic simulated annealing (GSA-BP).

5. Experimental Results and Analysis

5.1. Performance Analysis of Optimized Compensation Algorithm GSA-BP

The variables of the segmentation compensation function are the actual measured value of the humidity sensor and the ambient temperature value. GSA is a random search algorithm. The number of iterations is uncertain. The representative training process was compared with the BP neural network, which is not optimized.

It can be seen from Figure 6 that when the GSA-BP neural network evolves to 70 generations, its adaptation value reaches a minimum, the optimal weight and threshold of the BP neural network are found, and the number of termination iterations is 77. In Figure 7, the number of iterations of the BP neural network is 229. It can be seen that under the same conditions, the number of iterations of the BP neural network optimized by GSA is small, and the training speed is fast, which indicates that the BP neural network optimizes the operation efficiency significantly.

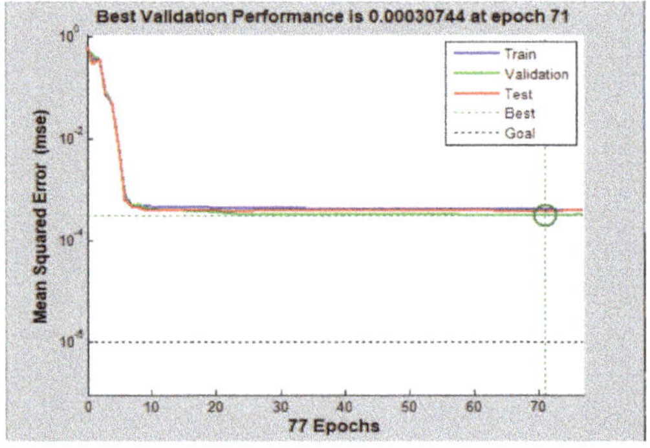

Figure 6. Training iterations of GSA-BP neural network.

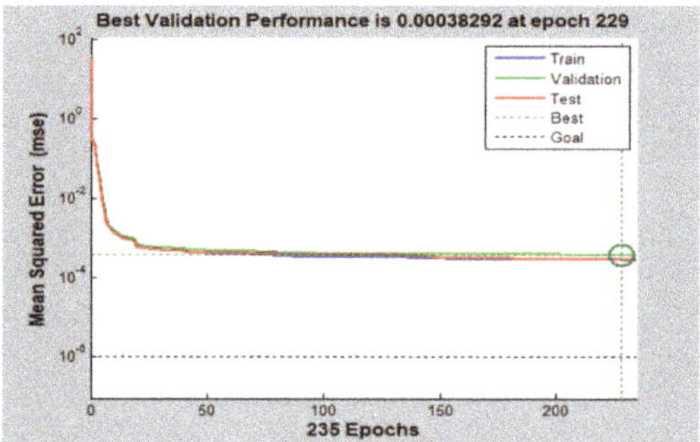

Figure 7. Training iterations of back propagation (BP) neural network.

Figure 8 is the humidity compensation effect curve of the GSA-BP neural network. It can be seen from the figure that the errors between the predicted output and the expected output are very small, and the BP neural network optimized by GSA has a good compensation effect.

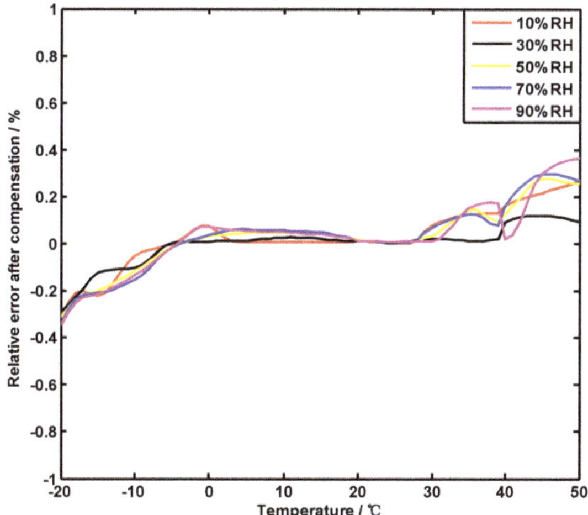

Figure 8. Compensation curve of GSA-BP neural network.

5.2. Segmentation Optimization Compensation Model

Through the above theoretical analysis and verification of experimental data, the influence of temperature on the humidity sensor can be seen. One hundred and fifty sets of data obtained by repeating experiments on the same humidity sensor were used as training samples, and 15 sets of data were used as network test samples. The process of establishing a humidity compensation model is as follows:

1. Read the measured data set and interpolate it. The temperature points set in the experiment have some discreteness. So the continuous function is added on the basis of the discrete data and that the continuous curve passes all the given discrete data points. The interpolated data will be compensated for later.

2. Read two groups of temperature and corresponding humidity in sequence from the temperature of 25 °C, use the binary linear regression function to regress the data after interpolation, and determine the regression effect according to the value of the criterion R^2. When $R^2 \leq 0.911$, the regression is stopped. The value of R^2 is 0.8468 when the experimental data stops returning. The

temperature value at this time is the temperature segmentation points T_1, T_2, which are 22.36 and 29.98, respectively.

3. Determine whether the interval of the temperature segmentation point is greater than 5. The experimental data satisfy this condition; therefore, segmentation compensation is made. The least squares method is used for linear fitting of the temperature range $[22.36, 29.98]$. Taking the straight line fitting effect of 30% RH as an example, the fitted straight line is: $\varepsilon = -0.2695T + 4.4723$, where ε is the compensation value, and T is the temperature value. When the value of measured humidity is 30% RH, input the value of temperature and obtain the corresponding humidity compensation value, then add the measured value to obtain the compensated value.

4. Use GSA to optimize the weight and threshold of the BP neural network, train the neural network, and compensate the data in the nonlinear interval.

5. The compensation function model is shown in Figure 9. The compensation effect diagram is shown in Figure 10. It can be seen from Figure 10 that the compensated humidity value has a good linear relationship with the standard humidity such that the measured value is closer to the true value.

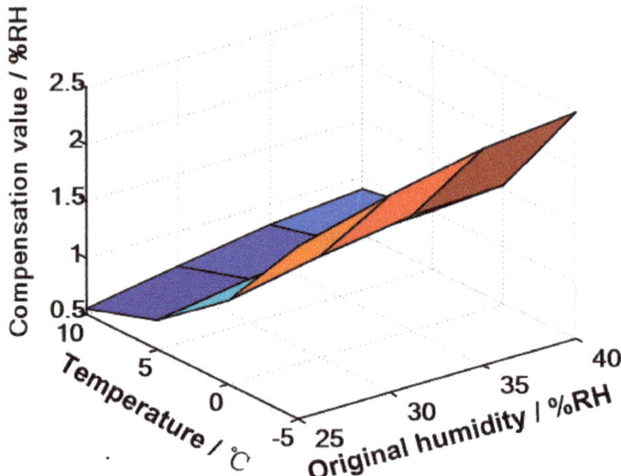

Figure 9. Compensation function model.

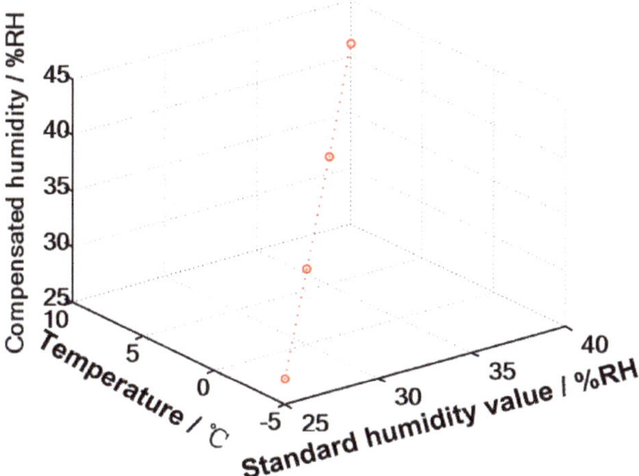

Figure 10. Compensation effect.

The original measurement data of the same humidity sensor were compensated by using reference 11, reference 13 and the method proposed in this paper. Reference 11 used a genetic algorithm (GA) improved BP neural network without segmentation compensation, where GA is a population-based optimization algorithm [19]. Reference 13 used a least squares and BP neural network to compensate. In this paper, the BP neural network improved by adaptive selection and combined with genetic simulated annealing was used for compensation. To avoid the randomness of the algorithm, the same group of data was run many times, and the compensation effect was the same; therefore, the average value of the same group of data after multiple runs was taken. The three methods compensate the data as shown in Table 1, and the corresponding error curve is shown in Figure 11.

Table 1. Error after compensation of three methods.

Environment Temperature (°C)	Standard Value of Relative Humidity (%RH)	Error of Different Methods of Compensation (%)		
		Reference 11	Reference 13	This Paper
−19.68	11.79	−0.34553	−0.3857	−0.2912
−14.98	11.56	−0.25367	−0.2678	−0.2198
−9.65	11.15	−0.24591	−0.1783	−0.1423
−4.78	9.26	0.0321	−0.0026	0.0035
3.45	9.61	0.0056	−0.0098	0.0102
20.01	23.55	−0.1785	0.0728	0.0128
24.58	25.05	−0.0148	−0.0118	0.0021
29.98	25.95	0.04726	0.0248	0.0219
38.89	27.8	0.0147	0.0975	0.0175
51.21	29.25	0.0947	0.0752	0.0642

It can be seen from Table 1 and Figure 11:

1. The overall trend of the compensation effects of the three methods is the same. The compensation error used in reference 11 is large, and the error at the segmentation node significantly increased. The compensation effect of the method in this paper is relatively stable. In particular, the curve between 0 and 40 °C tends to be gentle and close to zero.

2. The adaptive segmentation compensation method combines the simplicity and efficiency of the least squares method with the high precision of the GSA-BP neural network. The measurement error of humidity significantly improved over the entire temperature range. In the vicinity of the

segmentation point (22.36 °C, 29.98 °C) obtained by the adaptive calculation, the compensation effect is particularly significant.

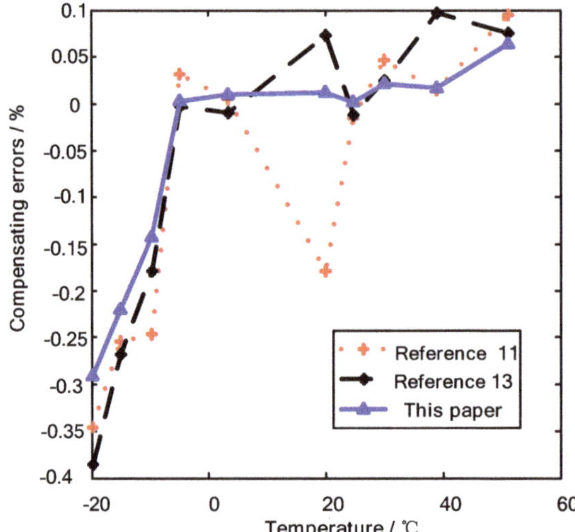

Figure 11. Error curve after compensation by three methods.

6. Conclusions

We use this method to compensate for humidity sensors of different temperature ranges and different individual temperatures. According to the error curve under different ambient temperatures, multiple linear regression analysis was used to determine the segmentation points, and different algorithms were used to compensate. The results show that the method has strong universality and is effective for different temperature ranges and individual measurements.

In addition, the introduction of independent component analysis method into the reconstruction of measured error might further improve the compensation effect [20]. It should be added that the initial temperature is 25 °C when determining the segmentation point in this method. This temperature is employed because the humidity error of the sensor used in this experiment is relatively linear at approximately 25 °C after multiple measurements. If other sensors are used, the initial values will be determined according to the humidity curves of those different sensors. However, the initial values will be not very strict, and the nearby values can be adaptively processed as initial values.

Author Contributions: Conceptualization, W.X. and X.F.; Methodology, W.X.; Software, X.F.; Validation, X.F. and W.X.; Data Curation, H.X.; Writing-Original Draft Preparation, X.F. and W.X.; Writing-Review & Editing, W.X.; Supervision, W.X.; Project Administration, H.X.; Funding Acquisition, W.X.

Funding: This work is supported by the National Key R&D Program of China (2018YFC1506102) and the National Natural Science Foundation of China (Grant NO. 41605121 and NO. 61671248).

Conflicts of Interest: The authors declare no conflicts of interest.

References

1. Weiwei, L.V.; Xiaohua, L.V.; Zuoyang, T.; Xiaohua, X.; Yong, X. Fault Analysis and Maintenance of DZZ Series of Automatic Weather Stations. *Meteorol. Environ. Res.* **2018**, *9*, 35–37.
2. Hai, M.; Wen, F.; Wang, C.; Li, X.J. Capacitive humidity sensing properties of CdS/ZnO sesame-seed-candy structure grown on silicon nanoporous pillar array. *J. Alloys Compd.* **2017**, *11*, 94–98.
3. Lob, V.; Geisler, T.; Brischwein, M.; Uhl, R.; Wolf, B. Humidity sensor using a single molecular transistor. *J. Appl. Phys.* **2015**, *118*, 135–171.

4. Wang, D.F.; Lou, X.; Bao, A.; Yang, X.; Zhao, J. A temperature compensation methodology for piezoelectric based sensor devices. *Appl. Phys. Lett.* **2017**, *111*, 083502. [CrossRef]

5. Ruirong, D.; Hongwei, Z.; Nan, S.; Bo, D.; Dengyue, W. Compensation and calibration of the high temperature and pressure downhole pressure sensor. *Chin. J. Sci. Instrum.* **2016**, *43*, 737–743.

6. Hsieh, C.; Hung, C.; Li, Y. Investigation of a Pressure Sensor with Temperature Compensation Using Two Concentric Wheatstone-Bridge Circuits. *Mod. Mech. Eng.* **2013**, *10*, 104–113. [CrossRef]

7. Zhou, G.; Zhao, Y.; Guo, F. A Smart High Accuracy Silicon Piezoresistive Pressure Sensor Temperature Compensation System. *Sensors* **2014**, *4*, 74–90. [CrossRef] [PubMed]

8. Li, T.; Liang, S.; Hong, Y.; Pan, L. Simulation of Temperature Compensation of Pressure Sensor Based on PCA and Improved BP Neural Network. *Adv. Mater. Res.* **2014**, *846–847*, 513–516. [CrossRef]

9. Li, Y.; Li, Y.; Li, F.; Zhao, B. The Research of Temperature Compensation for Thermopile Sensor Based on Improved PSO-BP Algorithm. *Math. Probl. Eng.* **2015**, *3*, 1–6. [CrossRef]

10. Zhu, L.; Xie, B.; Xing, Y.; Chen, D.; Wang, J. A Resonant Pressure Sensor Capable of Temperature Compensation with Least Squares Support Vector Machine. *Procedia Eng.* **2016**, *168*, 1731–1734. [CrossRef]

11. Jiwei, P.; Wenhua, L.V.; Hongyan, X.; Xiangjuan, W. Temperature compensation for humidity sensor based on improved GA-BP neural network. *Chin. J. Sci. Instrum.* **2013**, *34*, 153–160.

12. Guo, M.; Xing, H.; Zhang, D.; Zhang, L. Temperature Compensation for Humidity Sensor Based on the AFSA-BP Neural Network. *Instrum. Tech. Sens.* **2017**, *8*, 6–10.

13. Xing, H.Y.; Peng, J.W.; Lv, W.H. A fusion algorithm for humidity sensor temperature compensation. *Chin. J. Sens. Actuators* **2012**, *25*, 1711–1716.

14. Kutner, M.; Nachtsheim, C.; Neter, J. *Applied Linear Regression Models*; McGraw-Hill/Irwin Education: New York, NY, USA, 2004.

15. Dos Soares, T.S.; Mendes, D.; Rodrigues, T.R. Artificial neural networks and multiple linear regression model using principal components to estimate rainfall over South America. *Nonlinear Process. Geophys.* **2016**, *23*, 1317–1337.

16. Xu, W.; Li, F.; Liu, F. Optimality and Recursive Algorithm of General Least Squares Estimator of Seemingly Unrelated Linear Regression Models. In Proceedings of the 2010 International Conference on Computational Intelligence and Software Engineering, Wuhan, China, 10–12 December 2010.

17. Shidrokh, G.; Hassan, W.H.; Hossein, M.; Seyed, A.; Soleymani, A. MDP-Based Network Selection Scheme by Genetic Algorithm and Simulated Annealing for Vertical-Handover in Heterogeneous Wireless Networks. *Wirel. Pers. Commun.* **2017**, *2*, 399–436.

18. Örkcü, H.H. Subset selection in multiple linear regression models: A hybrid of genetic and simulated annealing algorithms. *Appl. Math. Comput.* **2013**, *5*, 11018–11028.

19. Wang, K.; Luo, X.; Shen, H.; Zhang, H. GSA-BP neural network model for back analysis of surrounding rock mechanical parameters and its application. *Rock Soil Mech.* **2016**, *37*, 631–638.

20. Chen, X.; Ma, D. Mode Separation for Multimodal Ultrasonic Lamb Waves Using Dispersion Compensation and Independent Component Analysis of Forth-Order Cumulant. *Appl. Sci.* **2019**, *9*, 555. [CrossRef]

 *electronics*

Article

An Image Compression Method for Video Surveillance System in Underground Mines Based on Residual Networks and Discrete Wavelet Transform

Fan Zhang [1,2,*], **Zhichao Xu** [1], **Wei Chen** [3,4], **Zizhe Zhang** [3], **Hao Zhong** [1], **Jiaxing Luan** [1] and **Chuang Li** [1]

1. School of Electrical and Information Engineering, China University of Mining and Technology (Beijing), Beijing 100083, China; generalxzc@hotmail.com (Z.X.); anatole_hao@163.com (H.Z.); sqt1800407113@student.cumtb.edu.cn (J.L.); zqt1800407138g@student.cumtb.edu.cn (C.L.)
2. Institute of Intelligent Mining and Robotics, China University of Mining and Technology (Beijing), Beijing 100083, China
3. School of Computer Science and Technology and Mine Digitization Engineering Research Center of the Ministry of Education, China University of Mining and Technology, Xuzhou 221116, China; chenwdavior@163.com (W.C.); z32000@126.com (Z.Z.)
4. School of Earth and Space Sciences, Peking University, Beijing 100871, China
* Correspondence: zf@cumtb.edu.cn

Received: 28 November 2019; Accepted: 13 December 2019; Published: 17 December 2019

Abstract: Video surveillance systems play an important role in underground mines. Providing clear surveillance images is the fundamental basis for safe mining and disaster alarming. It is of significance to investigate image compression methods since the underground wireless channels only allow low transmission bandwidth. In this paper, we propose a new image compression method based on residual networks and discrete wavelet transform (DWT) to solve the image compression problem. The residual networks are used to compose the codec network. Further, we propose a novel loss function named discrete wavelet similarity (DW-SSIM) loss to train the network. Because the information of edges in the image is exposed through DWT coefficients, the proposed network can learn to preserve the edges better. Experiments show that the proposed method has an edge over the methods being compared in regards to the peak signal-to-noise ratio (PSNR) and structural similarity (SSIM), particularly at low compression ratios. Tests on noise-contaminated images also demonstrate the noise robustness of the proposed method. Our main contribution is that the proposed method is able to compress images at relatively low compression ratios while still preserving sharp edges, which suits the harsh wireless communication environment in underground mines.

Keywords: underground mines; intelligent surveillance; residual networks; compressed sensing; image compression; image restoration; discrete wavelet transform

---

1. Introduction

1.1. The Image Compression Demand from Underground Mines

Coal is one of the major resources in China. In the foreseeable future, China will still be the largest consumer and the producer of coal [1]. Therefore, it is of great importance to research into technologies that contribute to the advancement in intelligent mine monitoring and safe mining practices.

One of the key components of intelligent mine monitoring is the video surveillance system since visual information plays a key role in how a human perceives the world. Because digital images usually require large storage, it is natural to think of transmitting images with high bandwidth channels, like cable networks. Although cable networks could potentially provide enough bandwidth, they are inflexible in that the cable networks are fixed and have to expand as the working surface expands. In favor of mobility, wireless networks are usually chosen as the information channel in mines. However, the bandwidth can be limited because of relatively limited narrow spaces, harsh environment diffraction, attenuation, and multi-path effect in underground mines. The problem can be especially serious when disasters such as explosion and collapse occur [2]. Therefore, it is necessary to investigate image compression methods in order to save the transmission bandwidth.

1.2. From Conventional Image Compressing to Compressed Sensing

There have been vast investigations into the field of image compression. Among the researches, JPEG (Joint Photographic Experts Group) [3] has been quite popular and influential. JPEG mainly employs discrete cosine transform (DCT) and entropy coding techniques to compress the images. While the JPEG compression method has gained widespread popularity, it does introduce visible artifacts including blurring, ringing and blocking [4]. JPEG2000 [5] is proposed forward to address the problems in JPEG. JPEG2000 adopts 2D wavelet transform and arithmetic coding to achieve higher compression efficiency.

Besides utilizing transforms and entropy coding techniques, a theory framework known as compressed sensing (CS) [6–8] was proposed to overcome the limitation that a signal must be sampled at the Nyquist sampling rate [9]. The CS theory has shed light on the problem of compression and reconstruction. Optimization techniques such as total variation (TV) minimization [10] and approximate message passing (AMP) [11] can be used in the recovery phase in the CS framework. TV minimization for image denoising was first introduced in [12]. TV minimization takes the advantage that it can better accurately preserve the edges or boundaries at certain compression ratios. In [13], the method "total variation minimization by augmented Lagrangian and alternating direction algorithms" (TVAL3) is proposed and has been used widely in image recovery problems. Comparisons in [14] suggest that the TVAL3 solver turns out to be fast and efficient so long as the reconstruction parameters are sufficient for a satisfying reconstruction. Meanwhile, based on the AMP [11] recovery algorithm, the D-AMP [15] algorithm is proposed to enhance CS recovery. In the scheme of D-AMP, the existing rich knowledge of signal denoiser is utilized to design the solver. Tests in [15] show that the D-AMP maintains a low computational footprint. Compressed sensing- based techniques have been explored in real-life scenarios like mine monitoring image compression [16] and landslide monitoring system [17]. The non-learning compressed sensing methods do achieve some success, but they struggle to produce sound recoveries at low compression ratios.

1.3. Data-driven Approaches

Due to the advancement of information technology, more data is within the reach of researchers. The data-driven approaches have found their way into various fields including signal processing [18], control systems [19–22] and especially vision tasks [23–27]. In particular, the deep learning-based method has stood out among the data-driven approaches. This section explores the recent development of deep learning-based image compression methods.

1.3.1. Convolution Neural Network based Image Compression

In more recent years, convolution neural networks (CNNs) has gained great attention due to the improvement of computing devices. As for image compression utilizing CNN, it generally involves designing image codecs with neural networks and constructing appropriate loss functions.

One genre of compression method combines the ideas of compressed sensing into CNN. For instance, the network DeepInverse proposed in [28] uses fully connected layers to simulate the compression process and stacks convolution layers for decompression. Back-propagation is applied to train the networks. This idea is extended further by ReconNet [29] which uses more convolution layers to attack the decompression problem. In [30], a deep residual reconstruction network is proposed to recover images more accurately. However, this series of methods are more likely to blur edges in the recovered image especially at low compression ratios, according to the results reported by their authors.

Another genre of CNN based compression methods utilize the semantic information in images, since preserving semantic information will render the recovered image more eye-pleasing. Ballé et al. introduce an end-to-end optimized CNN image compression network in [31]. The method is based on non-linear coding rather than linear coding used by JPEG. One important contribution of [31] is that the authors propose an method which simulates the quantizer in the training procedure to deal with the problem of zero derivatives due to quantization. Li et al. point out that in [4] it is inappropriate to allocate the same number of codes for each spatial position in an image. They propose the importance map to guide the spatially variant bit allocation. To further compress the data, they introduce the convolutional entropy encoder to compress the binary codes and the importance map. In [32], the authors combine the deep-learning-based image semantic analysis into image compression as well. Unlike [4] which focuses more on the edge of objects, the method in [32] emphasizes the semantic analysis of the whole region. Results in their experiments show the method can improve the visual quality under the same compression overhead. However, it can be quite complicated to adjust the compression ratios of this genre of methods. Moreover, these methods are rarely applied at very low compression ratios.

1.3.2. Recurrent Neural Network Based Image Compression

Unlike the feed-forward CNN, the recurrent neural network (RNN) is state-aware. The output of an RNN is not only related to current input, but also the previous input. Lyu et al. propose to combine the knowledge of block-sparsity recovery into RNN deep learning in [33]. Their method acquires the spatial correlations between nonzero elements of block-sparse signals. It is applied to not only images but also audio data. However, the method proposed in [33] requires the input data to be sparse, which limits its compression capability. In [34], Toderici et al. combine the scaled-additive coding framework into RNN-based image compression scheme. The highlight in [34] is that the architectures proposed can provide variable compression rates during deployment without retraining the network. In [35], Minnen et al. propose a spatially adaptive image compression framework with quality-sensitive bit rate adaptation. However, though their method outperforms JPEG, it is still inferior to JPEG2000 [36].

1.3.3. Generative Adversarial Network Based Image Compression

Generative adversarial network (GAN) is another promising deep learning method developed during recent years. In the GAN scheme, a generator network and a discriminator network are optimized simultaneously. The discriminator network is trained to determine whether a sample is generated by the generator network, while the generator network needs to fool the discriminator into wrong decisions. In regards of image compression utilizing GAN, Ripple and Bourdev in [37] propose an architecture of autoencoder featuring pyramidal analysis, an adaptive coding module, and regularization of the expected code length. It produces images 2.5 times smaller than JPEG and JPEG2000, while achieving realtime performance using GPU. Jia et al. in [38] propose a light filed image compression framework driven by a GAN-based sub-aperture image generation and a cascaded hierarchical coding structure. Their method outperforms the state-of-the-art learning-based light field image compression approach with on average 4.9% BD-rate [39] reductions. In [40], Agustsson et al. propose a GAN-based framework targeting extremely low bitrate compression. Their method pushes the bitrate below 0.1 bpp while still achieves eye-pleasing results.

1.4. The Objectives and the Organization of the Paper

Considering the demand of image compression at very low compression ratios in underground mines, in this paper, we propose an image codec network based on CNN and a new loss function based on discrete wavelet transform. The new loss function is dedicated to preserving edges in the images of underground mines. The remaining of the paper is organized as follows: Section 2 elaborates the proposed method by discussing the network architecture and the construction of the loss function. Section 3 provides experiments which demonstrate the performance and analysis of the proposed method. Section 4 concludes the paper with further discussion about the proposed method.

2. The Proposed Image Compression Method

2.1. Overview

Before introducing the network architecture, it is necessary to understand the workflow of the proposed compression method. As shown in Figure 1, a gray-scale image or one of the channels of an RGB color image is taken as the input. We view the input image as a matrix $\mathbf{x}$. For simplicity, we assume the input image is square, which means $\mathbf{x}$ has the same number of rows and columns. The image matrix $\mathbf{x}$ is "vectorized" into one vector $\mathbf{x}_v$ by concatenating each row of the matrix. The encoder module compresses $\mathbf{x}_v$ to a feature vector $\mathbf{y}$. Then the decoder module is applied to approximate $\mathbf{x}$ using the feature vector $\mathbf{y}$. The approximation of $\mathbf{x}$ is denoted as $\hat{\mathbf{x}}$. During training, both the recovered image and the original image are fed into the loss function. Back-propagation will try to minimize the value of loss by updating the weights in the encoder and decoder module.

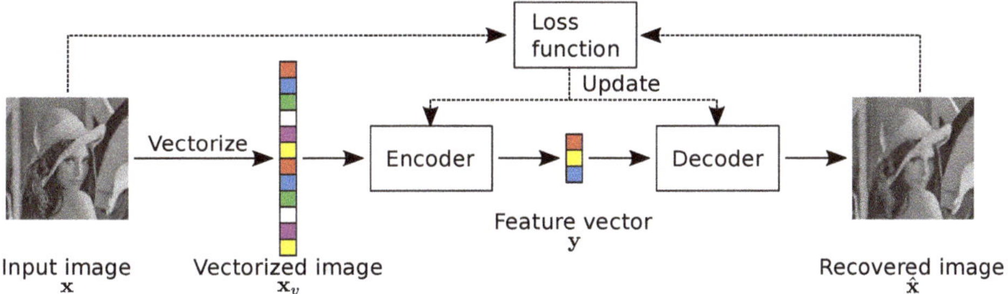

Figure 1. The workflow of the proposed method.

If there are N numbers in the image matrix $\mathbf{x}$ and M numbers in the feature vector $\mathbf{y}$, then we define the compression ratio r as

$$r = M/N. \tag{1}$$

In short, the encoder module is responsible for compressing the image and determining the compression ratio, while the decoder module takes care of the recovery process.

2.2. The Network Architecture

2.2.1. The Encoder Module

The weight matrix $\mathbf{W}$ of size $M \times N$ is multiplied by the "vectorized" image $\mathbf{x}_v$. Then the product is added by the bias vector $\mathbf{b}$ to derive the feature vector $\mathbf{y}$:

$$\mathbf{y} = \mathbf{W}\mathbf{x}_v + \mathbf{b}. \tag{2}$$

In Equation (2), both the weight matrix $\mathbf{W}$ and the bias vector $\mathbf{b}$ are parameters to be learnt during back-propagation. $\mathbf{W}$ is initialized using He initialization [41], while $\mathbf{b}$ is initialized with zeros.

2.2.2. The Decoder Module

The network architecture of the decoder module is illustrated in Figure 2. The feature vector **y** is first upsampled to **y**$'$ using nearest-neighbor interpolation [42]. The length of **y**$'$ is determined by Equation (3):

$$\text{length}(\mathbf{y}') = \lceil \sqrt{M} \rceil^{2}, \tag{3}$$

where M is the length of vector **y**. The symbol $\lceil z \rceil$ means rounding number z to the nearest integer more than or equal to z. The vector **y**$'$ is then reshaped into the initial feature map **F** using Equation (4):

$$\mathbf{F}[i,j] = \mathbf{y}'[(i-1) \times \lceil \sqrt{M} \rceil + j], 1 \leq i,j \leq \lceil \sqrt{M} \rceil. \tag{4}$$

Afterwards, the initial feature map **F** is convolved with 96 filters of size 3×3. We empirically add a batch-normalization [43] layer after the first convolution layer to accelerate training. Then the feature maps go through several residual units. Some residual units are followed by nearest-neighbor upsampling operation as in Figure 2. Finally, the feature maps are convolved with one filter of size 1×1 to derive the recovered image $\hat{\mathbf{x}}$.

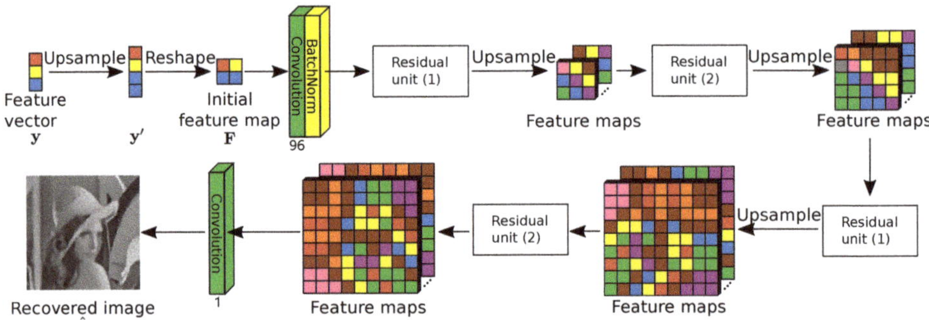

Figure 2. The network architecture of the decoder module.

The residual units. The introduction of residual units is inspired by [44]. As depicted in Figure 3, two types of residual units are used. Both types follow the two-branch connection pattern. The feature maps go through the two branches and add up at the output summator. The upper branches of the two types are identical. The lower branches differ in that residual unit (1) connects the input and the output with a stack of layers, but residual unit (2) connects the input and the output directly. Each convolution layer that appears in Figure 3 is composed of 96 filters of size 3×3. After each convolution layer, there is a batch-normalization layer [43]. Each batch-normalization layer is then followed by a Leaky ReLU activation layer [45] if the batch-normalization layer is not directly connected to the output summator.

The nearest-neighbor upsampling operations. If the input image **x** is of size $n \times n$, then the second, third, and fourth upsampling operation in Figure 2 resize the feature maps to size $\frac{1}{2}n \times \frac{1}{2}n \times 96$, $\frac{3}{4}n \times \frac{3}{4}n \times 96$, and $n \times n \times 96$, respectively.

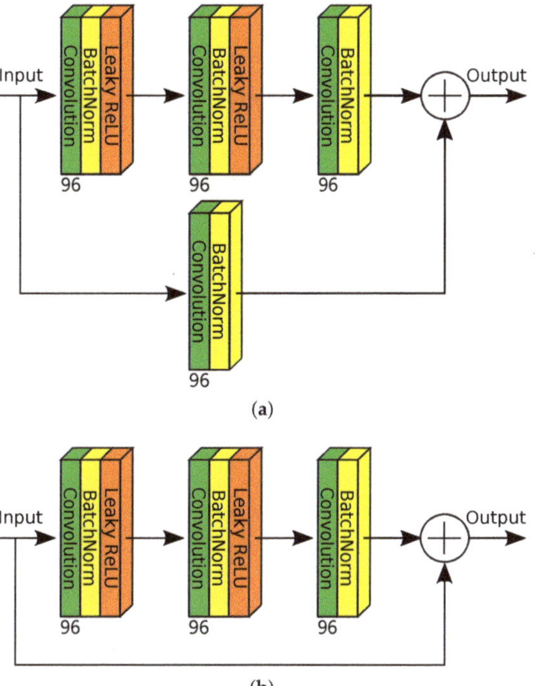

Figure 3. The residual units: (**a**) residual unit (1); (**b**) residual unit (2). All convolution layers in the two types of residual units employ filters of size 3×3.

2.3. The Proposed Loss Function

2.3.1. Combination of Two Types of Loss Functions

Image recovering problems are conventionally seen as optimization problems that minimize the l_2 loss between the recovered and original image. However, from the perspective of image recovery quality assessment, l_2 metric does not reflect every aspect of signal fidelity [46]. Therefore, it is necessary to combine other metrics that compensate for what is missing in l_2 loss when constructing the loss function.

In this section, we propose a metric termed discrete wavelet structural similarity (DW-SSIM) that focuses the recovery of edges of the images. Our loss function is the weighted sum of DW-SSIM loss and l_2 loss:

$$L(\mathbf{x}, \hat{\mathbf{x}}) = \sum_{\mathbf{x} \in \Omega} (\beta_1 L_F(\mathbf{x}, \hat{\mathbf{x}}) + \beta_2 L_S(\mathbf{x}, \hat{\mathbf{x}}))$$
$$\beta_1 + \beta_2 = 1 \tag{5}$$
$$0 \leq \beta_1, \beta_2 \leq 1,$$

where Ω represents a set of training image, $L_F(\mathbf{x}, \hat{\mathbf{x}})$ denotes the l_2 loss, $L_S(\mathbf{x}, \hat{\mathbf{x}})$ denotes the DW-SSIM loss, and $\beta_1 = 0.5$ and $\beta_2 = 0.5$ are weights. Both $L_F(\mathbf{x}, \hat{\mathbf{x}})$ and $L_S(\mathbf{x}, \hat{\mathbf{x}})$ are set up to fall in range $[0, 1)$. Section 2.3.2 will provide the expression of $L_F(\mathbf{x}, \hat{\mathbf{x}})$, while Section 2.3.3 will explain $L_S(\mathbf{x}, \hat{\mathbf{x}})$ in details.

2.3.2. l_2 Loss

We propose to use Frobenius norm in $L_F(\mathbf{x}, \hat{\mathbf{x}})$ to derive the l_2 loss:

$$L_F(\mathbf{x}, \hat{\mathbf{x}}) = \frac{\|\hat{\mathbf{x}} - \mathbf{x}\|_2^2}{\|\mathbf{x}\|_2^2}. \tag{6}$$

It is worth noting that the denominator of Equation (6) cannot be zero. However, since $\mathbf{x}$ is taken from the natural images instead of artificial generated matrices, it is impossible for $\mathbf{x}$ to be a zero matrix.

2.3.3. Discrete Wavelet Similarity (DW-SSIM) and DW-SSIM Loss

Inspired by structural similarity (SSIM) [47] and complex-wavelet structure similarity (CW-SSIM) [48], we propose to use two-dimensional discrete wavelet transform (2D-DWT) [49] to analyze the similarity between the recovered image and the original image. The similarity is termed DW-SSIM which stands for discrete-wavelet similarity.

2D-DWT. The 2D-DWT is able to decompose an image into different levels of subbands. The first level is the decomposition of the original image. Each level is composed of four subband images which can be referred to as low–low (LL), low–high (LH), high–low (HL) and high–high (HH). The LL image at each level can be further decomposed into the next level of subbands. The LH image represents the variation along the vertical direction, HL image the horizontal direction, and HH image the diagonal direction [49]. The high-frequency LH, HL, and HH subband images altogether form the details of the original image. As the decomposition level goes higher, the subband images become coarser, thus details of different scales can be analyzed. Figure 4 provides an example of a three-level 2D-DWT decomposition of an image.

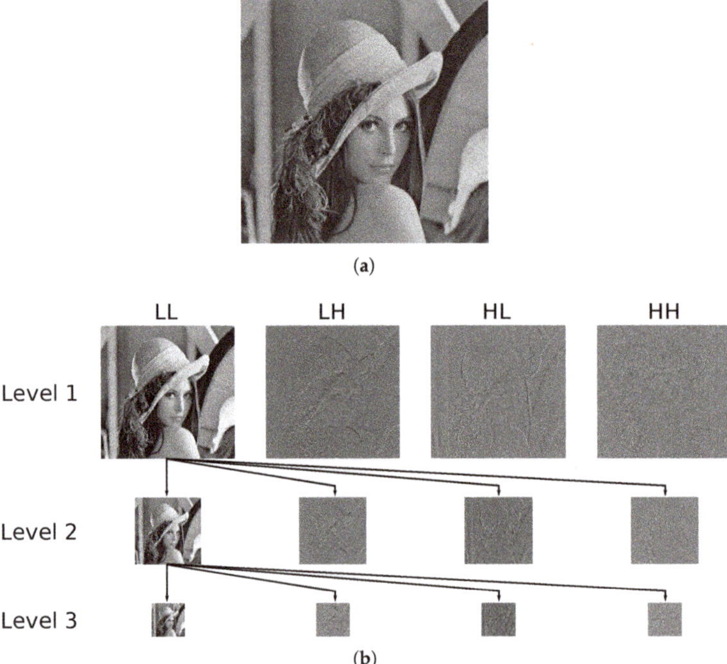

Figure 4. Illustration of 2D-discrete wavelet transform (DWT) image decomposition. (**a**) Original image. (**b**) Three-level decomposition of the image. For clarity, every intermediate low–low (LL) image is put in its place, yet DWT only preserves the LL image of the highest level.

DW-SSIM. We divide the calculation of DW-SSIM between the original image and the recovered image into two stages. The first stage involves figuring out the local DW-SSIM, where a "window" slides through the original image and the recovered image. 2D-DWT is performed on the image patches within the "window" to derive the decomposition. We define the local low frequency DW-SSIM $S_{L,t}$ and high frequency DW-SSIM $S_{H,t}$ of the image patches as

$$S_{L,t}(\mathbf{c}^{(1)}, \mathbf{c}^{(2)}) = \frac{2\left|\sum_u \sum_v \mathbf{c}_{LL}^{(1)}[u,v]\mathbf{c}_{LL}^{(2)}[u,v]\right| + K}{\sum_u \sum_v \left|\mathbf{c}_{LL}^{(1)}[u,v]\right|^2 + \sum_u \sum_v \left|\mathbf{c}_{LL}^{(2)}[u,v]\right|^2 + K}, \tag{7}$$

$$S_{H,t}(\mathbf{c}^{(1)}, \mathbf{c}^{(2)}) = \frac{1}{J}\sum_{j=1}^{J} \frac{2\left|\sum_i \sum_u \sum_v \mathbf{c}_i^{(1)}[u,v]\mathbf{c}_i^{(2)}[u,v]\right| + K}{\sum_i \sum_u \sum_v \left|\mathbf{c}_i^{(1)}[u,v]\right|^2 + \sum_i \sum_u \sum_v \left|\mathbf{c}_i^{(2)}[u,v]\right|^2 + K}, i \in \{LH_j, HL_j, HH_j\}. \tag{8}$$

In Equations (7) and (8), K is a small positive constant for arithmetic robustness and K is set to 0.01. $\mathbf{c}^{(1)}$ and $\mathbf{c}^{(2)}$ refer to the corresponding subband images of the original image patch and the recovered image patch after 2D-DWT, respectively. The wavelet function we use is the Haar wavelet. t is the patch index. $J = 3$ is the maximum decomposition level, and $c_{LH_j}, c_{HL_j}, c_{HH_j}$ are high frequency subband images at the j-th level.

To better understand Equation (7), one can ignore K, "vectorize" (as in Section 2.2.1) $\mathbf{c}$ into $\mathbf{c}_v$ and rewrite it as

$$\begin{aligned}
S_{L,t}(\mathbf{c}^{(1)}, \mathbf{c}^{(2)}) = S_{L,t}(\mathbf{c}_v^{(1)}, \mathbf{c}_v^{(2)}) &= 2\frac{|\mathbf{c}_v^{(1)} \cdot \mathbf{c}_v^{(2)}|}{\|\mathbf{c}_v^{(1)}\|_2^2 + \|\mathbf{c}_v^{(2)}\|_2^2} \\
&= 2\frac{\|\mathbf{c}_v^{(1)}\|_2\|\mathbf{c}_v^{(2)}\|_2}{\|\mathbf{c}_v^{(1)}\|_2^2 + \|\mathbf{c}_v^{(2)}\|_2^2}\left|\frac{\mathbf{c}_v^{(1)} \cdot \mathbf{c}_v^{(2)}}{\|\mathbf{c}_v^{(1)}\|_2\|\mathbf{c}_v^{(2)}\|_2}\right| \\
&= 2\left(\frac{\|\mathbf{c}_v^{(1)}\|_2}{\|\mathbf{c}_v^{(2)}\|_2} + \frac{\|\mathbf{c}_v^{(2)}\|_2}{\|\mathbf{c}_v^{(1)}\|_2}\right)^{-1}|\cos(\theta)|.
\end{aligned} \tag{9}$$

In Equation (9), the first term is determined by the energy of the subband images. It will reach its maximum value 1 only if $\|\mathbf{c}_v^{(1)}\|_2 = \|\mathbf{c}_v^{(2)}\|_2$. In the second term, $\cos(\theta) = \frac{\mathbf{c}_v^{(1)} \cdot \mathbf{c}_v^{(2)}}{\|\mathbf{c}_v^{(1)}\|_2\|\mathbf{c}_v^{(2)}\|_2}$ is the cosine similarity [50]. If $\mathbf{c}_v^{(1)}$ and $\mathbf{c}_v^{(2)}$ point to roughly the same direction, the cosine similarity will be close to 1. However, the cosine function falls in range $[-1, 1]$. Therefore we are taking the absolute value so that it falls in $[0, 1]$. The interpretation of Equation (8) is largely the same with that of Equation (7). Equation (8) additionally averages the contribution of each level of subband to the high frequency DW-SSIM in order to cope with the patterned noise in underground mine images. This can be better understood through the discussion in Section 3.3.

In the second stage, a weighted sum of $S_{H,t}$ and $S_{L,t}$ is figured out to form the final DW-SSIM S:

$$S(\mathbf{x}, \hat{\mathbf{x}}) = \frac{1}{T}\sum_{t=1}^{T}(\gamma_1 S_{L,t} + \gamma_2 S_{H,t}), \tag{10}$$

where T is the total number of image patches, γ_1 and γ_2 are parameters to adjust the weight of low frequency subband and high frequency subbands. Since we want to emphasize high frequency details such as edges and spikes in the image, we set $\gamma_1 = 0.2$ and $\gamma_2 = 0.8$.

The computation of DW-SSIM is summarized with Algorithm 1. The window length l in the proposed method is set to 15. The stride s that the window will move in each iteration is set to 8.

Algorithm 1: The procedure to compute discrete wavelet similarity (DW-SSIM).

Input: The original image img-ori and the recovered image img-rec of the same height H and width W ($H > 0, W > 0$); the decomposition level J; the stride s that the window will move in each iteration; the window length l; the weights γ_1 and γ_2 in Equation (10)

Output: The DW-SSIM similarity S between img-ori and img-rec

$S \leftarrow 0$;
$t \leftarrow 0$;
up $\leftarrow 0$;
left $\leftarrow 0$;
while up $< H$ **do**
    down $\leftarrow$ up $+ l$;
    while left $< W$ **do**
        $t \leftarrow t + 1$;
        right $\leftarrow$ left $+ l$;
        Get image patch patch-ori within the window [up, down, left, right] from img-ori;
        Get image patch patch-rec within the window [up, down, left, right] from img-rec;
        Derive $\mathbf{c}^{(1)}$ by performing 2D-DWT on patch-ori;
        Derive $\mathbf{c}^{(2)}$ by performing 2D-DWT on patch-rec;
        Derive $S_{L,t}$ from Equation (7);
        Derive $S_{H,t}$ from Equation (8);
        $S \leftarrow S + \gamma_1 S_{L,t} + \gamma_2 S_{H,t}$;
        left $\leftarrow$ left $+ s$;
    end
    up $\leftarrow$ up $+ s$;
end
$S \leftarrow S / t$;
return S;

DW-SSIM loss. The DW-SSIM defined in Equation (10) falls in range $(0, 1]$. The more the original image and the recovered image matches each other, the closer DW-SSIM S is to 1. However, the loss should be near 0 if the model has done a perfect recovery. Moreover, the loss should fall in range $[0, 1)$. Therefore, we define the DW-SSIM loss as:

$$L_S(\mathbf{x}, \hat{\mathbf{x}}) = 1 - S(\mathbf{x}, \hat{\mathbf{x}}). \tag{11}$$

2.4. Learning the Parameters

The encoder module and the decoder module can be trained in an end-to-end manner using the proposed network architecture and the proposed loss function. Mini-batch gradient descent is used to train the model with the batch size being 64. The Adam [51] optimizer is utilized as well. We set the initial learning rate to 5×10^{-4}. The learning rate is multiplied by 0.2 when the loss is not going down during training. The training is stopped if the learning rate drops below 1×10^{-6}.

3. Results

3.1. Overview

In order to generalize the recovery capability, the network of the proposed method is trained on both images from video images we have collected in underground mines and images from the COCO 2014 dataset [52]. We build the training set by extracting the 100×100 center-crop patches from the images, and converting them to grayscale images.

After the model is trained, test images (as in Figure 5) are passed to the model to perform the compression and recovery. We test our method on both standard images of Barbara, Fingerprint, and Lena to verify its effectiveness. In addition, we test the proposed method on images of coal cutter and tunnel boring machine (TBM) which are from real underground mines to evaluate the performance in the application-specific environment.

The recovery quality is quantitatively evaluated with peak-signal-to-noise ratio (PSNR) and structural similarity (SSIM) [46]:

$$\mathrm{PSNR}(\mathbf{x}, \hat{\mathbf{x}}) = 10 \log_{10} \frac{d^2}{\frac{1}{N} \sum_{i=1}^{N} (\mathbf{x}[i] - \hat{\mathbf{x}}[i])^2}, \tag{12}$$

$$\mathrm{SSIM}(\mathbf{x}, \hat{\mathbf{x}}) = \left(\frac{2\mu_\mathbf{x}\mu_{\hat{\mathbf{x}}} + C_1}{\mu_\mathbf{x}^2 + \mu_{\hat{\mathbf{x}}}^2 + C_1} \right) \cdot \left(\frac{2\sigma_\mathbf{x}\sigma_{\hat{\mathbf{x}}} + C_2}{\sigma_\mathbf{x}^2 + \sigma_{\hat{\mathbf{x}}}^2 + C_2} \right) \cdot \left(\frac{\sigma_{\mathbf{x}\hat{\mathbf{x}}} + C_3}{\sigma_\mathbf{x}\sigma_{\hat{\mathbf{x}}} + C_3} \right). \tag{13}$$

In Equation (12), d is the dynamic range of pixel intensities, and N is the number of pixels in the image. In Equation (13), $\mu_\mathbf{x}$ and $\mu_{\hat{\mathbf{x}}}$ are means of $\mathbf{x}$ and $\hat{\mathbf{x}}$, and $\sigma_\mathbf{x}^2$ and $\sigma_{\hat{\mathbf{x}}}^2$ are variances of $\mathbf{x}$ and $\hat{\mathbf{x}}$. $\sigma_{\mathbf{x}\hat{\mathbf{x}}}$ is the cross correlation of $\mathbf{x}$ and $\hat{\mathbf{x}}$. The small positive constants $C_1 = C_2 = C_3 = 0.01$ prevent numerical instability of each term.

To verify the effectiveness of the proposed method, the quantitative evaluation at compression ratios of 0.25, 0.20, 0.15, 0.10, 0.04 and 0.01 is carried out, with the compression ratio defined in Equation (1). In addition, the proposed method is compared to the algorithms of D-AMP [15], ReconNet [13] and TVAL3 [29] at different compression ratios. For simplicity, we do not re-implement the algorithms but use the demo code provided by the authors' websites instead.

Further, visual quality evaluation of recovery is presented at some specific compression ratios.

Finally, the robustness of the proposed method is tested by recovering images contaminated by different levels of Gaussian noise.

The proposed method was implemented with Pytorch [53] and pytorch_wavelet package (https://github.com/fbcotter/pytorch_wavelets). The training process is carried out on Ubuntu 18.04.2, with Nvidia Tesla K80 GPU and Intel Xeon CPU. More details about the implementation can be found in the code which we have made public on the Internet (https://github.com/y0umu/ResCSNet).

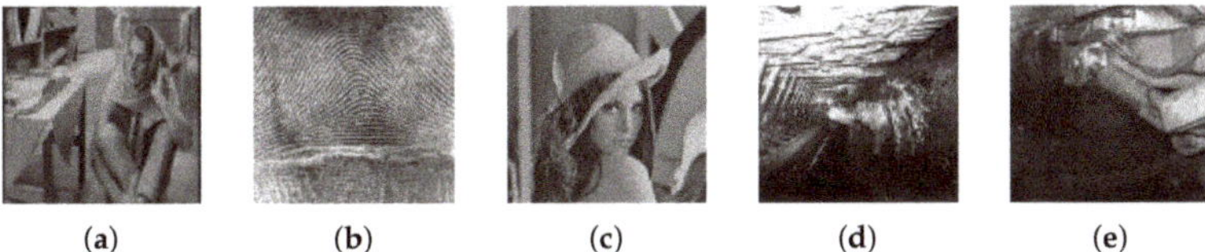

| (a) | (b) | (c) | (d) | (e) |

Figure 5. The test images: (**a**) Barbara; (**b**) Fingerprint; (**c**) Lena; (**d**) Coal cutter; (**e**) Tunnel boring machine (TBM).

3.2. Quantitative Evaluation

Tables 1 and 2 provide quantitative measurements of the proposed method and other algorithms at different compression ratios. As the compression ratio r decreases, all the algorithms being compared have PSNR and SSIM decreased. It can be interpreted from Table 1 that the proposed method is second only to D-AMP at compression ratio $r \geq 0.20$ for both standard test images and real underground mine images. Yet the proposed method achieves the highest PSNR compared to other algorithms at a compression ratio $r \leq 0.15$. It should be also noted that for the recoveries of images of coal cutter and TBM at compression ratios $r \leq 0.04$, the proposed method has an edge over other algorithms by a margin of at least 1.8 dB, indicating the potential of the application-specific usage in mines of the proposed method.

Table 1. Peak signal-to-noise ratio (PSNR) (in dB) comparison for different algorithms on test images. r is the compression ratio.

Image	Algorithm	r = 0.25	r = 0.20	r = 0.15	r = 0.10	r = 0.04	r = 0.01
Barbara	D-AMP	26.61	25.37	24.00	21.73	15.37	7.23
	ReconNet	25.14	22.80	21.41	21.79	19.74	16.20
	TVAL3	22.40	21.28	19.76	18.87	16.19	15.15
	DR^2-Net	25.43	21.64	19.86	20.99	18.34	16.08
	Proposed	**27.23**	**27.62**	**26.50**	**24.15**	**21.76**	**17.86**
Fingerprint	D-AMP	**20.99**	**20.64**	19.41	19.07	11.65	5.24
	ReconNet	17.56	17.20	17.25	16.68	16.10	15.55
	TVAL3	18.25	17.45	17.04	15.57	14.08	9.68
	DR^2-Net	18.30	16.57	15.98	17.16	16.26	15.20
	Proposed	19.76	19.80	**19.70**	**19.39**	**19.17**	**18.68**
Lena	D-AMP	**30.28**	28.40	26.57	24.38	11.71	6.57
	ReconNet	23.83	22.65	21.58	20.32	18.50	15.90
	TVAL3	21.26	20.68	19.51	17.81	16.37	15.17
	DR^2-Net	26.37	21.93	20.02	21.82	19.07	15.77
	Proposed	28.82	**29.01**	**28.44**	**25.48**	**24.08**	**19.47**
Coal cutter	D-AMP	**21.81**	**21.86**	20.90	19.10	14.36	8.14
	ReconNet	18.78	18.35	17.67	17.24	16.26	14.52
	TVAL3	12.52	10.94	9.87	8.17	10.48	12.50
	DR^2-Net	20.22	17.71	16.65	17.76	16.19	14.78
	Proposed	21.78	21.84	**21.40**	**20.05**	**18.08**	**17.34**
TBM	D-AMP	**29.68**	**28.02**	26.30	24.51	17.63	8.76
	ReconNet	23.89	22.95	22.13	21.21	19.24	17.65
	TVAL3	17.27	16.17	14.88	14.35	13.16	13.71
	DR^2-Net	25.65	22.11	20.87	22.04	19.50	17.53
	Proposed	27.67	27.27	**27.12**	**24.95**	**22.46**	**20.03**

Table 2. SSIM comparison for different algorithms on test images. r is the compression ratio.

Image	Algorithm	r = 0.25	r = 0.20	r = 0.15	r = 0.10	r = 0.04	r = 0.01
Barbara	D-AMP	0.8570	0.7781	0.7583	0.6189	0.0624	0.0129
	ReconNet	0.7449	0.7037	0.6062	0.5506	0.3805	0.2226
	TVAL3	0.7391	0.6834	0.6154	0.4692	0.3134	0.2281
	DR^2-Net	0.8165	0.7396	0.6774	0.6137	0.3947	0.2283
	Proposed	**0.8823**	**0.8950**	**0.8648**	**0.7832**	**0.6087**	**0.2859**
Fingerprint	D-AMP	**0.5530**	**0.4063**	0.2709	0.2288	0.1050	0.0029
	ReconNet	0.2438	0.2245	0.1890	0.1871	0.1412	0.0970
	TVAL3	0.3448	0.2884	0.2496	0.1948	0.1339	0.0774
	DR^2-Net	0.3030	0.2291	0.2115	0.2044	0.1484	0.0976
	Proposed	0.3464	0.3742	**0.3307**	**0.2871**	**0.2103**	**0.1498**
Lena	D-AMP	0.8867	0.8667	0.8174	0.7550	0.4343	0.0235
	ReconNet	0.7412	0.7084	0.6436	0.5997	0.4558	0.3181
	TVAL3	0.7420	0.7145	0.6596	0.5370	0.3735	0.2869
	DR^2-Net	0.8200	0.7771	0.7052	0.6597	0.5119	0.3352
	Proposed	**0.8930**	**0.9040**	**0.8879**	**0.8301**	**0.7437**	**0.4440**
Coal cutter	D-AMP	0.6854	0.6793	0.6376	0.5148	0.1735	0.0363
	ReconNet	0.5470	0.4947	0.4377	0.4267	0.3371	0.2431
	TVAL3	0.3830	0.3146	0.2574	0.1838	0.2110	0.1608
	DR^2-Net	0.6358	0.5482	0.4672	0.4899	0.3467	0.2634
	Proposed	**0.7320**	**0.7476**	**0.7049**	**0.6303**	**0.4923**	**0.3498**
TBM	D-AMP	0.8711	0.6793	0.8027	0.7187	0.2829	0.0634
	ReconNet	0.7728	0.7319	0.6771	0.6522	0.5460	0.4318
	TVAL3	0.5445	0.4868	0.5069	0.4127	0.3794	0.3372
	DR^2-Net	0.8184	0.7523	0.7171	0.6755	0.5714	0.4456
	Proposed	**0.8793**	**0.8764**	**0.8639**	**0.8058**	**0.6921**	**0.5359**

From Table 2, it can be learned that the proposed method achieves the highest SSIM at every compression ratio for all the images except the Fingerprint image. Since the SSIM metric describes structural similarity between the recovered and the original images, it can be drawn to the conclusion that the proposed method preserves specific characteristics of the images better.

3.3. Visual Quality Evaluation

Figures 6 and 7 illustrate the recovered images of the proposed method and the algorithms being compared. The green boxes zoom in the image patches within the red boxes so that the details can be viewed clearly. As can be seen in most of the pictures, the proposed method recovers sharper edges with less blurring compared to other algorithms. In Figure 7 where the compression ratio is relatively low, the edges can still be discerned in the recovered image of the proposed method, while other recoveries tend to be more blurred. Combined with Tables 1 and 2, it can be found that the characteristic which the proposed method preserves is the edges in the image.

Figures 6 and 7 also demonstrate an interesting phenomenon. In the recovery of the Fingerprint image, the proposed method fails to recover the details either at a compression ratio $r = 0.15$ or $r = 0.04$. This is intended behavior and actually the proposed method deliberately "blurs" dense patterns in the recovered images to cope with the noise which is often seen in underground mine images. To explain the rationale behind this, suppose we take the image patches of size 15×15 at the same location from the recovered image and the original image of Fingerprint. Then 3-level 2D-DWT is applied on both patches and it can be discovered that the level 2 or level 3 subband images are almost identical. The major difference of the subbands lies in the level 1 decomposition. Recall that in Equation (8) each level is given the same significance, the difference between the recovered and original patch in level 1 decomposition is in effect "averaged out". Therefore the DW-SSIM loss of the original dense patterned patch and the recovered blurred patch will be small, leading the proposed network to learn to blur the dense patterns.

3.4. Robustness against Noise

Since the tests in previous sections indicate that the proposed method takes an advantage when the compression ratio is low, we then test the noise robustness of the proposed method at a compression ratio $r = 0.04$ in this section. As depicted in Figures 8 and 9, Gaussian noise is added to the Lena and TBM test images to simulate the dusty environment in underground mines. The noise is zero-mean. The standard deviation σ of the noise is set to 5, 10, 15, 20, 25 and 30 to emulate different levels of noise. The noise-contaminated images are compressed at ratio $r = 0.04$. Then the similarity of the recovered images between the original test images is evaluated using the PSNR and SSIM measurement.

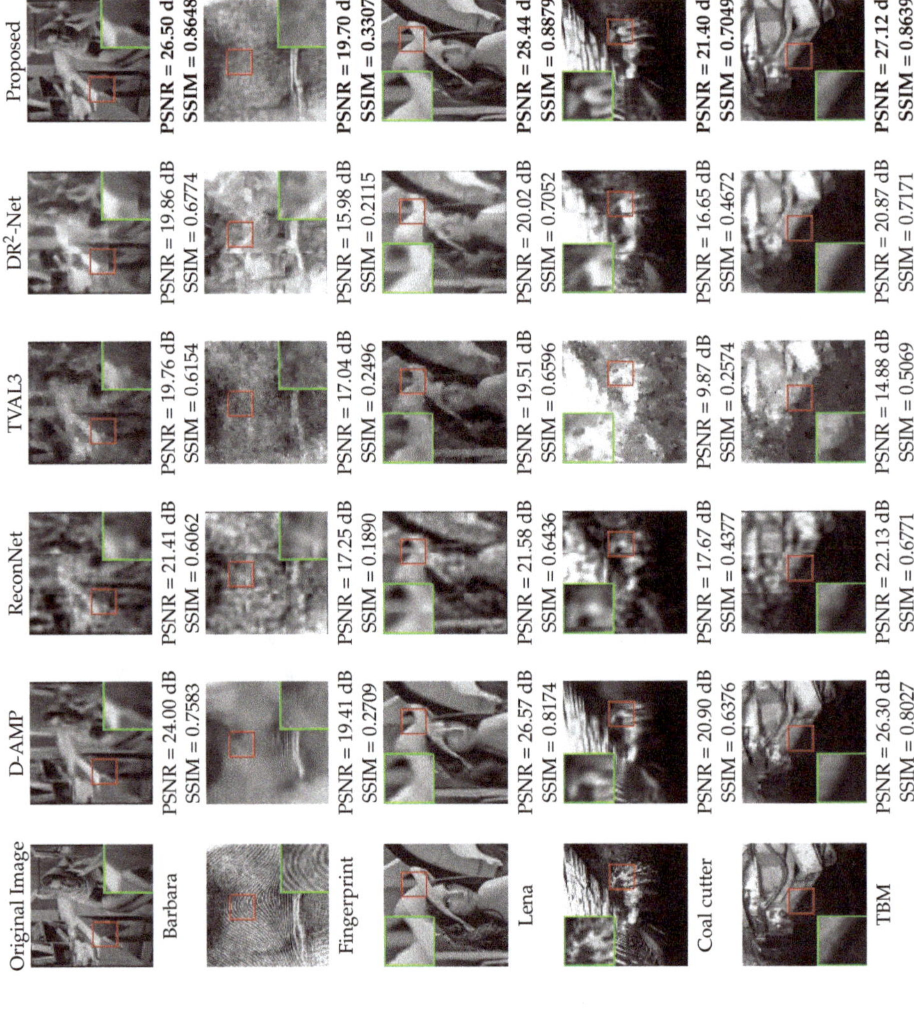

Figure 6. The recovered images at compression ratio $r = 0.15$.

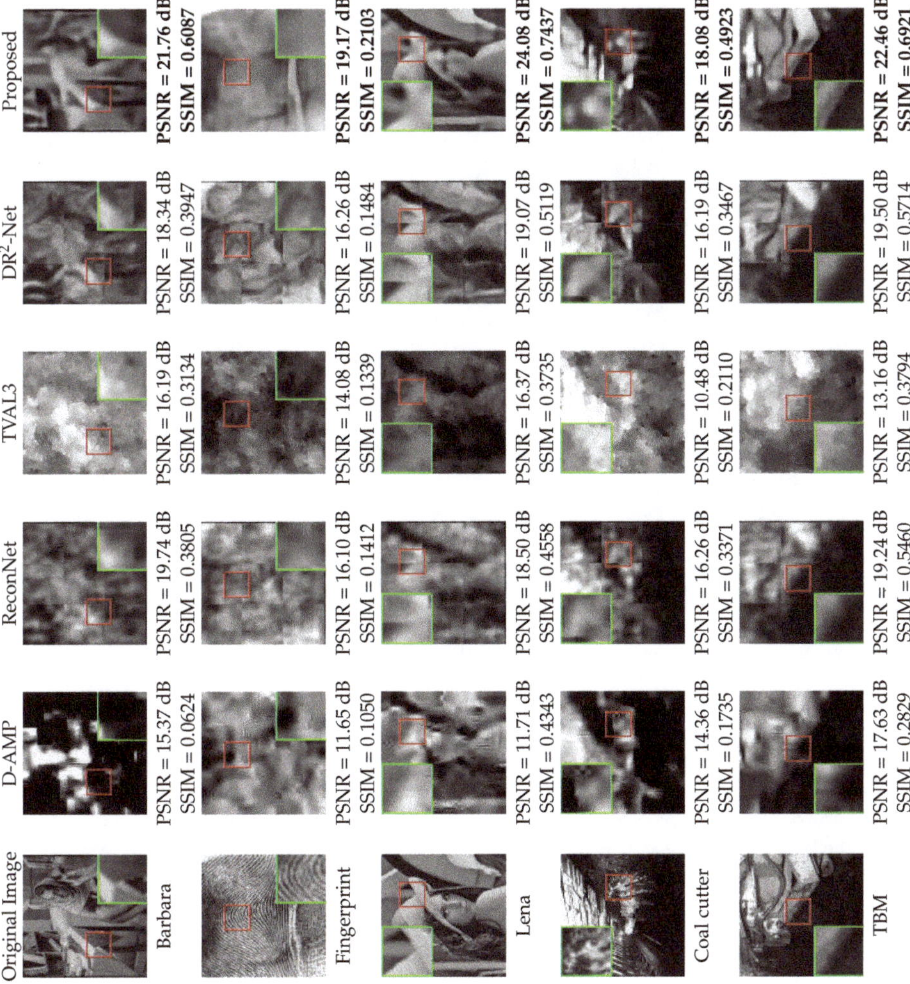

Figure 7. The recovered images at compression ratio $r = 0.04$.

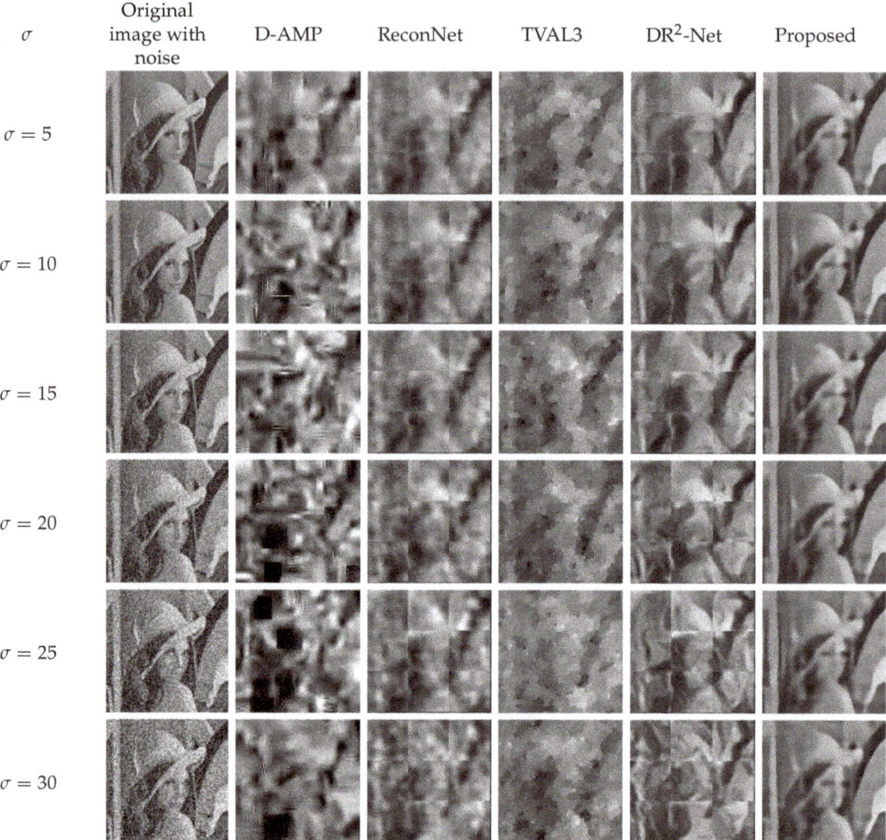

Figure 8. Comparison of recoveries of the Lena image at the presence of noise. σ denotes the standard deviation of the noise. The compression ratio r is 0.04.

As in Figure 8 and Figure 9, at all noise levels, fewer artifacts can be seen yet sharp edges are preserved in the recovered images of the proposed method. Further, Figure 10 plots the PSNR and SSIM curves as σ varies. The PSNR and SSIM of all algorithms drop as σ increases, yet PSNR and SSIM of the proposed method are higher than those of the algorithms being compared. As σ grows from 5 to 30, the decrease of PSNR and SSIM of the proposed method, which is no more than 1.6 dB and 0.11, is the least among the algorithms. Therefore, it can be concluded that the proposed method features noise robustness when the compression ratio is low.

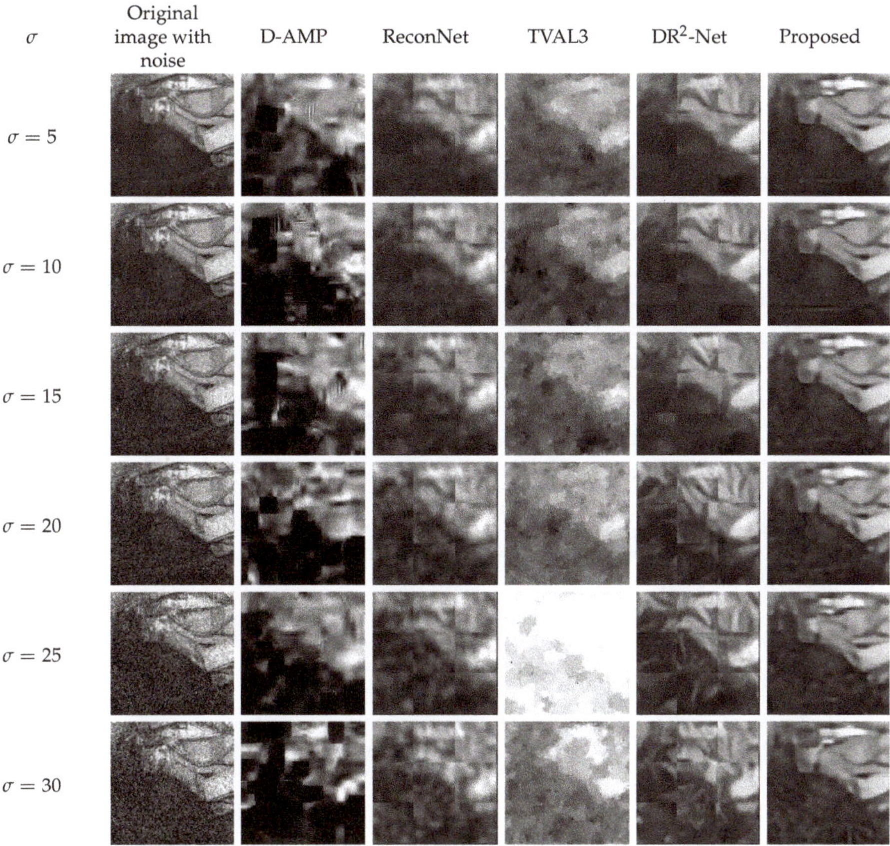

Figure 9. Comparison of recoveries of the TBM image at the presence of noise. σ denotes the standard deviation of the noise. The compression ratio r is 0.04.

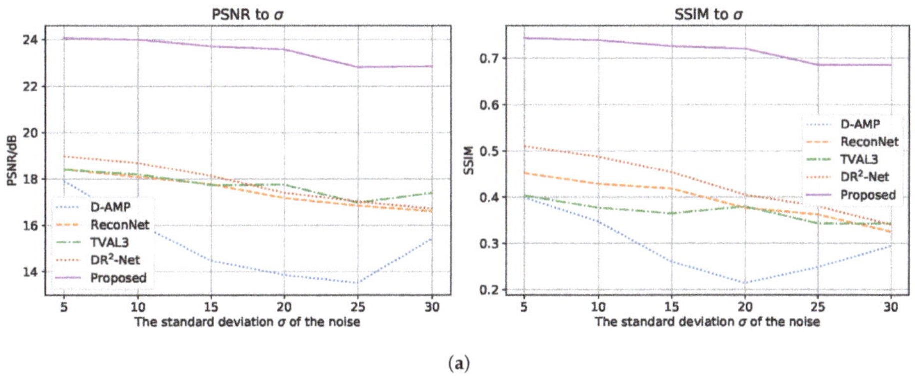

Figure 10. *Cont.*

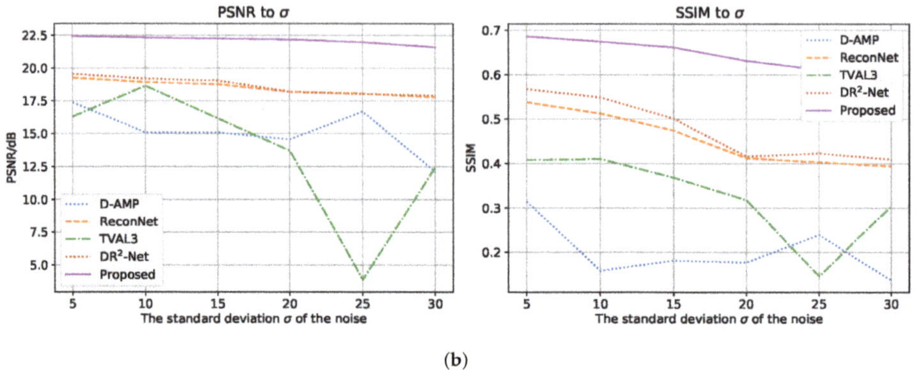

(b)

Figure 10. Plots of PSNR and SSIM against σ for the recoveries of noise contaminated (**a**) Lena image, (**b**) TBM image. The PSNR and SSIM are checked between the original test image (no noise added) and the recovered images. The compression ratio r is 0.04.

4. Conclusions

In this paper, we propose a CNN based image codec network which acts as the basis for the compression and recovery of images. We also propose a novel loss function that combines the knowledge of discrete wavelet transform to attack the problem of edge blurring in the recovered images. The proposed method is more suitable for the compression and recovery of underground mine images in that:

- The proposed method recovers sharp edges in the images. For underground mines, edges in the image are the key component to distinguish the foreground and background. By determining the boundaries of miners and equipment, it is possible for further image analysis to carry out.
- The proposed method features noise robustness. By blurring the dense patterns, the proposed method can filter out the noise especially seen in underground mines.
- Compared to other algorithms, the proposed method excels at low compression ratios. General image compression methods tend to strike a balance between the compression ratio and the recovery quality. They do not have to work at extremely low compression ratios as the transmission bandwidth available is comparably high. However, the proposed method is designed to work at low compression ratios to adapt to the harsh communication environment in underground mines.

In future work, we will combine other denoising techniques into the work presented in this paper is an attempt to achieve noise robustness without blurring the patterned areas. The current design of the DW-SSIM loss is not perfect in that the merits of cosine similarity is not fully preserved. Thus it is worth further investigating into the design of loss function. We will also train the model on other datasets in order to expand the application of the proposed method.

Author Contributions: Conceptualization, F.Z. and W.C.; data curation, Z.X.; formal analysis, F.Z. and Z.Z.; funding acquisition, F.Z. and W.C.; investigation, Z.X., H.Z. and J.L.; methodology, Z.X. and Z.Z.; project administration, F.Z. and W.C.; resources, F.Z.; software, Z.X., H.Z. and J.L.; Supervision, F.Z. and W.C.; validation, Z.X., W.C. and C.L.; visualization, Z.X.; writing—original draft, Z.X.; writing—review and editing, F.Z., Z.X., W.C. and C.L.

Funding: This research was funded by Foundation of the National Key Research and Development Program grant number 2016YFC0801800, National Natural Science Foundation of China grant number 51874300, National Natural Science Foundation of China and Shanxi Provincial People's Government Jointly Funded Project of China for Coal Base and Low Carbon grant number U1510115, and the Open Research Fund of Key Laboratory of Wireless Sensor Network and Communication, Shanghai Institute of Microsystem and Information Technology, Chinese Academy of Sciences grant numbers 20190902 and 20190913.

Conflicts of Interest: The authors declare no conflict of interest. The funders had no role in the design of the study; in the collection, analyses, or interpretation of data; in the writing of the manuscript, or in the decision to publish the results.

References

1. Dai, S.; Finkelman, R.B. Coal as a promising source of critical elements: Progress and future prospects. *Int. J. Coal Geol.* **2018**, *186*, 155–164, doi:10.1016/j.coal.2017.06.005. [CrossRef]
2. Dohare, Y.S.; Maity, T.; Das, P.S.; Paul, P.S. Wireless Communication and Environment Monitoring in Underground Coal Mines- Review. *IETE Tech. Rev.* **2015**, *32*, 140–150, doi:10.1080/02564602.2014.995142. [CrossRef]
3. Wallace, G. The JPEG still picture compression standard. *IEEE Trans. Consum. Electron.* **1992**, *38*, xviii–xxxiv, doi:10.1109/30.125072. [CrossRef]
4. Li, M.; Zuo, W.; Gu, S.; Zhao, D.; Zhang, D. Learning Convolutional Networks for Content-Weighted Image Compression. In Proceedings of the 2018 IEEE/CVF Conference on Computer Vision and Pattern Recognition, Salt Lake City, UT, USA, 18–23 June 2018; pp. 3214–3223, doi:10.1109/CVPR.2018.00339. [CrossRef]
5. Marcellin, M.W.; Gormish, M.J.; Bilgin, A.; Boliek, M.P. An overview of JPEG-2000. In Proceedings of the Data Compression Conference (DCC 2000), Snowbird, UT, USA, 28–30 March 2000; pp. 523–541, doi:10.1109/DCC.2000.838192. [CrossRef]
6. Candes, E.; Romberg, J.; Tao, T. Robust uncertainty principles: Exact signal reconstruction from highly incomplete frequency information. *IEEE Trans. Inf. Theory* **2006**, *52*, 489–509, doi:10.1109/TIT.2005.862083. [CrossRef]
7. Candès, E.J.; Romberg, J.K.; Tao, T. Stable signal recovery from incomplete and inaccurate measurements. *Commun. Pure Appl. Math.* **2006**, *59*, 1207–1223, doi:10.1002/cpa.20124. [CrossRef]
8. Donoho, D.L.; Tanner, J. Thresholds for the Recovery of Sparse Solutions via L1 Minimization. In Proceedings of the IEEE 2006 40th Annual Conference on Information Sciences and Systems, Princeton, NJ, USA, 22–24 March 2006; pp. 202–206, doi:10.1109/CISS.2006.286462. [CrossRef]
9. Joshi, A.M.; Sahu, C.; Ravikumar, M.; Ansari, S. Hardware implementation of compressive sensing for image compression. In Proceedings of the TENCON 2017—2017 IEEE Region 10 Conference, Penang, Malaysia, 5–8 November 2017; IEEE: Penang, Malaysia, 2017; Volume 2017-Decem, pp. 1309–1314, doi:10.1109/TENCON.2017.8228060. [CrossRef]
10. Chambolle, A. An Algorithm for Total Variation Minimization and Applications. *J. Math. Imaging Vis.* **2004**, *20*, 89–97, doi:10.1023/B:JMIV.0000011321.19549.88. [CrossRef]
11. Donoho, D.L.; Maleki, A.; Montanari, A. Message-passing algorithms for compressed sensing. *Proc. Natl. Acad. Sci. USA* **2009**, *106*, 18914–18919, doi:10.1073/pnas.0909892106. [CrossRef]
12. Rudin, L.I.; Osher, S.; Fatemi, E. Nonlinear total variation based noise removal algorithms. *Phys. D Nonlinear Phenom.* **1992**, *60*, 259–268, doi:10.1016/0167-2789(92)90242-F. [CrossRef]
13. Li, C.; Yin, W.; Jiang, H.; Zhang, Y. An efficient augmented Lagrangian method with applications to total variation minimization. *Comput. Optim. Appl.* **2013**, *56*, 507–530, doi:10.1007/s10589-013-9576-1. [CrossRef]
14. Kong, Q.; Gong, R.; Liu, J.; Shao, X. Investigation on Reconstruction for Frequency Domain Photoacoustic Imaging via TVAL3 Regularization Algorithm. *IEEE Photonics J.* **2018**, *10*, 1–15, doi:10.1109/JPHOT.2018.2869815. [CrossRef]
15. Metzler, C.A.; Maleki, A.; Baraniuk, R.G. From Denoising to Compressed Sensing. *IEEE Trans. Inf. Theory* **2016**, *62*, 5117–5144, doi:10.1109/TIT.2016.2556683. [CrossRef]
16. Zhao, X.; Shen, X.; Wang, K.; Li, W. A DCVS Reconstruction Algorithm for Mine Video Monitoring Image Based on Block Classification. *Preprints* **2018**, 2018070222, doi:10.20944/PREPRINTS201807.0222.V1. [CrossRef]
17. Wang, Y.; Liu, Z.; Wang, D.; Li, Y.; Yan, J. Anomaly detection and visual perception for landslide monitoring based on a heterogeneous sensor network. *IEEE Sens. J.* **2017**, *17*, 1, doi:10.1109/JSEN.2017.2704584. [CrossRef]
18. Qiao, X.; Yang, F.; Zheng, J. Ground Penetrating Radar Weak Signals Denoising via Semi-soft Threshold Empirical Wavelet Transform. *Ingénierie Des Systèmes d Information* **2019**, *24*, 207–213, doi:10.18280/isi.240213. [CrossRef]

19. Xie, S.; Imani, M.; Dougherty, E.R.; Braga-Neto, U.M. Nonstationary linear discriminant analysis. In Proceedings of the IEEE 2017 51st Asilomar Conference on Signals, Systems, and Computers, Pacific Grove, CA, USA, 29 October–1 November 2017; pp. 161–165, doi:10.1109/ACSSC.2017.8335158. [CrossRef]

20. Imani, M.; Ghoreishi, S.F.; Braga-Neto, U.M. Bayesian control of large MDPs with unknown dynamics in data-poor environments. In Proceedings of the Advances in Neural Information Processing Systems (NIPS), Montréal, QC, Canada, 3–8 December 2018, pp. 8146–8156.

21. Imani, M.; Ghoreishi, S.F.; Allaire, D.; Braga-Neto, U.M. MFBO-SSM: Multi-Fidelity Bayesian Optimization for Fast Inference in State-Space Models. In Proceedings of the AAAI Conference on Artificial Intelligence, Hilton Hawaiian Village, Honolulu, HI, USA, 27 January 27–1 February 2019; Volume 33, pp. 7858–7865, doi:10.1609/aaai.v33i01.33017858. [CrossRef]

22. Imani, M.; Dougherty, E.R.; Braga-Neto, U. Boolean Kalman filter and smoother under model uncertainty. *Automatica* **2020**, *111*, 108609, doi:10.1016/j.automatica.2019.108609. [CrossRef]

23. Ren, S.; He, K.; Girshick, R.; Sun, J. Faster R-CNN: Towards Real-Time Object Detection with Region Proposal Networks. *IEEE Trans. Pattern Anal. Mach. Intell.* **2017**, *39*, 1137–1149, doi:10.1109/TPAMI.2016.2577031. [CrossRef]

24. Redmon, J.; Divvala, S.; Girshick, R.; Farhadi, A. You Only Look Once: Unified, Real-Time Object Detection. In Proceedings of the 2016 IEEE Conference on Computer Vision and Pattern Recognition (CVPR), Las Vegas, NV, USA, 27–30 June 2016; Volume 2016-Decem, pp. 779–788, doi:10.1109/CVPR.2016.91. [CrossRef]

25. Liu, W.; Anguelov, D.; Erhan, D.; Szegedy, C.; Reed, S.; Fu, C.Y.; Berg, A.C. SSD: Single Shot MultiBox Detector. In *Lecture Notes in Computer Science (Including Subseries Lecture Notes in Artificial Intelligence and Lecture Notes in Bioinformatics)*; Springer International Publishing: Cham, Switzerland, 2016; Volume 9905 LNCS, pp. 21–37, doi:10.1007/978-3-319-46448-0_2. [CrossRef]

26. Chen, Y.; Zhu, K.; Zhu, L.; He, X.; Ghamisi, P.; Benediktsson, J.A. Automatic Design of Convolutional Neural Network for Hyperspectral Image Classification. *IEEE Trans. Geosci. Remote Sens.* **2019**, *57*, 7048–7066, doi:10.1109/TGRS.2019.2910603. [CrossRef]

27. Liu, Q.; Feng, C.; Song, Z.; Louis, J.; Zhou, J. Deep Learning Model Comparison for Vision-Based Classification of Full/Empty-Load Trucks in Earthmoving Operations. *Appl. Sci.* **2019**, *9*, doi:10.3390/app9224871. [CrossRef]

28. Mousavi, A.; Baraniuk, R.G. Learning to invert: Signal recovery via Deep Convolutional Networks. In Proceedings of the 2017 IEEE International Conference on Acoustics, Speech and Signal Processing (ICASSP), New Orleans, LA, USA, 5–9 March 2017; pp. 2272–2276, doi:10.1109/ICASSP.2017.7952561. [CrossRef]

29. Kulkarni, K.; Lohit, S.; Turaga, P.; Kerviche, R.; Ashok, A. ReconNet: Non-Iterative Reconstruction of Images from Compressively Sensed Measurements. In Proceedings of the 2016 IEEE Conference on Computer Vision and Pattern Recognition (CVPR), Las Vegas, NV, USA, 27–30 June 2016; pp. 449–458, doi:10.1109/CVPR.2016.55. [CrossRef]

30. Yao, H.; Dai, F.; Zhang, S.; Zhang, Y.; Tian, Q.; Xu, C. DR2-Net: Deep Residual Reconstruction Network for image compressive sensing. *Neurocomputing* **2019**, *359*, 483–493, doi:10.1016/j.neucom.2019.05.006. [CrossRef]

31. Ballé, J.; Laparra, V.; Simoncelli, E.P. End-to-end Optimized Image Compression. In Proceedings of the 5th International Conference on Learning Representations, ICLR 2017, Conference Track Proceedings, Toulon, France, 24–26 April 2017.

32. Wang, C.; Han, Y.; Wang, W. An End-to-End Deep Learning Image Compression Framework Based on Semantic Analysis. *Appl. Sci.* **2019**, *9*, 3580. [CrossRef]

33. Lyu, C.; Liu, Z.; Yu, L. Block-sparsity recovery via recurrent neural network. *Signal Process.* **2019**, *154*, 129–135. doi:10.1016/j.sigpro.2018.08.014. [CrossRef]

34. Toderici, G.; Vincent, D.; Johnston, N.; Jin Hwang, S.; Minnen, D.; Shor, J.; Covell, M. Full Resolution Image Compression with Recurrent Neural Networks. In Proceedings of the IEEE Conference on Computer Vision and Pattern Recognition (CVPR), Honolulu, HI, USA, 21–26 July 2017.

35. Minnen, D.; Toderici, G.; Covell, M.; Chinen, T.; Johnston, N.; Shor, J.; Hwang, S.J.; Vincent, D.; Singh, S. Spatially adaptive image compression using a tiled deep network. In Proceedings of the 2017 IEEE International Conference on Image Processing (ICIP), Beijing, China, 17–20 September 2017; pp. 2796–2800, doi:10.1109/ICIP.2017.8296792. [CrossRef]

36. Ma, S.; Zhang, X.; Jia, C.; Zhao, Z.; Wang, S.; Wanga, S. Image and Video Compression with Neural Networks: A Review. *IEEE Trans. Circuits Syst. Video Technol.* **2019**, *8215*, 1, doi:10.1109/TCSVT.2019.2910119. [CrossRef]

37. Rippel, O.; Bourdev, L. Real-time adaptive image compression. In Proceedings of the 34th International Conference on Machine Learning, ICML 2017, Sydney, Australia, 6–11 August 2017; Volume 6, pp. 4457–4473.

38. Jia, C.; Zhang, X.; Wang, S.; Wang, S.; Ma, S. Light Field Image Compression Using Generative Adversarial Network-Based View Synthesis. *IEEE J. Emerg. Sel. Top. Circuits Syst.* **2019**, *9*, 177–189, doi:10.1109/JETCAS.2018.2886642. [CrossRef]

39. Bjontegaard, G. Calculation of average PSNR differences between RD-curves. In Proceedings of the VCEG Meeting (ITU-T SG16 Q.6), Austin, TX, USA, 2–4 April 2001, pp. 2–4.

40. Agustsson, E.; Tschannen, M.; Mentzer, F.; Timofte, R.; Gool, L.V. Generative Adversarial Networks for Extreme Learned Image Compression. In Proceedings of the IEEE International Conference on Computer Vision (ICCV), Seoul, Korea, 27 October–2 November 2019.

41. He, K.; Zhang, X.; Ren, S.; Sun, J. Delving Deep into Rectifiers: Surpassing Human-Level Performance on ImageNet Classification. In Proceedings of the 2015 IEEE International Conference on Computer Vision (ICCV), Santiago, Chile, 7–13 December 2015; pp. 1026–1034, doi:10.1109/ICCV.2015.123. [CrossRef]

42. Miklós, P. Image interpolation techniques. In Proceedings of the 2nd Siberian-Hungarian Joint Symposium On Intelligent Systems, Subotica, Serbia and Montenegro, 1–2 October 2004; pp. 1–6.

43. Ioffe, S.; Szegedy, C. Batch Normalization: Accelerating Deep Network Training by Reducing Internal Covariate Shift. In Proceedings of the 32nd International Conference on Machine Learning, ICML 2015, Lille, France, 6–11 July 2015; Volume 1, pp. 448–456.

44. He, K.; Zhang, X.; Ren, S.; Sun, J. Deep Residual Learning for Image Recognition. In Proceedings of the 2016 IEEE Conference on Computer Vision and Pattern Recognition (CVPR), Las Vegas, NV, USA, 27–30 June 2016; pp. 770–778, doi:10.1109/CVPR.2016.90. [CrossRef]

45. Maas, A.L.; Hannun, A.Y.; Ng, A.Y. Rectifier nonlinearities improve neural network acoustic models. In Proceedings of the ICML Workshop on Deep Learning for Audio, Speech and Language Processing, Atlanta, GA, USA, 16 June 2013.

46. Wang, Z.; Bovik, A.C. Mean squared error: Love it or leave it? A new look at Signal Fidelity Measures. *IEEE Signal Process. Mag.* **2009**, *26*, 98–117, doi:10.1109/MSP.2008.930649. [CrossRef]

47. Wang, Z.; Bovik, A.C.; Sheikh, H.R.; Simoncelli, E.P. Image quality assessment: From error visibility to structural similarity. *IEEE Trans. Image Process.* **2004**, *13*, 600–612, doi:10.1109/TIP.2003.819861. [CrossRef] [PubMed]

48. Sampat, M.P.; Wang, Z.; Gupta, S.; Bovik, A.C.; Markey, M.K. Complex Wavelet Structural Similarity: A New Image Similarity Index. *IEEE Trans. Image Process.* **2009**, *18*, 2385–2401, doi:10.1109/TIP.2009.2025923. [CrossRef]

49. Mallat, S.G. A theory for multiresolution signal decomposition: The wavelet representation. *IEEE Trans. Pattern Anal. Mach. Intell.* **1989**, *11*, 674–693, doi:10.1109/34.192463. [CrossRef]

50. Kotu, V.; Deshpande, B. Classification. In *Data Science*; Elsevier: Amsterdam, The Netherlands, 2019; pp. 65–163, doi:10.1016/B978-0-12-814761-0.00004-6. [CrossRef]

51. Kingma, D.P.; Ba, J.L. Adam: A method for stochastic optimization. In Proceedings of the 3rd International Conference on Learning Representations, ICLR 2015 - Conference Track Proceedings, San Diego, CA, USA, 7–9 May 2015.

52. Lin, T.Y.; Maire, M.; Belongie, S.; Bourdev, L.; Girshick, R.; Hays, J.; Perona, P.; Ramanan, D.; Zitnick, C.L.; Dollár, P. Microsoft COCO: Common Objects in Context. In *Computer Vision—ECCV 2014*; Springer International Publishing: Cham, Switzerland, 2014; Volume 8693 LNCS, pp. 740–755 doi:10.1007/978-3-319-10602-1_48. [CrossRef]

53. Paszke, A.; Chintala, S.; Chanan, G.; Lin, Z.; Gross, S.; Yang, E.; Antiga, L.; Devito, Z.; Lerer, A.; Desmaison, A. Automatic differentiation in PyTorch. In Proceedings of the 31st Conference on Neural Information Processing Systems, Long Beach, CA, USA, 4–9 December 2017.

 *electronics*

Article

Trajectory Planning Algorithm of UAV Based on System Positioning Accuracy Constraints

Hao Zhou [1,4]**, Hai-Ling Xiong** [1,2,*]**, Yun Liu** [3]**, Nong-Die Tan** [1] **and Lei Chen** [1]

1   College of Computer and Information Science, Southwest University, Chongqing 400715, China;
    zhouhao19@email.swu.edu.cn (Z.H.); tnd1201@email.swu.edu.cn (N.-D.T.);
    chenlei0207@email.swu.edu.cn (L.C.)
2   Business College, Southwest University, Chongqing 402460, China
3   College of Artificial Intelligence, Southwest University, Chongqing 400715, China; yunliu@swu.edu.cn
4   College of Big Data and Artificial Intelligence, Chizhou University, Chizhou 247000, China
*   Correspondence: xionghl@swu.edu.cn; Tel.: +86-023-46752718

Received: 27 December 2019; Accepted: 29 January 2020; Published: 3 February 2020

Abstract: This paper describes a novel trajectory planning algorithm for an unmanned aerial vehicle (UAV) under the constraints of system positioning accuracy. Due to the limitation of the system structure, a UAV cannot accurately locate itself. Once the positioning error accumulates to a certain degree, the mission may fail. This method focuses on correcting the error during the flight process of a UAV. The improved genetic algorithm (GA) and A* algorithm are used in trajectory planning to ensure the UAV has the shortest trajectory length from the starting point to the ending point under multiple constraints and the least number of error corrections.

Keywords: unmanned aerial vehicle; UAV; trajectory planning; GA; A*; multiple constraints

1. Introduction

An unmanned aerial vehicle (UAV) is an aircraft capable of completing missions with the autonomous flight capability. UAVs are currently widely used in vision systems, such as cargo transfer [1], object detection [2–4], and vision-assisted navigation [5–7].

The idea for scientists to develop UAVs is to fly autonomously and accomplish specific tasks. In modern warfare, the air defense system is constantly improving, and air defense technology is becoming more and more advanced. Thus, improving the autonomy of unmanned aerial vehicles is an important trend for the future. Trajectory planning is a key technology to improve the autonomy of a UAV and an effective means to implement flight missions. It has important significance in both theoretical and practical applications. The aircraft path planning can effectively ensure the operational performance of a UAV, which provides technical support for a UAV to successfully complete the flight mission, realize the autonomous control of a UAV, and to complete the autonomous flight.

Trajectory planning refers to the planning of an optimal flight path of the aircraft between the starting point and the ending point, considering factors such as fuel consumption, maneuverability, arrival time, flight area, and threat level. Trajectory planning is an important guarantee for the successful completion of a UAV and one of the key technologies for mission planning systems.

Due to technical limitations, in the early 1980s, trajectory planning relied heavily on manual operations by technicians. With the continuous development and improvement of the prevention and control system and technology, the accuracy requirements of a UAV for planning the trajectory are getting higher and higher, and artificial path planning has become more and more difficult to meet the requirements. With the rapid development of communication technology, various methods for detecting flight environment information have emerged endlessly, which makes the information obtained by the trajectory planners more and more abundant. In order to improve flight accuracy, the planned trajectory should satisfy the requirements of terrain following, terrain avoidance, and threat avoidance while satisfying the performance constraints of the aircraft. Due to the complexity and variety of these constraints, it is difficult to handle and complete such complex tasks in manually.

In order to realize the autonomous trajectory planning under different planning environments, scientists have carried out in-depth research on the trajectory planning of an UAV, and proposed various algorithms, such as the Voronoi diagram method [8–11], A* algorithm [12–14], particle swarm optimization (PSO) algorithm [15–17], genetic algorithm [18–20], neural network algorithm [21,22], artificial potential field method [23–25], simulated annealing algorithm [26], and so on.

The process of finding the best trajectory includes consideration of constraints. The optimal path must produce an optimal objective function and satisfy multiple constraints. However, conventional methods can only handle one objective function at a time, and they cannot handle optimization problems involving two or more objective functions. Thus, a combined objective function is formed by mathematically aggregating two or more separate objective functions. The weighted values are introduced into the combined objective function formula to reflect its relative importance. In this work, we propose a trajectory path planning algorithm of a UAV based on genetic algorithm and A* algorithm under multiple constraints, and then satisfied the system positioning accuracy conditions.

The paper is organized as follows: Section 2 provides the necessary background: problem statement, model assumptions, and multiple constraints. Section 3 describes in detail an UAV trajectory planning Problem 1 based on positioning accuracy constraints, and uses an improved genetic algorithm to design a trajectory planning algorithm, and finally get the results of the trajectory planning. Section 4 describes in detail an UAV trajectory planning Problem 2 based on Problem 1, and uses an improved A* algorithm to design a trajectory planning algorithm, and finally get the results of the trajectory planning. Section 5 presents the performance comparison of the proposed algorithm with the traditional swarm intelligence algorithm. Section 6 concludes the paper and gives the further work.

2. Background

2.1. Problem Statement

In practical applications, the planning scope is often up to one million square kilometers, and the planning area environment is very complicated. In addition to the topographical factors, the planning process needs to consider various constraints such as aircraft maneuverability, penetration requirements, and flight missions. This research is a simplified version of the actual problem, without considering the aircraft's maneuverability, penetration requirements, threats, and many other factors. However, there are multiple constraints.

In this research, we mainly consider the trajectory error correction problem of an UAV based on system positioning accuracy constraints. Due to system structure limitations, the positioning system of such aircraft cannot accurately locate itself. Once the positioning error accumulates to a certain extent, the task may fail. Therefore, correcting the positioning error during flight is an important task in UAV trajectory planning. The trajectory planning problem of an UAV is a complex multi-constrained optimization problem. UAV trajectory planning refers to considering the positioning error during the flight due to a series of factors such as the environment and weather during the flight. Therefore, some safe positions are assumed in the flight area (called correction points) for safety correction. In order to enable a UAV to follow the original trajectory planning from the starting point to the ending point, several correction points are needed in the flight area to correct the trajectory error of an UAV.

Under the premise of many constraints, the correct point is selected to correct the error of a UAV, and the best flight path from the starting point to the ending point is planned for an UAV, so that the number of times a UAV is corrected by the correct points during the flight is as small as possible, and the trajectory length is as small as possible. How to ensure that a UAV meets the various constraints and the optimal trajectory requirements, and to quickly and accurately obtain the true flight trajectory is the problem that needs to be solved urgently. In the aircraft trajectory planning problem, the standard planning problem is usually to establish a flight trajectory with the optimal cost function value through a pre-set cost function.

2.2. Model Assumptions and Multiple Constraints

The aim of this study is to find the best optimal trajectory planning for an UAV under the limit of the positional accuracy of a UAV system, which is a multi-constraint combinatorial optimization problem.

Assume that the flight area of an UAV and the safe positions is as shown in Figure 1. The starting point is A and the ending point is B. The constraints of its trajectory are as follows:

(1) A UAV needs real-time positioning during flight, and its positioning error includes vertical error and horizontal error. For every 1m flight, the vertical error and horizontal error will be increased by δ dedicated units, respectively. The vertical error and horizontal error should be less than θ units when reaching the ending point, and for the sake of simplification, assume that when the vertical error and the horizontal error are both less than θ units, an UAV can still follow as the planned trajectory to fly.

(2) A UAV needs to correct the positioning error during flight. There are some safety positions in the flight area (called correction points) can be used for error correction. The type of the correction point includes horizontal and vertical correction points. If a UAV reaches the correct point, the error can be corrected based on the type of the correct points, assuming that the safety positions in the flight area (i.e., the position of the correction points) is determined before flight trajectory planning. Figure 1 is a schematic diagram of a certain trajectory. If the vertical error and the horizontal error can be corrected in time, a UAV can fly according to the predetermined trajectory, and finally reaches the destination.

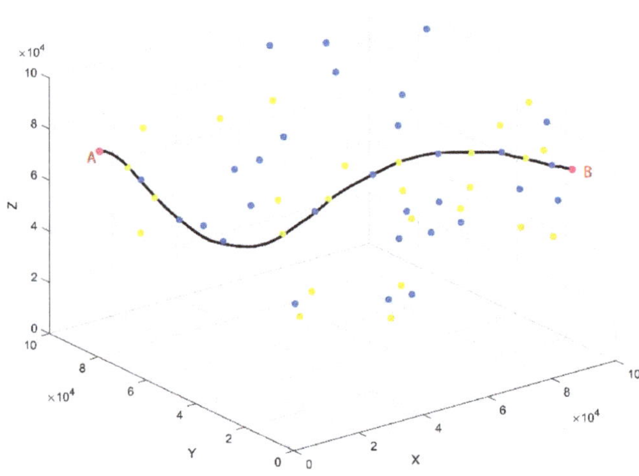

Figure 1. Schematic diagram of an UAV trajectory planning area. The yellow points are the horizontal error correction point, the blue points are the vertical error correction point, the starting point is point A, and the destination is point B. The black curve represents a trajectory.

(3) At the starting point A, the vertical and horizontal error of an UAV are both zero.

(4) After the correction of the vertical error correction point, the vertical error will become zero and the horizontal error will remain unchanged.

(5) After the correction of the horizontal error correction point, the horizontal error will become zero and the vertical error will remain unchanged.

(6) Vertical error correction can be performed when the vertical error of the aircraft is not greater than α_1 units and the horizontal error is not greater than α_2 units.

(7) Horizontal error correction can be performed when the vertical error of the aircraft is not greater than β_1 units and the horizontal error is not greater than β_2 units.

(8) An UAV is limited by the structure and control system during the turn and cannot complete the immediate turn (i.e., the direction of an UAV cannot be changed abruptly), if the minimum turning radius of an UAV is 200 m [27].

3. Problem 1

To plan a trajectory for an UAV from point A to point B, for the above-mentioned conditions (1) to (7), and comprehensively consider the following optimization goals: (A) the trajectory length is as small as possible; (B) the number of corrections through the correct points is as small as possible.

If the above-mentioned parameters of the data are:

$\alpha_1 = 25, \alpha_2 = 15, \beta_1 = 20, \beta_2 = 25, \theta = 30, \delta = 0.001$.

A data set contains the location and type of correction points. Table 1 shows some data of the data set. The unit of the coordinate is meter. There are 613 points in the data set. The number 0 is the starting point, the number 612 is the ending point, and the rest is correction points. The spatial position of each correction point is determined by the three-coordinate information of x, y, and z. For the type of the correction points, 0 represents the horizontal error correction point and 1 represents the vertical error correction point.

Table 1. Some data of the data set [27].

The Number of the Correction Points	X-Coordinate	Y-Coordinate	Z-Coordinate	The Type of the Correction Points
0	0.00	50,000.00	5000.00	The Starting Point
1	33,070.83	2789.48	5163.52	0
2	54,832.89	49,179.22	1448.30	1
3	77,991.55	63,982.18	5945.82	0
4	16,937.18	84,714.34	5360.29	0
5	339.69	14,264.46	3857.85	1
6	3941.93	74,279.86	9702.92	1
7	45,474.01	26,849.48	6411.72	1
8	86,806.90	5351.31	4409.85	0
9	23,602.88	68,460.10	88.47	0
... ...	... ...	... ...	... ...	... ...
610	14,870.60	95,939.17	8248.84	0
611	93,009.57	4549.33	7882.61	1
612	100,000.00	59,652.34	5022.00	The Ending Point

3.1. Multi-Constraints Optimization Problem

The focus of this research is to plan a trajectory for an UAV from the starting point to the ending point, and finding the optimal trajectory satisfied with the multi-constraints in Problem 1, so we build mathematical models of the problem.

3.1.1. Correction Area

The trajectory planned by a UAV is three-dimensional, and it is a space with many correction points. (x, y, z) is defined as the coordinates of correction point in the correction area, where x is the error in the horizontal direction, y is the error in the vertical direction, and z indicates the altitude. The physical space of the trajectory planning can be expressed as a set:

$$S = \{(x, y, z) \mid 0 \leq x \leq X_{max}, 0 \leq y \leq Y_{max}, 0 \leq z \leq Z_{max}\} \tag{1}$$

where $X_{\max}, Y_{\max}$ and $Z_{\max}$ refer to the maximum values of the corresponding coordinates, which can be obtained from the data set. Moreover, from the data set, we can dram the vertical error correction point and the horizontal error correction point in three-dimensional space. Figure 2 is the result we have drawn with MATLAB.

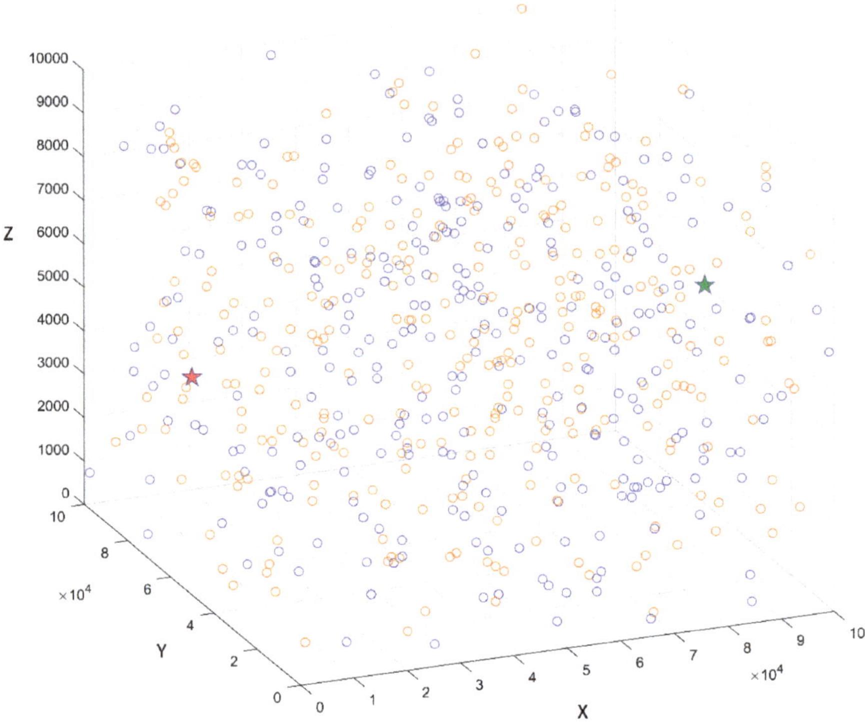

Figure 2. The physical space and the correct point. The blue point is the vertical error correction point, the yellow point is the horizontal error correction point, the red pentagram indicates starting point, the green pentagram indicates ending points. The range of axes comes from the data set.

3.1.2. Objective Function

Two objective optimization functions are proposed for minimizing the trajectory length of a UAV and the number of times an UAV has been corrected:

$$F_1 = \min(\sum_{i=1}^{m}(\eta_i + \lambda_i + h_i)) \tag{2}$$

$$F_2 = \min(\sum_{i=1}^{m}(a_i + b_i)) \tag{3}$$

where F_1 represents the trajectory length of an UAV from the starting point A to the ending point B, F_2 represents the number of times an UAV has been corrected through the correction area, m represents the total number of corrections of an UAV throughout the flight, η_i represents the distance of an UAV's i-th flight, λ_i represents the correction distance of the i-th flight to the vertical error correction point, h_i represents the correction distance of the i-th flight to the horizontal error correction point, a_i represents an UAV's i-th flight reaches the vertical error correction point, and b_i represents a UAV's i-th flight at the horizontal error correction point.

3.1.3. Multi-Constraints

The vertical and horizontal errors will increase by δ units when a UAV flies one meter each time. The vertical and horizontal errors should be less than θ units when the ending point is reached. Therefore, the total value of the vertical and horizontal errors of an UAV during the entire flight must be less than θ, and the distance η_i of each flight of a UAV will not exceed the displacement l_{AB} of the entire trajectory:

$$\sum_{i=1}^{m} a_i \eta_i \delta < \theta \tag{4}$$

$$\sum_{i=1}^{m} b_i \eta_i \delta < \theta \tag{5}$$

$$0 \le \eta_i \le l_{AB} \tag{6}$$

For the correction distance λ_i meters of a UAV's i-th flight to the vertical error correction point, when a UAV reaches the vertical error correction point, it is assumed that a UAV's i-th flight, η_i meters will increase its vertical error by $\delta\eta_i$ units:

$$\lambda_i = a_i \eta_i \delta, i = 1, \dots, m \tag{7}$$

For the correction distance h_i meters of an UAV's i-th flight to the vertical error correction point, when a UAV reaches the horizontal error correction point, it is assumed that a UAV's i-th flight, η_i meters will increase its horizontal error by $\delta\eta_i$ units:

$$h_i = b_i \eta_i \delta, i = 1, \dots, m \tag{8}$$

During the flight, a UAV is either corrected after reaching the vertical error correction point or reaching the horizontal error correction point. In order to perform error correction more accurately according to the type of error correction of the correction point, for the i-th flight, the binary variables a_i and b_i are introduced. The vertical error correction can be performed when a UAV's vertical error is not greater than α_1 units and the horizontal error is not greater than α_2 units. The horizontal error correction can be performed when a UAV's vertical error is not greater than β_1 units and the horizontal error is not greater than β_2 units. The binary variables a_i and b_i can be expressed as follows:

$$a_i = \begin{cases} 1, & \textit{when UAV reaches the vertical error correction point,} \\ 0, & \textit{when UAV does not reach the vertical error correction point,} \end{cases} \tag{9}$$

$$b_i = \begin{cases} 1, & \textit{when UAV reaches the horizontal error correction point,} \\ 0, & \textit{when UAV does not reach the horizontal error correction point,} \end{cases} \tag{10}$$

Then, the objective function should also meet the following constraints, which must be satisfied when an UAV performs vertical error correction:

$$\lambda_i \le \alpha_1, h_i \le \alpha_2 \tag{11}$$

Further, the objective function must be satisfied when an UAV performs horizontal error correction:

$$\lambda_i \le \beta_1, h_i \le \beta_2 \tag{12}$$

where λ_i is the corrected distance of an UAV to the vertical correction point on the i-th flight, h_i is the corrected distance of an UAV to the horizontal correction point on the i-th flight, α_1, β_1 is the vertical error of an UAV, and α_2, β_2 is the horizontal error of an UAV.

For Problem 1, the objective function is to make the times of corrections through the correction area as small as possible while the trajectory length is as small as possible.

In summary, we get a two-objective optimization model with multiple constraints:

$$F_1 = \min\left(\sum_{i=1}^{m}(\eta_i + \lambda_i + h_i)\right)$$

$$F_2 = \min\left(\sum_{i=1}^{m}(a_i + b_i)\right)$$

$$s.t.\begin{cases} \sum_{i=1}^{m}a_i\eta_i\delta < \theta \\ \sum_{i=1}^{m}b_i\eta_i\delta < \theta \\ 0 \le \eta_i \le l_{AB} \\ \lambda_i = a_i\eta_i\delta, \\ h_i = b_i\eta_i\delta, \\ a_i, b_i = 0, 1, \\ a_i + b_i = 1, \\ \lambda_i \le \alpha_1, h_i \le \alpha_2, \\ \lambda_i \le \beta_1, h_i \le \beta_2, \end{cases} \quad, i = 1, \ldots, m \tag{13}$$

3.2. Trajectory Planning Algorithm for the Problem 1

In order to solve the two-objective optimization model, it should be transformed into a single-objective programming model. In order to weigh the optimization goals of F_1 and F_2, we can choose a combination that takes into account both of them. The two are given a weight $\alpha(\alpha \le 1)$, and α is called a preference coefficient. So, the objective function F can be expressed as:

$$F = \min\left(\alpha\sum_{i=1}^{m}(\lambda_i + \eta_i + h_i) + (1 - \alpha)\sum_{i=1}^{m}(a_i + b_i)\right), \alpha \le 1 \tag{14}$$

We have successfully established the objective function and constraints for Problem 1, and we will begin to solve it. The trajectory planning problem is a nonlinear programming problem with multiple constraints, which needs to consider the overall optimization, and to minimize the calculation amount while avoiding falling into a local optimum, and our study is to find the optimal flight path of an UAV from the starting point to the destination in three-dimensional space. Then we use the genetic algorithm. The basic idea of a genetic algorithm is to simulate the evolutionary process of biological genetics. Based on the principles of "survival of the fittest", with the help of selection, crossing, and mutation, the problem to be solved approaches the optimal solution step by step from the initial solution. In trajectory planning, each chromosome (individual) of a genetic algorithm represents the trajectory of a UAV. The coding method of genes is also the coding method of trajectory nodes. The fitness function is changed by the cost function.

In this research, a UAV trajectory planning problem is combined with the idea of the genetic algorithm, and the real number gene coding method and specific genetic operator are used to meet the flight path trajectory of various constraint parameters to achieve the approximate optimal solution. The specific operation process is as follows:

The first step: using the real number gene coding. Use the fixed-length real number gene coding method showed in Figure 3 to convert the position information of an UAV in three-dimensional space into the chromosomal gene structure (as shown in Figure 4). Each gene of the chromosome contains the three-dimensional spatial coordinate (x, y, z) information of the gene, which records the following information of the spatial gene: one is whether the gene is feasible, and the other is whether the connecting line segment between the gene and the next gene is feasible. This gene sequence is only feasible if the above conditions are acceptable.

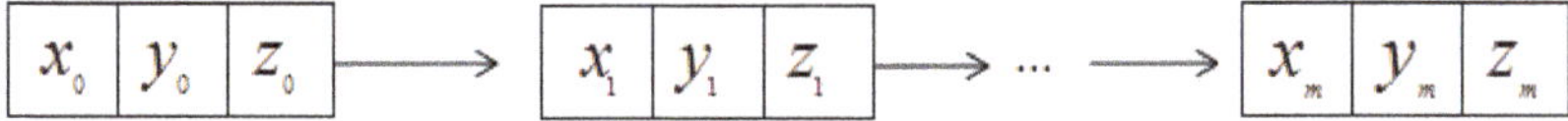

Figure 3. Fixed-length real-valued gene.

Figure 4. Chromosome structure.

As can be seen from Figure 3, fixed-length real number gene coding is used to convert the position information of a UAV in three-dimensional space into the chromosomal gene structure, and the advantages of the fixed-length real number gene coding method are: one is to avoid increasing the number of the error correction points which let the coding length is too long, and the second is that in the genetic process, high frequency encoding and decoding operations are not required, the calculated amount is reduced, and the search efficiency of the algorithm is improved.

As for a UAV trajectory planning problem, each chromosome represents a complete sequence of trajectory points of a UAV, and this point sequence may or may not be feasible. Therefore, from the chromosome structure diagram of Figure 4, each chromosome contains information on the start and end positions, plus the gene composition $H_1, H_2, H_3, \ldots, H_n$ that does not repeat.

The second step: the initialization of the population. During the conversion of trajectory nodes to genetic algorithm chromosomes, in order to make the generated random nodes as close as possible to the planning area, the generated nodes can be evenly and effectively distributed in the planning space. In this research, we use a specific initialization method to complete the initialization work. It mainly uses a starting point and an ending point as the center of symmetry as a rectangular area, and the length of the rectangle is the length of the line connecting the starting point and the ending point, and the width is the length of the rectangle, and the rectangular area is the planning area. The planning area is evenly divided into m grids, and the trajectory nodes are uniformly generated in the planning area using a loop statement as shown in Figure 5:

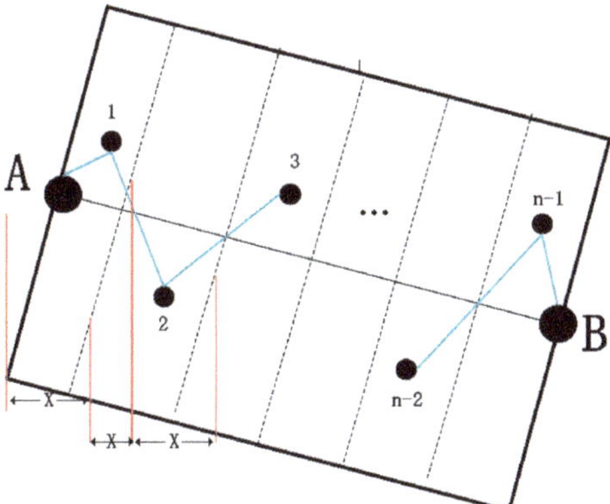

Figure 5. Trajectory initialization principle.

Figure 5 shows the generation process of the trajectory node when $i = 0$: the starting point A to the ending point B can determine the slope of the line, and then find the function equation of the four sides of the area, use the Random function to generate the random number x of the specific area, and then in the three cases, the Random function is used again to combine the functions of the four sides to generate y, and then the height z of a UAV is generated within a specific range, thereby obtaining the first gene of the trajectory, which is calculated in turn, and will be obtained at that time. For the last gene of the trajectory, we save these genes one by one and form the gene sequence of the trajectory, then the gene sequence is a chromosome. In the next work, this method is used to cycle through, thus completing the entire trajectory initialization work.

The third step: determine the trajectory fitness function. Suppose that A can determine that the trajectory of B is represented by z_1, and its cost function is composed of optimization item $f_1(z_1)$ and penalty item $f_2(z_1)$, so:

$$f(z_1) = f_1(z_1) + f_2(z_1) \tag{15}$$

Among them, the optimization term of the trajectory cost function:

$$f_1(z_1) = \omega_1[g_1(z_1)] + \omega_2[g_2(z_1)] \tag{16}$$

where $g_1(z_1)$ denotes the cost function value of the horizontal error on the arc of the trajectory A to B, $g_2(z_1)$ denotes the cost function value of the vertical error on the arc of the trajectory A to B, ω_1 represents the weight of the horizontal error in the total cost of the horizontal error and the vertical error, w_2 represents the weight of the vertical error in the total cost of the horizontal error and the vertical error, they should satisfy $\omega_1 + \omega_2 = 1$ and the weight distributions of ω_1 and ω_2 are similar.

The penalty of the trajectory cost function:

$$f_2(z_1) = \omega_1[p_1(z_1)] + \omega_2[p_2(z_1)] \tag{17}$$

where $p_1(z_1)$ denotes the trajectory which the flight height of an UAV exceeds the maximum flight altitude of an UAV, $p_2(z_1)$ denotes the trajectory which an UAV trajectory length is greater than the maximum range of an UAV, p_i denotes a penalty function for the i-th term, and ω_i denotes a penalty factor for the i-th term, $i = 1, 2$.

Take the trajectory fitness function:

$$fitness(z_1) = \frac{1}{f(z_1)} \tag{18}$$

where $f(z_1)$ represents the cost of the trajectory chromosome z_1, that is to say, the trajectory planning problem will be transformed from the cost minimization problem to the trajectory fitness maximization problem in genetic evolution.

The fourth step: Solve the fitness value of the formula of the fitness function, as shown in Figure 6.

Figure 6 shows the process of solving the fitness value of the trajectory population. First, each trajectory in the population is extracted in turn, and the trajectory is segmented. Each trajectory is equally divided by 10 points, and the actual length of the trajectory segment and all penalty items is calculated. Ten points respectively ask for the penalty value and are accumulated for averaging. Then, calculate the length, vertical error, and horizontal error of the trajectory segment, and combine the constraints determined by the adjacent trajectory nodes to meet the requirements of each segment. Then, calculate the cost function and punishment for the trajectory segment. The values are accumulated. If the current trajectory is calculated, we need to save their value and penalty value separately, then calculate each trajectory in turn, and finally, when all the trajectories are calculated, find the fitness, which is the maximum value of the cost. The given constraints are used to determine if the trajectory is feasible and to save it.

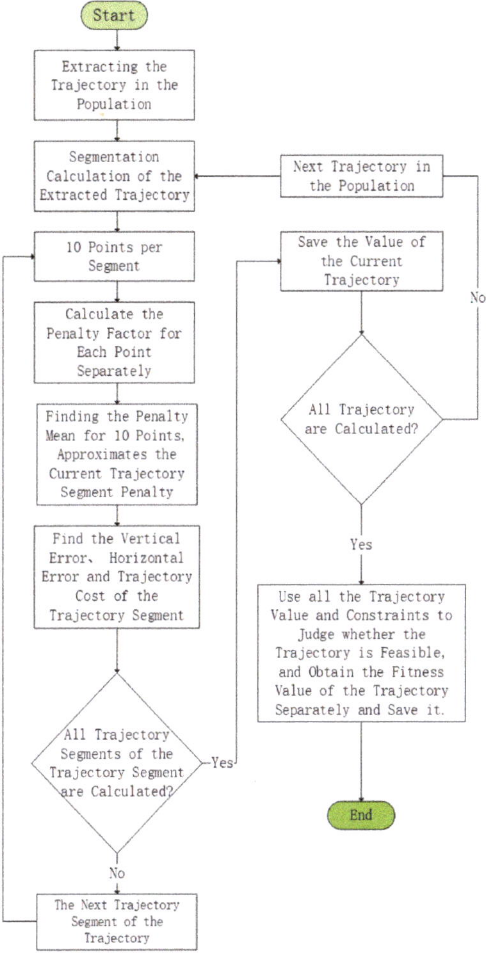

Figure 6. Solution process of trajectory fitness value.

The fifth step: determination of crossover probability and mutation probability. To prevent the genetic algorithm from converging prematurely, an improved adaptive genetic algorithm is used to solve the optimal flight path. The improved adaptive genetic algorithm adjusts the corresponding cross-probability P_r and genetic probability P_m according to the change of fitness during the evolution process. In the improved adaptive genetic algorithm, P_r and P_m are adaptively adjusted as follows:

$$P_r = \begin{cases} P_{r1} - \dfrac{(P_{r1} - P_{r2})\left(f^* - \overline{f}\right)}{f_{\max} - \overline{f}}, & f^* \geq \overline{f} \\ P_{r1}, & f^* \leq \overline{f} \end{cases} \tag{19}$$

$$P_m = \begin{cases} P_{m1} - \dfrac{(P_{m1} - P_{m2})\left(f - \overline{f}\right)}{f_{\max} - \overline{f}}, & f \geq \overline{f} \\ P_{m1}, & f \leq \overline{f} \end{cases} \tag{20}$$

where $f_{\max}$ represents the maximum value of the population fitness, $\overline{f}$ represents the average value of the current population fitness, f^* represents the fitness value of the larger of the two individuals currently used to cross, and f represents the fitness value of the individual which needs to perform mutation operation. The improved adaptive genetic algorithm mainly makes the P_r, P_m of the optimal individual in the early stage of evolution not zero, and it is not easy to fall into the local optimal solution, as shown in Figure 7:

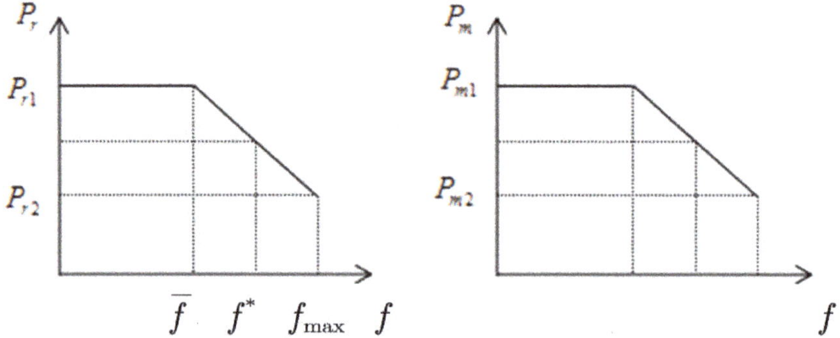

Figure 7. Improved adaptive crossover probability and mutation probability.

Because adaptive adjustment may cause the population to fall into a local optimal solution, we adopt an improved method to improve P_r, P_m so that the optimal individuals P_r, P_m in the early evolution stage are not zero. The specific improvement process is to increase the population's maximum fitness value P_r to P_{r1} and P_m to P_{m1}. As shown in Figure 6, this will make the individual with the largest fitness value in the initial stage of evolution in a constant state, and check and mutate according to the size of P_{r1}, P_{m1}.

The above five steps are an UAV trajectory planning algorithm process of Problem 1. The overall algorithm flow can be expressed in Figure 8.

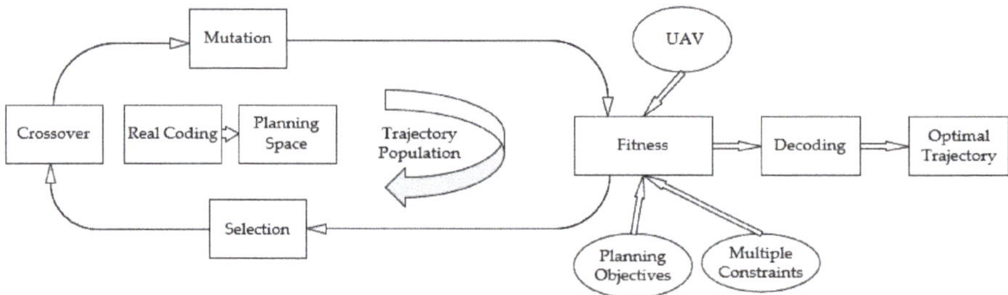

Figure 8. Flow chart of an UAV trajectory planning algorithm based on an improved genetic algorithm.

Figure 8 shows the combination of genetic algorithm and trajectory planning and the entire planning process. In Problem 1, we used the real number coding method to transform the trajectory planning problem into a genetic algorithm chromosome problem. Under multiple constraints and driven by the fitness function, the evolution operation of each generation is completed through operations such as genetic crossover and mutation, and finally decoded to obtain the required track node and complete the solution of the optimal solution for the trajectory planning.

3.3. Simulation Results and Analysis of the Trajectory Planning Algorithm for the Problem 1

According to the above algorithm, the results of the Problem 1 are calculated by using the data set.

For the parameters of the data set, the vertical error and horizontal error will increase by δ dedicated units for each flight of 1 m; when a UAV reaches the ending point, the vertical error and horizontal error should be less than θ units; when vertical error correction is performed, the vertical error of an UAV is not more than α_1 units, and the horizontal error is not more than α_2 units; when the horizontal error correction is performed, the vertical error of an UAV is not more than β_1 units, and the horizontal error is not more than β_2 units. Under these parameters, we calculate the shortest path of the trajectory under the given constraint condition, and calculate the minimum number of times an UAV corrected by the correct area. The trajectory planning path map of the data set is shown in Figure 9.

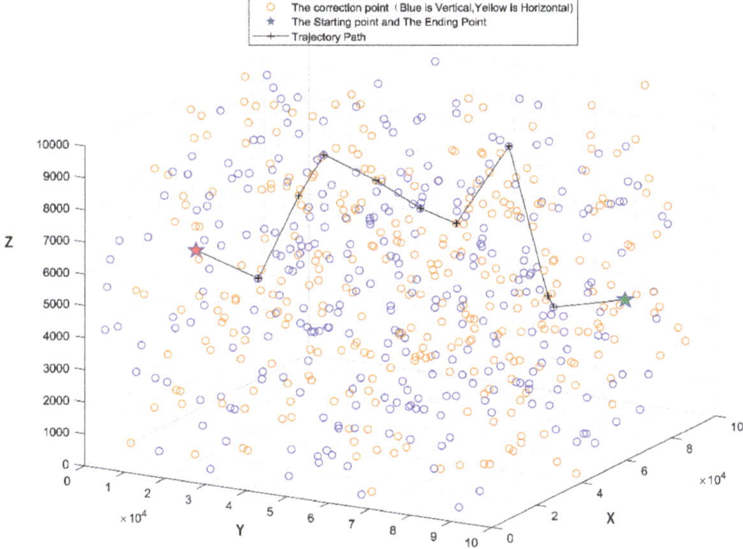

Figure 9. Trajectory planning path for the Problem 1.

Figure 9 shows the trajectory planning path of the data set. Firstly, the data of the data set is filtered and processed, and imported into MATLAB software. The discrete point maps of these three-dimensional spaces are drawn according to the type of correction points of an UAV. The mathematical model is established by given multiple constraints and designed objective function, and then the idea of genetic algorithm is used to analyze the feasibility of the horizontal error correction point and the vertical error correction point in the discrete point respectively. After the screening, through the fitness function, some fitness values are determined in turn, and the trajectory formed by the fitness function values is the optimal trajectory length.

Before determining the above-mentioned trajectory planning path and the number of times an UAV through the correct area for correction, the position of the corrected vertical and horizontal errors can be determined, and the error correction point number and the path planning of the error before the correction are obtained. The optimal trajectory length is 16,972.695304 m. The results are shown as Table 2.

It can be seen from Table 2 that the shortest path to the trajectory set of the data set is according to the process of the above algorithm, that is, to initialize a gene, save these genes and form the gene sequence of the trajectory, then the gene sequence is a chromosome. Using this method to cycle back and forth, we can get a lot of chromosomes, then get the appropriate function value from the determined moderate function, and use the given constraints to determine whether the chromosome is feasible, to get the optimal trajectory length. On this basis, the number of times of correction of the correction area is calculated to be nine, and the error correction point number is obtained from the starting point.

Table 2. Corrects result for Problem 1.

The Number of the Correction Points	The Type of the Correction Points
0	The Starting Point
504	1
295	0
124	1
76	0
34	1
12	0
404	1
595	0
398	1
612	The Ending Point

4. Problem 2

In addition to the constraints including Problem 1, Problem 2 including the above-mentioned condition (8) in Section 2. That is to say, the condition that an UAV is restricted by the structure and control system when turning is unable to complete the instant turn. And, comprehensively consider the following optimization goals: (A) the trajectory length is as small as possible; (B) the number of corrections through the correct points is as small as possible.

If the above-mentioned parameters of the data are:

$\alpha_1 = 25, \alpha_2 = 15, \beta_1 = 20, \beta_2 = 25, \theta = 30, \delta = 0.001.$

4.1. Multi-Constraints Optimization Problem

4.1.1. Objective Function

The focus of this research is to plan a trajectory for a UAV from the starting point to the ending point, and finding the optimal trajectory satisfied with the multi-constraints in Problem 2, so we build mathematical models of the problem.

$$F_1 = \min\left(\sum_{i=1}^{m} (\eta_i + \lambda_i + h_i + \sum_{1 \le j \le i} l_j)\right) \tag{21}$$

$$F_2 = \min\left(\sum_{i=1}^{m} (a_i + b_i)\right) \tag{22}$$

where F_1 represents the trajectory length of an UAV from the starting point A to the ending point B, F_2 represents the number of times an UAV has been corrected through the correction area, m represents the total number of corrections of an UAV throughout the flight, η_i represents the distance of an UAV's *i*-th flight, λ_i represents the correction distance of the *i*-th flight to the vertical error correction point, h_i represents the correction distance of the *i*-th flight to the horizontal error correction point, l_j represents the distance that an UAV turns during the *j*-th flight, a_i represents an UAV's *i*-th flight reaches the vertical error correction point, and b_i represents an UAV's *i*-th flight reaches the horizontal error correction point.

4.1.2. Multi-Constraints

The above objective function should satisfy the constraints of Problem 1 and the following constraints:

(1). When a UAV is flying from position H_i to position H_{i+1}, due to the limitation of the UAV's maneuverability, it will move in an arc segment $H_i\,H_{i+1}$. At this time, the circle is centered on the intersection of the mid-permanent line of the distance from position H_i to position H_{i+1} and the ideal trajectory of H_i to H_{i+1}, so the radius r_i can be expressed as:

$$r_i = \frac{\eta_i}{2}\cos\phi, i = 1,\ldots,m \tag{23}$$

where η_i represents the distance of an UAV which flies from position H_i to position H_{i+1}, r_i represents the turning radius of an UAV's i-th flight, and ϕ is the maximum yaw angle that an UAV is allowed to fly. During the flight of a UAV, it will be limited by the structure and control system, and it will not be able to complete the instant turn, so the turning distance of the j-th flight of an UAV can be obtained by:

$$l_j = \pi r_j - 2r_j, 1 \le j \le i \tag{24}$$

(2). Due to the limitation of the UAV's own maneuverability, a UAV can only turn within a certain range of yaw angle, that is, its yaw angle is less than or equal to the maximum allowed yaw angle before it can fly to the next trajectory point. The yaw angle limit is the minimum turning radius limit. The smaller the turning angle, the more smoothly a UAV can fly. Therefore, suppose the horizontal projection of the i-th trajectory segment is $\gamma_i = (x_i - x_{i-1}, y_i - y_{i-1})$, and the maximum yaw angle allowed by a UAV to fly is ϕ, then the constraint condition can be expressed as:

$$\cos\phi \le \frac{\gamma_i^T\gamma_{i+1}}{\|\gamma_i\|\|\gamma_{i+1}\|}, \quad i = 1, 2, \ldots, m \tag{25}$$

(3). A UAV is limited by the structure and control system during the turn and cannot complete the immediate turn. The minimum turning radius of an UAV is 200 m in Section 2.1, then the constraint condition can be expressed as:

$$200 \le r_j \le \frac{\eta_j}{2}\frac{\gamma_i^T\gamma_{i+1}}{\|\gamma_i\|\|\gamma_{i+1}\|}, \quad j = 1,2,\ldots,i \tag{26}$$

In summary, the model for Problem 2 can be established as:

$$
\begin{aligned}
&F_1 = \min\left(\sum_{i=1}^{m}\left(\eta_i + \lambda_i + h_i + \sum_{1\le j\le i} l_j\right)\right)\\
&F_2 = \min\left(\sum_{i=1}^{m}(a_i + b_i)\right)\\
&s.t.\begin{cases}
\sum_{i=1}^{m} a_i\eta_i\delta < \theta\\
\sum_{i=1}^{m} b_i\eta_i\delta < \theta\\
0 \le \eta_i \le l_{AB}\\
\lambda_i = a_i\eta_i\delta, i = 1,\ldots,m\\
h_i = b_i\eta_i\delta, i = 1,\ldots,m\\
a_i, b_i = 0, 1, i = 1,\ldots,m\\
a_i + b_i = 1, i = 1,\ldots,m\\
\lambda_i \le \alpha_1, h_i \le \alpha_2, i = 1,\ldots,m\\
\lambda_i \le \beta_1, h_i \le \beta_2, i = 1,\ldots,m\\
l_j = \pi r_j - 2r_j,\\
\cos\phi \le \frac{\gamma_i^T\gamma_{i+1}}{\|\gamma_i\|\|\gamma_{i+1}\|}, \quad i = 1, \ldots, m\\
200 \le r_j \le \frac{\eta_j}{2}\frac{\gamma_i^T\gamma_{i+1}}{\|\gamma_i\|\|\gamma_{i+1}\|}, \quad j = 1,\ldots,m
\end{cases}
\end{aligned}
\tag{27}
$$

4.2. Trajectory Planning Algorithm for the Problem 2

Problem 2 has a constraint on the minimum turning radius. Therefore, we adopt an improved sparse A* algorithm, which is an effective method to avoid useless trajectory nodes in space and reduce the time to search for feasible successor nodes. Using trajectory constraints to search only the effective space reduces the search space and speeds up the search. So, for Problem 2, we propose a trajectory planning problem based on the improved sparse A* algorithm. First, we plan the division of space. Attention should be paid to reducing the number of divided cells as much as possible, thereby reducing the amount of calculation and improving the algorithm's convergence speed, so that the algorithm can plan the feasible trajectory of a UAV in the shortest time. Then, there is the determination of the cost function.

If $g(n)$ represents the actual cost of a UAV at the current trajectory node n in space:

$$g(n) = q_1 R_n(P) + q_2 S_n(P) \tag{28}$$

where R,S represents the flight cost of a UAV respectively, the horizontal and vertical error correction costs of a UAV, and q_1, q_2 each represent the weight coefficient of R,S.

$$R_n = \sum_{i=1}^{m} \eta_i \tag{29}$$

$$S_n = \sum_{i=1}^{m} (a_i + b_i)\eta_i\delta \tag{30}$$

where η_i represents the distance of an UAV's i-th flight, a_i represents an UAV's i-th flight to the vertical correction point, b_i represents an UAV's i-th flight to the horizontal correction point, and δ indicates that the horizontal error and vertical error increase by δ units each time an UAV flies 1 metre. Let $h(n)$ represent the Euclidean distance from the current trajectory node to the target trajectory node $B(x_B, y_B, z_B)$, and q_3 represent the weight coefficient of $h(n)$, then:

$$h(n) = \sqrt{(x_n - x_B)^2 + (y_n - x_B)^2 + (z_n - x_B)^2} \tag{31}$$

Therefore, the improved cost function $f(n)$ can be determined as:

$$f(n) = q_1 R_n(P) + q_2 S_n(P) + q_3 h(n) \tag{32}$$

A UAV's online real-time trajectory planning task requires an improved sparse algorithm to complete the trajectory planning in a short time and a small memory space. The node can be expanded by an improved cost function, nodes can be expanded with an improved cost function, using two linked lists, of which the open table stores the nodes to be expanded, and the close table stores the nodes already. The specific algorithm flow is as follows and can be seen in

Step 1: The search space is divided according to the requirements of the aircraft for the accuracy of the trajectory.

Step 2: Put the starting point A into the open table, and the closing table is initially empty.

Step 3: If the open table is empty, it means that the search failed, and the algorithm exited.

Step 4: Remove the least costly trajectory node in the open table as the current trajectory node, and store the point in the closing table.

Step 5: If the current node is midpoint B, the search is ended, and the algorithm exits successfully. Starting from point B and going back to point A, we get the optimal path from point A to point B. Otherwise, go to the next step.

Step 6: The successor node of the node with the least expansion cost. Point the parent pointer of the succeeding node to the node with the least cost.

Step 7: Calculate the generation value of the successor node according to the improved cost function formula, and insert the successor node into the open table according to the generation value.

Step 8: Return to step 3. Figure 10:

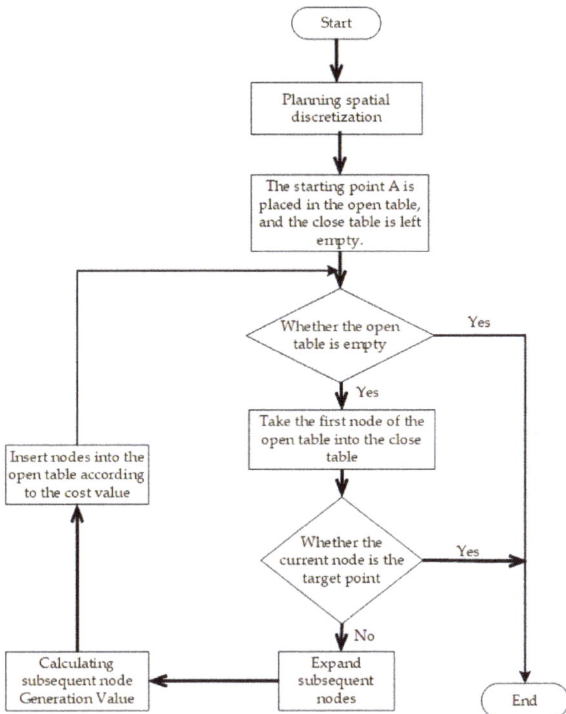

Figure 10. Flow chart of the improved sparse A* algorithm.

Step 6 is to classify the successor nodes of the node with the least expansion point cost as follows:

The first type: If the successor node already exists in the open table, the successor node is discarded; the node is added to the node table with the least cost; and then calculate and compare the actual cost g value of the point, determine whether the parent node of the point needs to be relocated, and if so, update the generation value;

The second type: If the subsequent node is in the closing table, call the Step 7 to calculate the path cost. If it is smaller, update the total generation value f and actual generation value g of the parent node. If find that there is a better path to reach this point in the path search, extend this step to the subsequent nodes of this point;

The third type: If the successor node is not in the open table and the closing table, then the node is placed in the open table.

4.3. Simulation Results and Analysis of the Trajectory Planning Algorithm for the Problem 2

According to the above algorithm, the results of the Problem 2 are calculated by using the data set. As the Problem 2 is based on the Problem 1, considering that an UAV is limited by the structure and control system when turning, it cannot complete the instant turn. According to the Problem 2, the minimum turning radius of the aircraft is 200 m. Then, use the improved A* algorithm to calculate its maximum yaw angle, and use MATLAB software to make a path planning path map of the data set. Figure 11 shows the trajectory planning path map of an UAV.

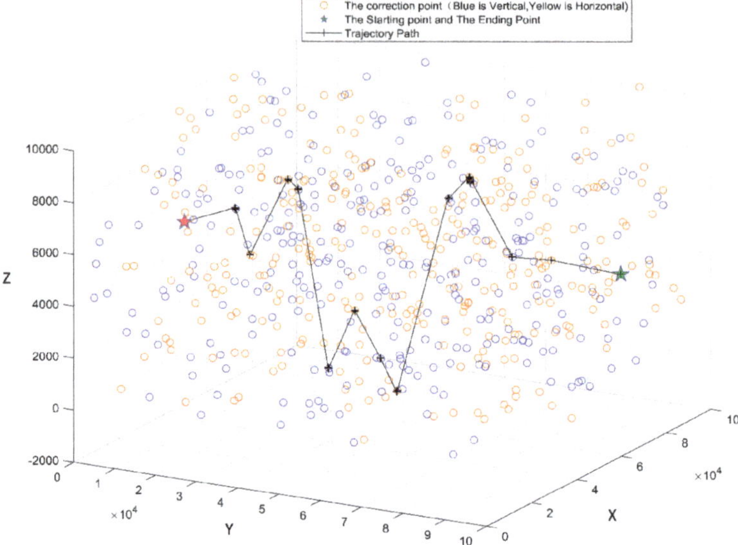

Figure 11. Trajectory planning path for the Problem 2.

Figure 11 shows that both the horizontal error correction point and the vertical error correction point are evenly distributed in the three-dimensional space. This question considers that when a UAV turns from the current trajectory node in the horizontal direction to the next trajectory node, due to the limitations of the UAV's own performance, it can only turn within a certain yaw angle range. Therefore, through the above algorithm, the feasible correction point is searched in turn. Connect these feasible correction points, that is, get a trajectory planning path, use this method to cycle back and forth, which constitutes the optimal trajectory length in Figure 11.

Under the constraint condition that a UAV is turning within a certain yaw angle, the position of the trajectory planning path and the number of times a UAV passes the correct area for correction can be determined before the correction of the vertical and horizontal errors, and a UAV starts from the starting point. The error correction point number and the trajectory planning result table of the error before correction are shown in Table 3.

Table 3. Corrects result for Problem 2.

The Number of the Correction Points	The Type of the Correction Points
0	The Starting Point
346	1
200	0
294	0
136	1
108	1
74	0
462	1
543	0
369	1
457	0
388	1
436	0
612	The Ending Point

Obtaining the optimal trajectory length from starting point to ending point is 14957.842315 m. From the table, the number of times of correction of the correction area is calculated 12 times, and the error correction point number passing through the starting point from the starting point is obtained.

5. Performance Comparison of the Proposed Algorithm with Traditional Swarm Intelligence Algorithm

5.1. For the Problem 1

For the data set, we compare the proposed improved GA algorithm with the traditional GA algorithm in trajectory planning. Figure 12 shows the fitness change of traditional GA algorithm:

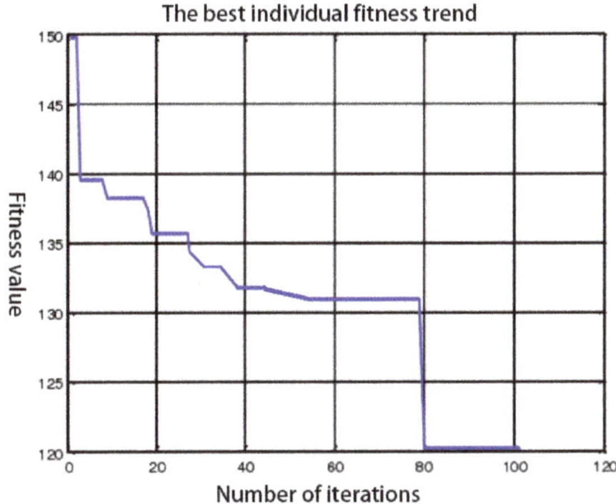

Figure 12. Fitness curve of traditional GA algorithm for the Problem 1.

Figure 13 shows the fitness change of the proposed improved GA algorithm:

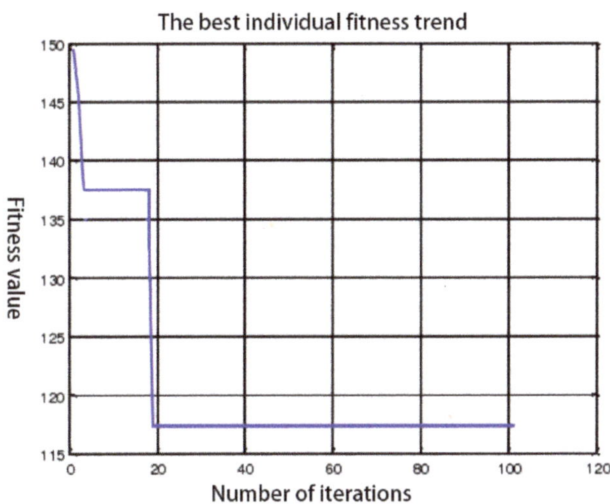

Figure 13. Fitness curve of the proposed improved GA algorithm for the Problem 1.

In order to compare the performance of the two algorithms more clearly, draw the fitness curves of Figures 12 and 13 into a unified coordinate system, as shown in Figure 14:

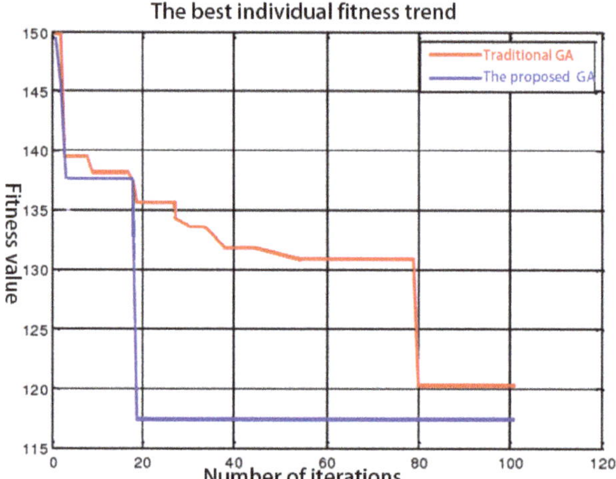

Figure 14. Two algorithms for fitness.

It can be seen from Figure 14 that: (1) the proposed improved GA algorithm converges significantly faster than the traditional GA algorithm, (2) the proposed improved GA algorithm has a lower overall cost for the optimal trajectory search, (3) during the search of the proposed improved GA algorithm, the shock is small. Therefore, the proposed improved GA algorithm in the paper is significantly better than the traditional GA algorithm.

5.2. For the Problem 2

Experiments were performed using the basic A* algorithm and the improved A* algorithm, and the superiority of the algorithm was verified by comparing the path length of the trajectory planning and the number of correction points passed. Using the data set, the two weight coefficients q_1, q_2 and q_3 of Formula (32) are set, respectively, as shown in Table 4.

Table 4. The weight coefficient of the improved cost function $f(n)$.

Number	q_1	q_2	q_3
1	0.3	0.2	0.5
2	0.3	0.5	0.2

In the course of trajectory selection, the smaller the value of the cost function, the better the trajectory. Table 5 compares the experimental performance of the proposed improved sparse A* algorithm and the basic A* algorithm in the same environment.

Table 5. Performance comparison of A* algorithm and the proposed improved sparse A* algorithm.

Number	Algorithm	Number of Correction Points Passed	Trajectory Length
1	Before improvement	31	35,436.632176
	After improvement	12	14,957.842315
2	Before improvement	39	41,258.581392
	After improvement	15	19,513.452647

In Table 5, before improvement means that the basic A* algorithm, and after improvement means that the proposed improved sparse A* algorithm in this paper is used. Numbers 1 and 2 represent the weight coefficients of numbers 1 and 2 in Table 4. It can be seen from Table 5 that the larger q_2 is, the larger correction points are passed, and the larger q_3 is, the shorter the trajectory length is. Therefore, the values of these weighting factors can be adjusted from different needs. Furthermore, it can be seen that the performance of the improved sparse A* algorithm is significantly better than the basic A* algorithm.

6. Conclusions

The proposed method allows for multiple constraints optimal trajectory planning using the improved genetic algorithm (GA) and A* algorithm. The conclusions of this research work are: (1) The numerical results of the experiment proved that the improved genetic algorithm (GA) and A* algorithm are good for optimal trajectory planning. (2) All essential constraints like the correction and turning radius of a UAV are satisfied with all solutions obtained from the trajectory planning algorithm.

Future work should take into account more complex and realistic constraints, such as threats to the ground or in the air, the avoidance of obstacles, and other constraints of the UAV itself (such as its own battery life), improve the algorithm optimization model and objective functions (such as considers flight time [28]), and ensure that UAVs are able to adapt to more complex environments.

Author Contributions: Optimization problem software development, H.Z.; supervision, H.-L.X.; data curation, N.-D.T. and L.C.; resources, Y.L.; writing, H.Z. All authors have read and agreed to the published version of the manuscript.

Funding: This research was funded by the Key Projects of Natural Science Research of Universities in Anhui of China (KJ2019A0864), and by the Chongqing Technology Innovation and Application Development Project (cstc2019jscx-gksbX0103), and by the Fundamental Research Funds for the Central Universities under Project (SWU119044).

Conflicts of Interest: The authors declare no conflict of interest. The funders had no role in the design of the study; in the collection, analyses, or interpretation of data; in the writing of the manuscript, or in the decision to publish the results.

References

1. Zhao, S.Y.; Hu, Z.Y.; Chen, B.M.; Lee, T.H. A Robust Real-Time Vision System for Autonomous Cargo Transfer by an Unmanned Helicopter. *IEEE Trans. Ind. Electron.* **2015**, *62*, 1210–1218. [CrossRef]
2. Shen, Y.F.; Rahman, Z.; Krusienski, D.; Li, J. A vision-based automatic safe landing-site detection system. *IEEE Trans. Aerosp. Electron. Syst.* **2013**, *49*, 294–311. [CrossRef]
3. Huh, S.; Cho, S.; Jung, Y.; Shim, D.H. Vision-based sense-and-avoid framework for unmanned aerial vehicles. *IEEE Trans. Aerosp. Electron. Syst.* **2015**, *51*, 3427–3439. [CrossRef]
4. ElMikaty, M.; Stathaki, T. Car Detection in Aerial Images of Dense Urban Areas. *IEEE Trans. Aerosp. Electron. Syst.* **2018**, *54*, 51–63. [CrossRef]
5. Zhao, L.; Wang, D.; Huang, B.; Xie, L. Distributed Filtering-Based Autonomous Navigation System of UAV. *Unmanned Syst.* **2015**, *3*, 17–34. [CrossRef]
6. Carrillo, L.R.G.; Fantoni, I.; Rondon, E.; Dzul, A. Three-dimensional position and velocity regulation of a quad-rotorcraft using optical flow. *IEEE Trans. Aerosp. Electron. Syst.* **2015**, *51*, 358–371. [CrossRef]
7. Kaiser, M.K.; Gans, N.R.; Dixon, W.E. Vision-Based Estimation for Guidance, Navigation, and Control of an Aerial Vehicle. *IEEE Trans. Aerosp. Electron. Syst.* **2010**, *46*, 1064–1077.
8. Khatib, O. Real-time obstacle avoidance for manipulators and mobile robots. *Int. J. Robot. Res.* **1986**, *5*, 90–99. [CrossRef]
9. Chazelle, B.; Edelsbrunner, H. An improved algorithm for constructing kth-order Voronoi diagrams. *IEEE Trans. Comput.* **1987**, *100*, 1349–1354. [CrossRef]
10. Li, Q.Q.; Zhe, Z. A Voronoi-based hierarchical graph model of road network for route planning. In Proceedings of the 11th Conference Intelligent Transportation Systems, Beijing, China, 12–15 October 2008; pp. 599–604.

11. Sugihara, K. Approximation of generalized Voronoi diagrams by ordinary Voronoi diagrams. *Cvgip Graph. Models Image Process.* **1993**, *55*, 522–531. [CrossRef]
12. Meng, B.B.; Gao, X.G. UAV Path Planning Based on Bidirectional Sparse A* Search Algorithm. In Proceedings of the 2010 International Conference on Intelligent Computation Technology and Automation (ICICTA), Changsha, China, 11–12 May 2010.
13. Szczerba, R.J.; Galkowski, P.; Glicktein, I.S.; Ternullo, N. Robust algorithm for real-time route planning. *IEEE Trans. Aerosp. Electron. Syst.* **2000**, *36*, 869–878. [CrossRef]
14. Chen, X.; Chen, X.M.; Zhang, J. The Dynamic Path Planning of UAV Based on A* Algorithm. *Appl. Mech. Mater.* **2014**, *494–495*, 1094–1097. [CrossRef]
15. Wang, H.; Zhou, H.; Yao, H. Research on autonomous planning method based on improved quantum Particle Swarm Optimization for Autonomous Underwater Vehicle. In Proceedings of the OCEANS 2016 MTS/IEEE, Monterey, CA, USA, 19–23 September 2016.
16. Zhang, H.; Liu, Z. 3D path planning for micro air vehicles based on quantum-behaved particle swarm optimization algorithm. *J. Cent. South Univ.* **2013**, *44*, 58–62.
17. Tokgo, M.; Li, R. Estimation Method for Path Planning Parameter Based on a Modified QPSO Algorithm. In Proceedings of the International Conference on Artificial Intelligence: Methodology, Systems, and Applications, Varna, Bulgaria, 11–13 September 2014.
18. Shorakaei, H.; Vahdani, M.; Imani, B.; Gholami, A. Optimal cooperative path planning of unmanned aerial vehicles by a parallel genetic algorithm. *Robotica* **2016**, *34*, 823–836. [CrossRef]
19. Men, J.Z.; Qun, E.; Yao, K.M. Study on the Route Planning for Anti-Submarine Ship-Based UAV Based on Genetic Algorithm. *Adv. Mater. Res.* **2012**, *433–440*, 4823–4826. [CrossRef]
20. Zhang, D.Q.; Zhao, J.F.; Lei, G.; Wang, S.H.; Zheng, X.L. Hurry Path Planning Based on Adaptive Genetic Algorithm. *Appl. Mech. Mater.* **2014**, *446–447*, 1292–1297. [CrossRef]
21. Shiri, H.; Park, J.; Bennis, M. *Massive Autonomous UAV Path Planning: A Neural Network Based Mean-Field Game Theoretic Approach*; Cornell University: Ithaca, NY, USA, 2019.
22. Hao, Z.; Cao, C.; Xu, L.; Gulliver, T.A. AN UAV Detection Algorithm Based on an Artificial Neural Network. *IEEE Access* **2018**, *6*, 24720–24728.
23. Chen, Y.B.; Luo, G.C.; Mei, Y.S.; Yu, J.Q.; Su, X.L. UAV path planning using artificial potential field method updated by optimal control theory. *Int. J. Syst. Sci.* **2016**, *47*, 1407–1420. [CrossRef]
24. Xia, C.; Jing, Z. The Three-Dimension Path Planning of UAV Based on Improved Artificial Potential Field in Dynamic Environment. In Proceedings of the International Conference on Intelligent Human-Machine Systems & Cybernetics, Hangzhou, China, 26–27 August 2013.
25. Liu, J.Y.; Guo, Z.Q.; Liu, S.Y. The Simulation of an UAV Collision Avoidance Based on the Artificial Potential Field Method. *Adv. Mater. Res.* **2012**, *591–593*, 1400–1404. [CrossRef]
26. Meng, H.; Xin, G. UAV route planning based on the genetic simulated annealing algorithm. In Proceedings of the IEEE International Conference on Mechatronics and Automation, Xi'an, China, 4–7 August 2010; pp. 788–793.
27. China Ministry of Education Degree and Graduate Education Development Center. Available online: https://cpipc.chinadegrees.cn/ (accessed on 19 September 2019).
28. Kwasniewski, K.K.; Gosiewski, Z. Genetic Algorithm for Mobile Robot Route Planning with Obstacle Avoidance. *Acta Mech. Autom.* **2018**, *12*, 151–159. [CrossRef]

Article

OTL-Classifier: Towards Imaging Processing for Future Unmanned Overhead Transmission Line Maintenance

Fan Zhang [1,2], Yalei Fan [2,3,*], Tao Cai [2,3], Wenda Liu [2,3], Zhongqiu Hu [2,3], Nengqing Wang [2,3] and Minghu Wu [1,2,*]

[1] Hubei Collaborative Innovation Center for High-efficiency Utilization of Solar Energy, Hubei University of Technology, Wuhan 430068, China; zhangfan@mail.hbut.edu.cn

[2] Hubei Power Grid Intelligent Control and Equipment Engineering Technology Research Center, Hubei University of Technology, Wuhan 430068, China; ct2437686482@163.com (T.C.); darwinliu1994@gmail.com (W.L.); huzhongqiu_hzq@163.com (Z.H.); WangNengQing_WNQ@163.com (N.W.)

[3] Hubei Key Laboratory for High-efficiency Utilization of Solar Energy and Operation Control of Energy Storage System, School of Electrical and Electronic Engineering, Hubei University of Technology, Wuhan 430068, China

* Correspondence: FanYalei_FYL@163.com (Y.F.); wuxx1005@mail.hbut.edu.cn (M.W.)

Received: 19 September 2019; Accepted: 29 October 2019; Published: 1 November 2019

Abstract: The global demand for electric power has been greatly increasing because of industrial development and the change in people's daily life. A lot of overhead transmission lines have been installed to provide reliable power across long distances. Therefore, research on overhead transmission lines inspection is very important for preventing sudden wide-area outages. In this paper, we propose an Overhead Transmission Line Classifier (OTL-Classifier) based on deep learning techniques to classify images returned by future unmanned maintenance drones or robots. In the proposed model, a binary classifier based on Inception architecture is incorporated with an auxiliary marker algorithm based on ResNet and Faster-RCNN(Faster Regions with Convolutional Neural Networks features). The binary classifier defines images with foreign objects such as balloons and kites as abnormal class, regardless the type, size, and number of the foreign objects in a single image. The auxiliary marker algorithm marks foreign objects in abnormal images, in order to provide additional help for quick location of hidden foreign objects. Our OTL-Classifier model achieves a recall rate of 95% and an error rate of 10.7% in the normal mode, and a recall rate of 100% and an error rate of 35.9% in the Warning–Review mode.

Keywords: smart grid; foreign object; binary classification; convolutional network

---

1. Introduction

Nowadays, people's daily life and industrial facilities are highly dependent on electric power. Therefore, research on electric power facilities inspection and maintenance is very important for ensuring a stable power supply. A lot of overhead transmission lines have been installed to distribute energy across long distances in the world. It is meaningful to prevent sudden wide-area outages caused by foreign objects suspended on uninsulated overhead transmission lines.

At present, foreign objects could be detected by foot patrol, piloted helicopter patrol, drones inspection, and transmission line robots inspection. Foot patrol is risky or unable to pass through complex areas such as highways, rivers, and mountains. Helicopter inspection is expensive and also limited by the shortage of pilots. Though unmanned drones and specialized robots are still not used in practice for various limitations such as path planning, law, and regulations. However, they are still highly considered by the electric power field for future maintenance of smart grids.

The main challenges of using drones for UAV (Unmanned Aerial Vehicles) maintenance are automatic pilot, flight time and communication bandwidth. The authors in [1] aim at solving automatic transmission line tracking problems. The authors in [2,3] face path planning and routing challenges when UAVs are flying along power transmission lines. The authors in [4,5] study wireless charging techniques for the increasing of drone's flight time. The authors in [6] focus on UAV communication toward 5G, which supports high-speed camera data transmission.

The PTL (Power Transmission Line, PTL) maintenance robot equipped with cameras can walk through transmission line for inspection. It is possible to perform inspection and maintenance work at a low cost in the future. Recently, transmission lines have been built with bundled conductors because of the increasing power demand. However, conventional robots can only inspect a line while traveling along it [7]. Thus, most research focused on developing new robot architectures for smart navigation over bundle transmission lines [8–10].

With the rapid hardware development of smart drones and PTL maintenance robots, the demand of automatic data processing for transmission line inspection will increase quickly. A number of research works have been carried out to extract transmission lines, insulators, and foreign objects from aerial images automatically. The authors in [11] use Robot LiDAR data for cable inspection, [12] extracts power lines based on Markov Random Field theory, foreign objects are detected with a morphology-based approach [13] and a motion compensation based method [14], and all of them use traditional algorithms. As reviewed in [15], the potential of deep learning in power line inspection is promising. For example, the automated inspection of insulator [16], transmission towers [17], and transmission lines [18] based on deep learning have already been carried out. The detection of foreign objects on transmission lines based on Faster-RCNN and YOLO (You Only Look Once) were also studied in [19,20] respectively. However, the foreign object image used for the experiment are images with foreign objects by default, so the algorithm does not have the classification function. In addition, the amount of data they use for experiments is very small, and the number of images in our dataset is more than 10 times that of them. There are also some detection algorithms for the detection of insulators. It is more challenging to detect foreign objects without a fixed shape compared to insulators with regular shapes.

Enlightened by image classification and object detection architectures based on deep learning (i.e., VGG [21], ResNet [22], Inception [23,24], Faster RCNN [25], and SSD (Single Shot MultiBox Detector) [26]), a two-stage

approach is proposed for automated image processing, which detects and marks foreign objects in the image. The model is trained, fine-tuned, and tested with images collected by electric maintenance departments. The reminder of the article is organized as follows: Section 1 reviews related work. Section 2 presents the methodology of the proposed model. Section 3 describes the preparation of data set. In Section 4, the experiment is analyzed and discussed. Finally, conclusions and contributions of this work are drawn in Section 5.

2. Methodology

2.1. Problem Statement

Detecting foreign objects on overhead transmission lines is a very important work regarding power system maintenance. Overhead transmission lines are a primary method for transmitting high-voltage power across long distances. The high energy of transmission lines requires very thick insulation to prevent the insulating material from catching fire itself. If they are insulated, the insulation would make power distribution lines too costly and very heavy and thus unlikely to set up in air. Thus, unlike low-voltage cable, overhead transmission lines don't have insulation, they are insulated by air. During high-wind events, foreign objects such as plastic greenhouses, kites, and balloons blew onto overhead transmission lines, thus prone to short-cuts or electrical sparks, causing power trips during humid seasons or wildfires during dry seasons.

In this study, we collected and sorted out the images that were retained during the manual cleaning of foreign objects in the transmission line, as shown in Figure 1. In addition, in the classification and marking, whether it is balloons, kites, or plastic greenhouses, we are uniformly classified as one class foreign object.

Figure 1. *Cont.*

(e)

Figure 1. Sample images in the data set. (**a**) Balloons, kites and agricultural plastic which are the main foreign objects hanging on overhead transmission lines. (**b**) Images without foreign objects which have been used as negative samples of the classification task.These negative sample images also contain high-voltage towers and transmission lines, as well as daily images collected by inspection equipment. (**c**) Foreign objects of different sizes. Some tiny foreign objects are not easy to detect. (**d**) Data we collected included not only colorful balloons and kites, but also translucent plastic and black agricultural greenhouses. (**e**) The contrast between the translucent plastic and the sky is not so obvious, and the black plastic shed is easily confused with the trees.

2.2. Warning–Review Strategy

In the first part, the algorithm workflow that constitutes the whole 'foreign object image classification-warning-personnel review' is introduced. We also introduce the framework used in the image classification algorithm and the target detection algorithm. In the second part, the preparation and division process of the data used in the experiment are described.

All of the test images in references [19,20] are images with foreign objects, which is equivalent to artificially removing the interference image without foreign objects before their foreign object detection algorithm detects it.

However, the images collected by the current intelligent inspection equipment contain a large number of images without foreign objects. In order to get closer to the real inspection situation, in this paper, the images used for training and testing algorithms are composed of the image with foreign objects and the image without foreign objects. The whole process is shown in Figure 1, which is the biggest difference between the research work of this paper and the previous research. After the mixed inspection image passes the classification algorithm of the first stage, the image with a foreign object may be marked by the classification algorithm, thereby alerting the power grid staff and prompting the staff to review the image with the alarm.

In the first stage, this paper focuses on the 100% recall rate algorithm and trains and tests SVM (Support Vector Machine), InceptionV3-retrain, InceptionV3-fine-tuning, and InceptionV4-fine-tuning. In the second stage, the inspection image marked as the presence of foreign object is sent to the target detection algorithm, and the region where the foreign object exists in the image is located and marked with a rectangular frame. The significance of this step is to assist the staff to quickly determine the type of foreign object and locate the position of the suspended foreign object. This paper has trained and tested SSD, Faster-RCNN , Faster-RCNN, and Faster-RCNN in this section. All the algorithm structures are concentrated in Figures 2 and 3. Please note that the rectangular block in the network structure is only indicative and does not represent the true size of a layer in the actual network.

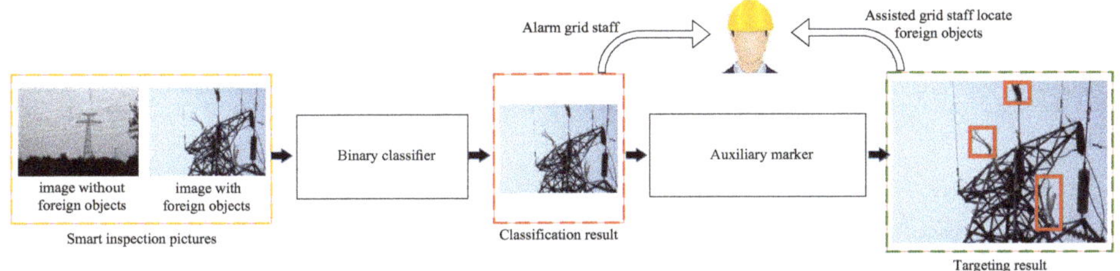

Figure 2. Warning–review strategy.

When we test the algorithm, all of the 753 images in the testing set are first input into the classifier of the first stage, and then according to the set classification threshold, a part of the images in the testing set are marked as "images with foreign objects" by the classifier. Finally, only the image marked by the classifier is sent to the foreign object indicator of the second stage for foreign object detection. The entire algorithm flow is shown in Figure 3.

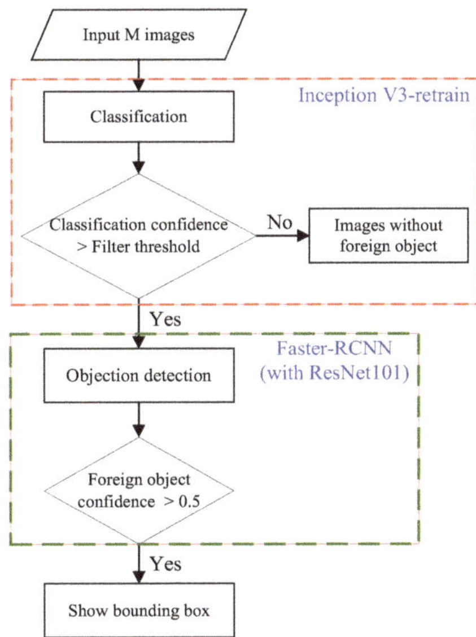

Figure 3. Two-stage foreign object detection flow chart.

2.3. Binary Classification Module

2.3.1. SVM

The SVM algorithm flow can be represented as the first part of Figure 4. SVM is a generalized linear classifier that classifies data according to the supervised learning method. Its decision boundary is the maximum margin hyperplane for solving learning samples. The purpose of the SVM is to find a hyperplane to divide the samples into two categories with the largest interval. The ω obtained by the algorithm represents the coefficient of the hyperplane that the algorithm needs to find. In mathematical terms, it can be described as Label (1),

$$\max \frac{1}{\|\omega\|}, s.t., y_i(\omega^T x_i + b) \geq 1, i = 1, \ldots, n, \tag{1}$$

where $y_i \in \{-1, 1\}$. The larger the score obtained by y_i $(\omega^T x_i + b)$, the greater the probability of predicting the category. Each image input into the SVM is compressed into a matrix of $[1 \times 3072]$. There are two categories in the power line image classification. Therefore, the size of the matrix ω is $[2 \times 3072]$, and the size of the matrix b is $[2 \times 1]$.

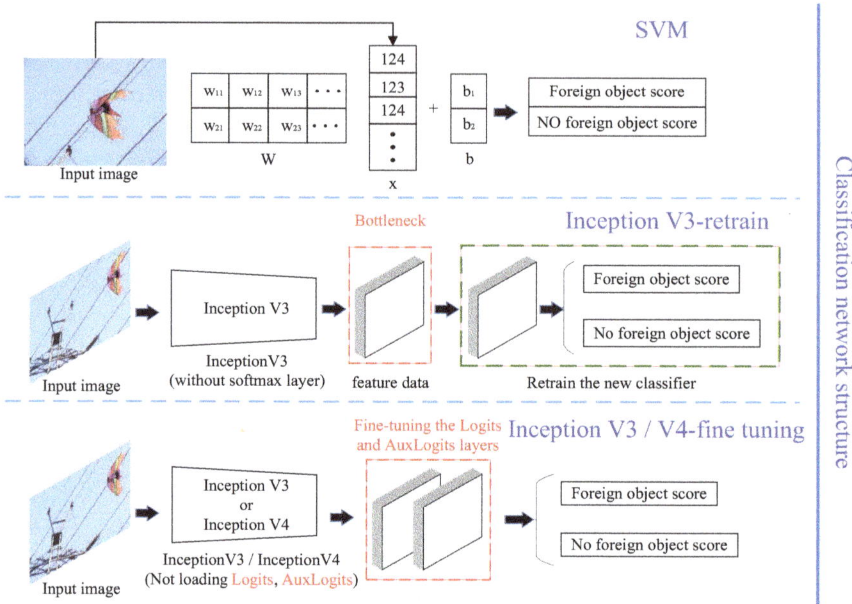

Figure 4. Classification network structure.

2.3.2. Inception Retrain

The InceptionV3-retrain algorithm flow can be represented as the second part of Figure 4. This method using InceptionV3-retrain as a foreign object classification model for power lines mainly utilizes the PB file derived from the InceptionV3 model based on ImageNet image training. The last softmax layer of the model is changed to the foreign object image classifier, and the Bretleneck feature of the pre-training network is used as training data for the new classifier. When the new two-class network is trained by the retrain method, the parameters of all the layers in the previous part of the network are fixed. The program first loads the pre-training model of InceptionV3, and then stores the Bottleneck file generated after the training data set is input into the network. Then, use the feature data stored in the Bottleneck file to train the modified softmax layer as a classifier for the new task. Except for the last layer, the parameters of the other layers are all solidified and cannot be updated, so the training speed is faster and less time-consuming.

2.3.3. Inception Fine-Tuning

The InceptionV3/V4-fine-tuning algorithm flow can be expressed as the third part of Figure 4, and the foreign object classification model is fine-tuned under the InceptionV3 and InceptionV4 models provided by Google. The fine-tuning mode is to use a CKPT (checkpoint) file, which derived from the InceptionV3 or InceptionV4 model based on ImageNet image training. During the training process, the parameters of the entire network can be modified accordingly, not only limited to the replaced softmax layer. The fine-tuning for InceptionV3/V4 is done by loading the pre-trained model without loading the parameters of the Logits layer and AuxLogits layer, and then fixing the parameters of all layers before. The foreign object training data set only trains the newly created Logits layer and AuxLogits layer.

When fine-tuning the model, restoring checkpoint weights requires attention. In particular, when a new task is fine-tuned with an output tag different from the number of ImageNet detection tasks, the final classification layer cannot be restored. Therefore, this paper uses the $'checkpoint\_exclude\_scopes'$ flag, which prevents certain variables from being loaded. For example, if the ImageNet trained model is fine-tuned on the foreign object classification dataset, the pre-trained logits layer has dimensions $[2048 \times 1001]$, but the new logits layer has dimensions $[2048 \times 2]$. Therefore, the flag indicates that the TF-Slim avoids loading these weights from the checkpoint.

2.4. Auxiliary Marking Module

2.4.1. SSD with VGG16

The SSD (with VGG16) algorithm flow can be represented as the first part of Figure 5. When training the target detection algorithm, the training data used are manually labeled foreign object images. The VGG-16 (Visual Geometry Group Network 16) model has sixteen convolutional layers and five pooled layers and three fully connected layers connected to a softmax layer. Conv4, Conv7, Conv8, Conv9, Conv10, and Conv11 are extracted separately in SSD as the feature layer of classification and box regression. In the SVM model experiment, each image is scaled to a size of 300×300, and the number of predicted classifications (ClassesNum) of the six feature layers according to Formula (2) is 2,

$$ClassesNum = ObjectNum + 1, \tag{2}$$

where *ObjectNum* represents the number of manually labeled categories, and one represents an additional background classification.

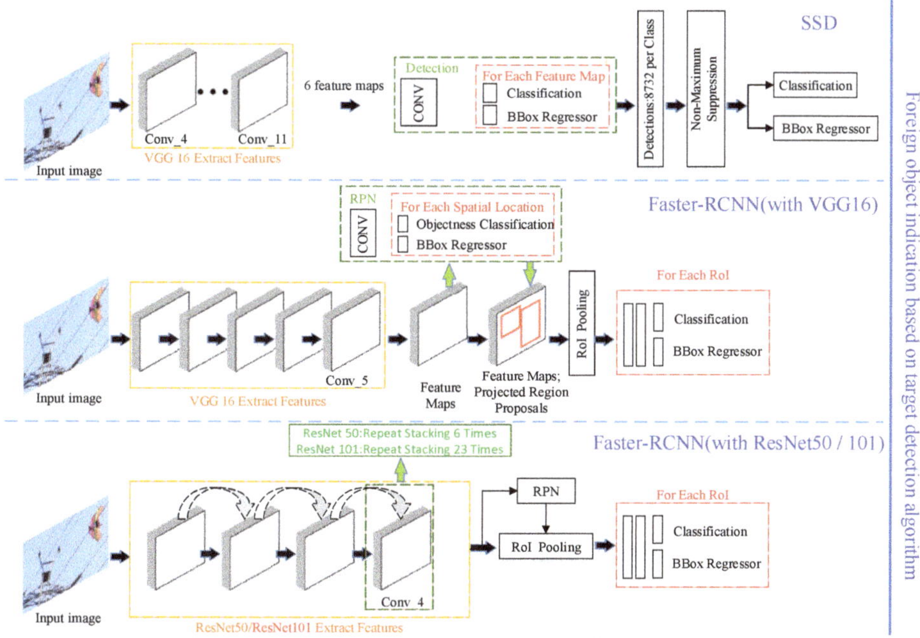

Figure 5. Foreign object indication based on a target detection algorithm.

2.4.2. Faster-RCNN with VGG16

The Faster-RCNN (with VGG16) algorithm flow can be represented as the second part of Figure 5. In the framework of the Faster-RCNN algorithm, the input image is extracted by the convolutional network of the feature extraction layer, and the feature map output by the specified convolution layer is used as the input of the RPN. The feature map obtained under different convolutional structures has different characterization capabilities for input images. In this experiment, the Faster-RCNN structure using VGG16 as the feature extraction network is first tested. This paper standardizes the image size to 1000×600 as input, which is consistent with the original author's parameter settings in [18]. The feature map output by the convolution layer Conv5 is used as the input of the RPN, and nine different size anchor boxes are generated according to the regulations at each anchor point. All bounding boxes with high confidence in the anchor box are selected as a region proposal and sent to the full convolution layer through ROI (Region of Interest) Pooling to obtain the category confidence and regression box of the detected image.

2.4.3. ResNet

The Faster-RCNN (with ResNet50/ResNet101) algorithm flow can be represented as the third part of Figure 5. The overall algorithm flow of Faster-RCNN (with ResNet50/ResNet101) is basically the same as the previous one. The original VGG16 network is replaced by the ResNet network in the feature extraction layer. At the same time, considering the performance of the experimental server, the input image is adjusted. The image size of the input algorithm is 500×300, which enables the experimental server to completely load a large network such as ResNet in a relatively low hardware configuration.

3. Data Set Preparation

When we divide the data set, the ratio of positive and negative samples in the training set is about 1:1.6. This division belongs to a balanced division mode, which helps the algorithm to learn more key features of the classification in the learning phase. However, the ratio of positive and negative samples on the test set is about 1:6, which is to simulate the real situation that the foreign object accident is a low-frequency high-risk electric accident. Most of the real inspection images are images without foreign objects.

As shown in Table 1, in the training of the classification algorithm, 305 images with foreign objects are used as positive samples, and 500 images without foreign objects are used as negative samples. Because we prepare training and testing data for SVM based on the cifar-10 data format, in this paper, only the amount of data used by SVM has been trimmed. There are 300 images with foreign objects and 500 images without foreign objects in the SVM training set; 100 images with foreign objects and 500 images without foreign objects in the SVM testing set. In this paper, the image is processed according to the rules into a dictionary format that Python can read quickly, as shown in Figure 6.

Table 1. Division of training and testing dataset.

	Foreign Object	No Foreign Object
Training set	305	500
Testing set	101	652
Total	406	1152

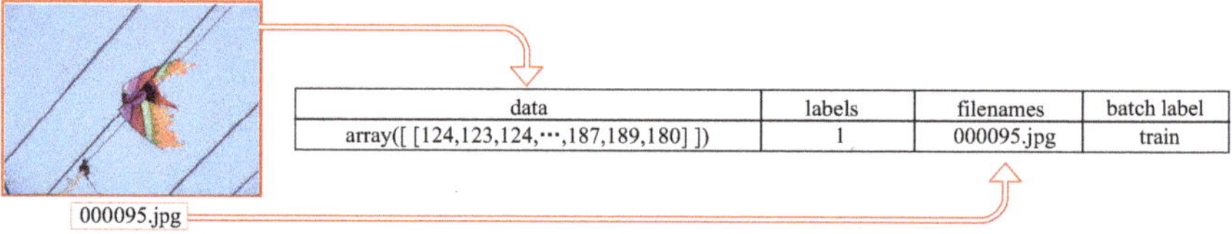

Figure 6. Cifar-10 data storage rules. The data stored in the $[32 \times 32 \times 3]$ matrix are the compressed data of the original image. Labels store image tags, image with foreign objects is 1, and image without foreign objects is 0. File names are stored in filenames. The batch label stores the image as being divided into training or test tags.

For the training of the target detection algorithm in the foreign object indicator, this paper only used 305 foreign object images, and the images were manually labeled. This is because the target detection algorithm uses anchors for training. The positive samples during training are taken up by the proposal region with the artificially labeled ground truth IOU (Intersection over Union) value being the largest or larger than the set threshold. The negative sample is assumed by the proposal region with the ground truth, whose IOU value is less than the set threshold. Therefore, there can be no negative sample input during training. The experimental environment of the above experiment is one server, and its specific parameters are as follows: The Tensorflow experimental framework is the Linux and Windows10 environment. The software environment is Python3.6, CUDA9.0, cuDNN7.0, and Tensorflow1.8; and the hardware environment is a Lenovo ZHENGJIUZHE REN7000 desktop PC produced in China, it is equipped with Intel i7-8700 Core , 8 GB memory and NVIDIA GeForce GTX 1060 GPU.

4. Results

4.1. Calculation of Various Evaluation Indicators

4.1.1. Calculation Formula for Classifier Evaluation

After the above test, the paper analyzes the classification results of the first stage, and draws the receiver operating characteristic (ROC) curves of the SVM, InceptionV3-retrain, InceptionV3-fine-tuning and InceptionV4-fine-tuning models, as shown in Figure 7. The abscissa of the ROC curve is false positive rate (FPR), the ordinate is true positive rate (TPR), and the area under curve (AUC) is defined as the area under the ROC curve. When a positive sample and a negative sample are randomly selected, the probability that the current classification algorithm ranks the positive example before the negative example based on the calculated score is the AUC value. Therefore, the larger the value of AUC, the more likely the current classification algorithm is to sort the positive samples before the negative samples, which enables better classification. The above values can be calculated using Equations (3), (4) and (5), respectively,

$$TPR = \frac{TP}{TP + FN}, \tag{3}$$

$$FPR = \frac{FP}{FP + TN}, \tag{4}$$

$$AUC = \int_0^1 TPRdFPR, \tag{5}$$

where TP : *True positive*; TN : *True negative*; FP : *False positive*; TN : *True negative*. At the same time, the point selected according to the Youden index is plotted in Figure 7. The ROC curve is often used as an evaluation curve for medical diagnosis. When a comprehensive evaluation of the diagnosis results is required, the sensitivity and specificity can be given the same weight in the medical field. It is characterized by the same significance of the missed diagnosis rate and the misdiagnosis rate of the research object. The larger the Youden index, the better the screening ability. See (6) for the calculation method. The x-axis of the ROC curve is $(1 - specificity)$, so the final formula can be simplified to Label (7):

$$Youden\ index = max(sensitivity + specificity - 1), \tag{6}$$

$$Youden\ index = max[sensitivity - (1 - specificity)] = max(TPR - FPR). \tag{7}$$

4.1.2. Calculation Formula for Evaluation of Foreign Object Indicator

The foreign object indicator performance index based on the target detection algorithm is the recall rate and accuracy rate, where $TP + FN = 126$. $Average Precision (AP)$ is the integral of the PR-curve, which is the area under the curve (8)–(10):

$$Precision = \frac{TP}{TP + FP},\tag{8}$$

$$Recall = \frac{TP}{TP + FN},\tag{9}$$

$$Average\ Precision = \int_0^1 Precision(Recall)dRecall.\tag{10}$$

4.2. Classification Performance Evaluation

In Figure 7, the blue dashed line is referred to as the "random chance", which means that the probability of the sample being classified as a positive or negative sample is random. In the ROC coordinate system, the classification threshold at the $(0,0)$ point is the largest, and the classification threshold at the $(1,1)$ point is the smallest. According to the ROC curve, it can be clearly observed that when the classification recall rate is 100% (corresponding to the ordinate value is 1), the InceptionV3-retrain model has the lowest error rate. At the same time, the InceptionV3-retrain model reduced the AUC value by 0.4% compared to the InceptionV3-fine-tuning model, but reduced the error rate by 8%.

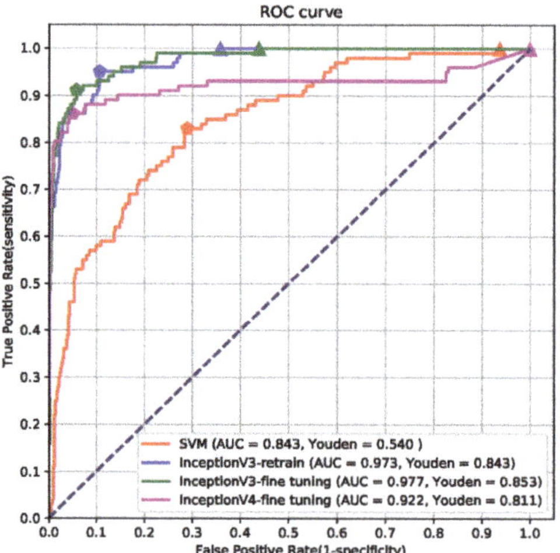

Figure 7. ROC of the classifier. (1) in the case of a 100% recall. The threshold points of the classifier are marked with triangles, indicating that all images with foreign objects can be screened when the images are classified by the threshold corresponding to the points. The abscissa value corresponding to the threshold point is the misclassification rate, which means that the triangle mark closer to the left side corresponds to the lower classification error rate, and the performance of the algorithm is better. (2) We calculate the optimal classification threshold by using the Yoden index. The optimal threshold point of the classifier is marked by a pentagon, which means that only a part of the foreign object image can be selected when the image is classified by the threshold value of the mark point, and the classification error rate is also decreased. (3) InceptionV3-fine-tuning has AUC = 0.977 as the maximum value and SVM has AUC = 0.843 as the minimum value.

In the foreign object image detection task, the paper should focus more on finding out all the foreign objects in the image. InceptionV3-retrain, as the first stage classifier, can maintain the lowest error rate among the four classification algorithms under the premise of 100% recall. Although the error rate of InceptionV3-retrain is reduced by 25% at the optimal threshold classification point, the algorithm cannot classify all the foreign object images, and there is a security risk in the scene of the foreign object inspection of the transmission line.

4.3. Automate Marking Performance Evaluation

In Table 2 this paper, the InceptionV3-retrain algorithm is selected as the classifier when the recall rate is 100%, and the corresponding classification threshold is 0.102. The remaining 233 sheets are all misclassified images with an error rate of 36%. In addition, 334 classified images are used as the input of the second stage target detection algorithm. The target detection algorithm PR curve is shown in Figure 8.

Table 2. Classification algorithm data.

		SVM	InceptionV3-Retrain	InceptionV3-Fine-Tuning	InceptionV4-Fine-Tuning
Recall 100%	recall rate	100%	**100%**	100%	100%
	error rate	94%	**35.9%**	43.9%	100%
	threshold	0.264	0.102	0.092	0.0
Optimal threshold	recall rate	83%	**95%**	91%	86%
	error rate	29%	**10.7%**	5.8%	5%
	Yoden index	0.54	0.843	0.853	0.811
	threshold	0.428	0.546	0.503	0.4

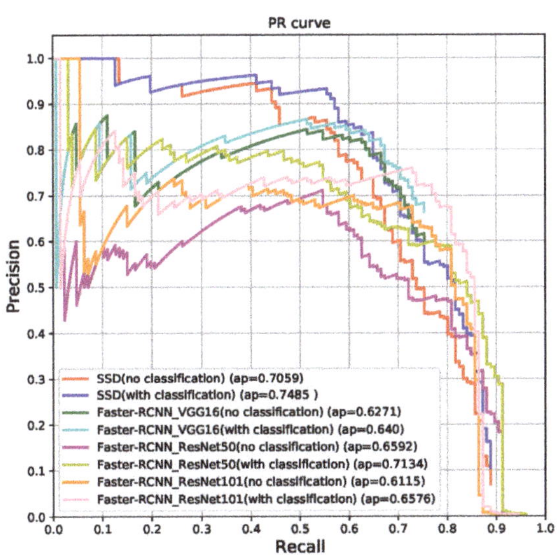

Figure 8. Target detection algorithm PR (Precision-Recall) curve. (1) In the figure, each target detection algorithm draws two PR curves. One is a PR curve that is directly input with 753 sheets as an algorithm without going through the classification process. The other is the PR curve drawn with 334 images as the algorithm input. (2) After using the two stages framework, each algorithm has a different degree of increase in the AP value, which is since the first stage classifier filters out 60% of the negative samples. (3) The SSD has the highest 0.7485 AP value, while the Faster-RCNN (with VGG16) AP is the lowest 0.64 of all experiments.

When 0.5 is used as the display threshold, only the bounding box whose confidence is higher than this threshold is displayed on the test picture. The specific values are shown in Table 3. The total box indicates the total number of bounding boxes that the algorithm ultimately presents to the power grid staff, and the TP number indicates the number of targets that are correctly found. The missed target indicates the number of bounding boxes that the algorithm missed. Target detection precision and Target detection recall are calculated according to Equations (8) and (9), respectively.

Table 3. Target detection algorithm data.

	Total Box	TP Number	Missed Target	Target Detection Precision	Target Detection Recall
ResNet50	218	106	20	48.62%	84.13%
ResNet101	183	106	20	57.92%	**84.13%**
VGG16	122	90	36	73.77%	71.43%
SSD	150	93	33	62.0%	73.81%

5. Conclusions

In this paper, we introduce an OTL-Classifier, a binary classifier with an auxiliary automate marker module. Compared to recent research, our method is much more application oriented. We have three main differences:

- Our **OTL-Classifier module can classify images with and without foreign objects**. However, recent research only processes images with foreign objects; they focused on detecting the type and location of the foreign objects in the abnormal images. However, aerial images return by drones and robots inspection include much more normal images than abnormal images. Searching abnormal images manually is not only time-consuming, but also has poor precision due to attention feature of human. Therefore, it is much more important to design a module which could automatically extract abnormal images directly from original images returned by unmanned vehicles.
- During the evaluation phase, we consider recall rate as more important than precision in our application. A sudden wide-area outage caused by even one undetected foreign object will affect people's lives and industrial production seriously and may lead to a lot of economic loss. Therefore, we think it is very critical to have a recall rate of 100%, so no abnormal images will be missed during classification.
- Most recent research evaluated detection speed. For example, RCNN4SPL module spends 230 ms per frame, YOLOv3 based module is 46 ms in average, Morphology based module is 95.8 ms in average, and Motion compensation-based module is 64 ms. We didn't test execution time because it is highly dependent on the hardware. In addition, in our application, we don't have a very high timing requirement as path planning for automatic drive.

In this article, we have evaluated the classification performance of SVM and three Inception variants, and the marking performance of SSD, Faster-RCNN with VGG, and ResNet. Experiments shows our module based on Inceptionv3-retrain, and Faster-RCNN with ResNet101 achieves best performance on the data set we collected from electric maintenance departments.

We summarize our contributions as follows:

- We proposed an OTL-Classifier module; it can classify images with and without foreign objects. It can work in either Warning-Review mode or Normal mode.
- In the normal mode, the OTL-Classifier works the same as most common classification tasks, the module uses optimal parameters that balances recall rate and error rate. It can achieve a recall rate of 95% and an error rate of 10.7%.
- In the Warning-Review mode, the OTL-Classifier achieves a recall rate of 100% and an error rate of 35.9%. It has a two-stage workflow. In the first stage, the binary classifier module provides the

warning. In the second stage, the automated marker module helps electric workers review the image quickly. This strategy can prevent outage caused by foreign objects and save more than half of the time on image checking. Our future work will focus on decreasing the error rate with a recall rate of 100%.

Author Contributions: Conceptualization, F.Z. and Y.F.; methodology, F.Z.; software, Y.F.; validation, Y.F., Z.H. and N.W.; formal analysis, Y.F.; investigation, F.Z.; resources, Y.F.; data curation, Y.F., T.C., W.L., Z.H. and N.W.; writing–original draft preparation, Y.F.; writing–review and editing, F.Z. and Y.F.; visualization, Y.F.; supervision, F.Z.; project administration, M.W.; funding acquisition, F.Z. and M.W.

Funding: This work was supported by the National Natural Science Foundation of China (Grant No. 11605051 and A050508), Excellent Young and Middle-aged Science and Technology Innovation Team Project in Higher Education Institutions of Hubei Province (T201805), Major Technological Innovation Projects of Hubei(No. 2018AAA028)

Acknowledgments: The authors would like to thanks to all the reviewers who helped us in the review process of our work.Moreover special thanks to the Jiangsu Electric Power staff for providing the authors with the opportunity to collect experimental image data.

References

1. Menéndez, O.; Pérez, M.; Auat Cheein, F. Visual-Based Positioning of Aerial Maintenance Platforms on Overhead Transmission Lines. *Appl. Sci.* **2019**, *9*, 165. [CrossRef]
2. Baik, H.; Valenzuela, J. Unmanned Aircraft System Path Planning for Visually Inspecting Electric Transmission Towers. *J. Intell. Robot. Syst.* **2019**, *95*, 1097–1111. [CrossRef]
3. Campbell, J.F.; Corberán, Á.; Plana, I.; Sanchis, J.M. Drone arc routing problems. *Networks* **2018**, *72*, 543–559. [CrossRef]
4. Lu, M.; Bagheri, M.; James, A.P.; Phung, T. Wireless Charging Techniques for UAVs: A Review, Reconceptualization, and Extension. *IEEE Access* **2018**, *6*, 29865–29884. [CrossRef]
5. Citroni, R.; Di Paolo, F.; Livreri, P. A Novel Energy Harvester for Powering Small UAVs: Performance Analysis, Model Validation and Flight Results. *PubMed* **2019**, *19*, 1771. [CrossRef] [PubMed]
6. Li, B.; Fei, Z.; Zhang, Y. UAV Communications for 5G and Beyond: Recent Advances and Future Trends. *IEEE Internet Things J.* **2019**, *6*, 2241–2263. [CrossRef]
7. Richard, P.L.; Pouliot, N.; Morin, F.; Lepage, M.; Hamelin, P.; Lagac, M.; Sartor, A.; Lambert, G.; Montambault, S. LineRanger: Analysis and Field Testing of an Innovative Robot for Efficient Assessment of Bundled High-Voltage Powerlines. In Proceedings of the 2019 International Conference on Robotics and Automation (ICRA), Montreal, QC, Canada, 20–24 May 2019; pp. 9130–9136.
8. Zhang, Y.; Li, J.; Li, C.; Tao, Q.; Xiong, X. Development of foreign matter removal robot for overhead transmission lines. In Proceedings of the 2nd International Conference on Mechanical, Electric and Industrial Engineering, Hangzhou, China, 25–27 May 2019.
9. Disyadej, T.; Promjan, J.; Poochinapan, K.; Mouktonglang, T.; Grzybowski, S.; Muneesawang, P. High Voltage Power Line Maintenance & Inspection by Using Smart Robotics. In Proceedings of the 2019 IEEE Power and Energy Society Innovative Smart Grid Technologies Conference (ISGT), Washington, DC, USA, 18–21 February 2019; doi:10.1109/isgt.2019.8791584. [CrossRef]
10. Seok, K.H.; Kim, Y.S. A State of the Art of Power Transmission Line Maintenance Robots. *J. Electr. Eng. Technol.* **2016**, *9*, 1412–1422. [CrossRef]
11. Qin, X.; Wu, G.; Lei, J.; Fan, F.; Ye, X.; Mei, Q. A Novel Method of Autonomous Inspection for Transmission Line based on Cable Inspection Robot LiDAR Data. *Sensors* **2018**, *18*, 596. [CrossRef] [PubMed]
12. Zhao, L.; Wang, X.; Yao, H.; Tian, M.; Jian, Z. Power Line Extraction From Aerial Images Using Object-Based Markov Random Field With Anisotropic Weighted Penalty. *IEEE Access* **2019**, *7*, 125333–125356. [CrossRef]
13. Cao, Z.; Ma, J.; Lin, P.; Peng, Z. Morphology-Based Visual Detection of Foreign Object on Overhead Line Tower. In Proceedings of the 2018 3rd IEEE International Conference on Image, Vision and Computing, ICIVC 2018, Chongqing, China, 27–29 June 2018; pp. 468–472.
14. Jiao, S.; Wang, H. The Research of Transmission Line Foreign Body Detection Based on Motion Compensation. In Proceedings of the 2016 First, International Conference on Multimedia and Image Processing (ICMIP), Bandar Seri Begawan, Brunei, 1–3 June 2016; pp. 10–14.

15. Nguyen, V.N.; Jenssen, R.; Roverso, D. Automatic autonomous vision-based power line inspection: A review of current status and the potential role of deep learning. *Int. J. Electr. Power Energy Syst.* **2018**, *99*, 107–120. [CrossRef]

16. Miao, X.; Liu, X.; Chen, J.; Zhuang, S.; Fan, J.; Jiang, H. Insulator Detection in Aerial Images for Transmission Line Inspection Using Single Shot Multibox Detector. *IEEE Access* **2019**, *7*, 9945–9956. [CrossRef]

17. Michalski, P.; Ruszczak, B.; Lorente, P.J.N. The Implementation of a Convolutional Neural Network for the Detection of the Transmission Towers Using Satellite Imagery. In Proceedings of the 40th Anniversary International Conference on Information Systems Architecture and Technology, Wrocław, Poland, 15–17 September 2019; pp. 287–299.

18. Dong, J.; Chen, W.; Xu, C. Transmission line detection using deep convolutional neural network. In Proceedings of 2019 IEEE 8th Joint International Information Technology and Artificial Intelligence Conference, Chongqing, China, 24–26 May 2019; pp. 97–980.

19. Xia, P.; Yin, J.; He, J. Neural Detection of Foreign Objects for Transmission Lines in Power Systems. *J. Phys. Conf. Ser.* **2019**, *1267*, 012043. [CrossRef]

20. Zhang, W.; Liu, X.; Yuan, J. RCNN-based foreign object detection for securing power transmission lines (RCNN4SPTL). In Proceedings of the International Conference on Identification, Information and Knowledge in the Internet of Things (IIKI), Beijing, China, 19–21 October 2018; pp. 331–337.

21. Simonyan, K.; Zisserman, A. Very deep convolutional networks for large-scale image recognition. *arXiv* **2015**, arXiv:1409.1556v6.

22. He, K.; Zhang, X.; Ren, S.; Sun, J. Deep residual learning for image recognition. In Proceedings of the IEEE Conference on Computer Vision and Pattern Recognition (CVPR), Las Vegas, NV, USA, 26 June–1 July 2016; pp. 770–778.

23. Szegedy, C.; Vanhoucke, V.; Ioffe, S. Rethinking the inception architecture for computer vision. In Proceedings of the IEEE Conference on Computer Vision and Pattern Recognition (CCVPR), Las Vegas, NV, USA, 26 June–1 July 2016; pp. 2818–2826.

24. Szegedy, C.; Ioffe, S.; Vanhoucke, V. Inception-v4, Inception-Resnet and the impact of residual connections on learning. In Proceedings of the Thirty-First, AAAI Conference on Artificial Intelligence (AAAI-17), San Francisco, CA, USA, 4–9 February 2017; pp. 4278–4284.

25. Ren, S.; He, K.; Girshick, R.; Sun, J. Faster r-cnn: Towards Real-Time Object Detection with Region Proposal Networks. In Proceedings of the Advances in Neural Information Processing Systems (NIPS), Montreal, QC, Canada, 7–12 December 2015; pp. 91–99.

26. Liu, W.; Anguelov, D.; Erhan, D. SSD: Single Shot MultiBox Detector. In Proceedings of the European Conference on Computer Vision (ECCV), Amsterdam, The Netherlands, 8–16 October 2016; Springer: Cham, Switzerland, 2016; pp. 21–37.

 *electronics*

Review

Model Update Strategies about Object Tracking: A State of the Art Review

Deyu Wang [1,2], **Weidong Fang** [3], **Wei Chen** [1,2,4,*], **Tongfeng Sun** [1,2] and **Tingjie Chen** [1,2]

[1] School of Computer Science and Technology, China University of Mining Technology, Xuzhou 221000, China; hnsdwdy@126.com (D.W.); suntf@cumt.edu.cn (T.S.); 2390@cumt.edu.cn (T.C.)
[2] Mine Digitization Engineering Research Center of the Ministry of Education, China University of Mining and Technology, Xuzhou 221116, China
[3] Key Laboratory of Wireless Sensor Network & Communication, Shanghai Institute of Micro-System and Information Technology, Chinese Academy of Sciences, Shanghai 201800, China; weidong.fang@mail.sim.ac.cn
[4] School of Earth and Space Sciences, Peking University, Beijing 100871, China
* Correspondence: chenwdavior@163.com

Received: 24 September 2019; Accepted: 17 October 2019; Published: 23 October 2019

Abstract: Object tracking has always been an interesting and essential research topic in the domain of computer vision, of which the model update mechanism is an essential work, therefore the robustness of it has become a crucial factor influencing the quality of tracking of a sequence. This review analyses on recent tracking model update strategies, where target model update occasion is first discussed, then we give a detailed discussion on update strategies of the target model based on the mainstream tracking frameworks, and the background update frameworks are discussed afterwards. The experimental performances of the trackers in recent researches acting on specific sequences are listed in this review, where the superiority and some failure cases on each of them are discussed, and conclusions based on those performances are then drawn. It is a crucial point that design of a proper background model as well as its update strategy ought to be put into consideration. A cascade update of the template corresponding to each deep network layer based on the contributions of them to the target recognition can also help with more accurate target location, where target saliency information can be utilized as a tool for state estimation.

Keywords: visual tracking; update occasion; update mechanism; background model; network layer contribution; saliency information

---

1. Introduction

With the progress of computer vision technology, moving target tracking is being increasingly popularly researched, which has become a challenging topic in the area of smart application. As the development of computerÿ hardware devices and rapid progress of machine learning and deep learning techniques, researches on each respect of moving target tracking has been endowed with great essence. Object tracking has been greatly related to many applications in modern life, i.e., player identification, vehicle monitor, smart human-computer interactions [1]. The mechanism of tracking a moving target is that the target, which is distinguishable from the background, is separated out and marked by a bounding box, which is usually regarded as a classification issue that target samples and background ones should be from different classes. Nowadays, lots of frameworks of image classifiers, i.e., support vector machine (SVM) [2], extreme learning machine (ELM) [3], Integrated Circulant Structure Kernels (ICSK) [4], etc., have been widely utilized for researches of visual tracking. Furthermore, deep learning is getting more and more popularly concerned, trackers using which framework have gained more excellent performances due to the development of neuroscience.

Diverse variations regarding the target usually occur in the process of tracking, i.e., variations arise from changes of the outside environment, such as view angle, camera orientation, environmental illumination, etc., and inherent changes of the object, such as self-rotation, self-deformation, and self-variation of target appearance; therefore, a tracker with more robust capacity has to be designed, whose framework structure and sample learning strategy are of key importance, which guarantees its real-time and accuracy. Consequently, researching an update strategy with higher robustness and efficiency has been of greater essence.

Object tracking framework can be usually typed into two categories: generative frameworks and discriminative ones, where, for the former framework, i.e., particle filter, sparse coding, linear predictions [5,6], Kalman filter, etc., target and background models are established at the beginning and the features of them are extracted for the search of similar target or background features in succeeding frame images to iteratively locate the target; The latter, i.e., deep neural networks, correlation filter, random forest, feature bagging [7], etc., gets the object location by drawing candidate target patches within a region and then select one that is distinguished from given background patches. With the progress of researches on machine learning and deep learning tracking frameworks, the model update has become a widely concerned part in recent researches. A good update mechanism is a crucial respect measuring the reliability of a tracker. On the one hand, template models of the target and background should be constantly updated to catch up with the their variation, which is a fundamental requirement of model adaptation. On the other hand, the parameter model must be adjusted with the same pace of the variations of the samples to satisfy the real-time requirement. Generally speaking, when and how to update make the major parts of the update task. In general, a less-frequent update cannot make sure that the target model can catch up with the change of target appearance, which gives rise to tracking failure, while much too frequent update makes it excessively adaptive to new characteristics of targets but neglects the influences of historical ones, which leads to background drift after a sudden occlusion comes across, thus incurring fatal errors. Up to now, specific update methods are designed to deal with tracking under irregular situations, such as occlusion and background clutter. For instance, more attention will be given to the background analysis when partial occlusion occurs. Although model update technology of visual tracking is gaining rapid progress and has obtained substantial break through at present, there are too few reviews about it compared to other works of tracking, as most reviews still focus on model construction and mathematical algorithms. This review will provide discussions on recently-proposed model update mechanisms and talk about the merits and drawbacks of them. Measures of improvement based on the superiority of existed update strategies and the remaining challenging tracking problems are proposed at the end of this paper. The remaining part of this paper is organized, as follows:

In Section 2, target update occasions in recent researches are talked about, in which three common tools—occlusion detection, response map, and similarity judgement—and two complementary update occasions—conservative update and long-short-term update (LST)—are respectively discussed in detail. In Section 3, the update strategies of target models are illustrated, where recent strategies under four commonly used frameworks—correlation filter (CF), dictionary sparse coding, bag-of-words (BoW), and deep neural network—are respectively analyzed in detail. Background update mechanisms are then illustrated in Section 4, where a new background update framework, called tracking with background estimation (TBE), is briefed. In Section 5, tracking experiment performances of recent trackers are listed, afterwards superior performances under several challenge factors of each typical tracker and some failure cases are exampled and analyzed. Specific conclusions regarding the update mechanisms are drawn from the testing statistics, and improvement measures of model update are briefly summarized in Section 6.

2. Review on Target Model Update Occasions

Determination of the model update occasion is a key part of the update process. Low-frequency update makes it difficult for a tracker to adapt to variations of target appearance, while too frequent update might make the target model introduce too much newest bounding box information that increases the probability of background drift, meanwhile datum calculation burden grows, which cuts down the tracker's efficiency. In general, update occasions often embodies the types below:

(1)   update frame-by-frame;
(2)   update for every certain amount of frames;
(3)   update when the target response is higher than a threshold; and,
(4)   update when the target becomes less distinguished from the background.

Generally, the method that to merely update for every certain period neglects the distinction of the target variation and its response, as well as the consideration of dealing with wrong updates, which makes the tracker update too frequently when the target appearance remains stable for quite a long period or update less frequently if the target constantly changes it appearance, which gives rise to error accumulation that leads to tracking drift. Therefore, trackers with this kind of update method have less robustness. Though update frame-by-frame, i.e., correlation filter, might well make the model tightly pace with the variation of the target, this kind of update unavoidably brings about calculation burden, thus lengthening the datum processing time, incurring unnecessary troubles to some extent. Accordingly, to speed up the calculation, Fast Fourier Transmission (FFT) and Kernelized Correlation Filter (KCF) have been recently proposed that are usually combined with the traditional correlation filter method for image procession. For the construction of a more robust tracker that can pace with target appearance variation as well as avoid error accumulation that is caused by improper update and decrease calculation burden, mere frame-by-frame update or updating with a fixed time interval is rarely adopted in recent researches, hence lots of target response assessment mechanisms, i.e., response maps, foreground and background histogram, multiple-class dictionaries, etc., are proposed. Once the tracked target in a frame is regarded as responsible, target the model update is then enabled, otherwise the tracked object has less responsibility and model update is temporarily stopped.

2.1. Update Using Occlusion Detection

Occlusion is one of the most challenging factors in the process of tracking. It is unavoidable that information of the occluding background part will be integrated into the target model if mere frame-by-frame or fixed-time-interval update is adopted, which makes the tracker mistakenly detect the occluding background part as the target, thus the bounding box stops at the occluding part [8]. Therefore, occlusion detection is required for judging whether the target has been occluded. Occlusion comprises of partial occlusion and full occlusion. In the latter case, almost all of the pixels in the view are background, which means that the target has temporarily disappeared. It is hardly possible to observe the variation of the target's appearance, so target model update is usually stopped when full occlusion happens. However, when the target is partial occluded, only a part of it is visible, hence part of the pseudo target information can be mixed with the target one in the target model if the regular update mode is still used in this case. A special update mode should be utilized in the case of partial occlusion.

There are increasing researches dealing with occlusions in recent year. Although it is easy for the tracker to identify whether the target is under full occlusion, partial occlusion or no occlusion, in quite a few researches, the update is only enabled when there is no occlusion, while it is disabled if partial occlusion happens. For instance, several small patches will be drawn within and around the bounding box after the target is located in a frame in [9] and the patches are classified into three types, where the patches from class #A do not overlap with the bounding box at all, while those from class #B overlap with the bounding box with higher target response and class #C with lower target response. The target is regarded to be occluded if the number of patches from class #C reaches the threshold, thus the target model is prohibited. Conventional correlation filter model update method is adopted in [10], where the fixed learning rate is used for target appearance model update when there is no occlusion; otherwise, the appearance model remains unmodified. Similar strategy is utilized in [11] for target occlusion detection, in which the occluding coefficient of each patch is calculated after the target is located. An update is disabled when the sum of the coefficients is above a given threshold. Complementary features, histogram of oriented gradient (HoG) and Hue, Saturation, Value (HSV), are used in [12] for tracking, where templates that are related to HoG and HSV are respectively established. Background pixel masking is carried out when there is occlusion and target's accurate position and scale is further calculated when partial occlusion happens. Still, the update of two feature templates is enabled only if the target undergoes no occlusion. The Bhattacharyya Distance between the candidate filters and the template in [13] has been used to identify occlusion in this research. Occlusion happens if the distance is above a threshold and thus the template is no updated.

Although the conservative update strategy that target model update is prohibited when the target is partial occluded can well prevent background patches from contaminating target templates, the probability of target appearance variation in each frame never equals to zero, even if the target is in the status of occlusion, therefore if the appearance model of the target is not properly updated at this stage, the tracker might also be unable to pace with the change of the target, thus losing the tracking before the target completely disappears. Local patterns are commonly used in some works to solve the problem of target model update under partial occlusion. In the framework of local patterns, a target model is departed into multiple non-overlapped patches, each of which is respectively tracked to alleviate the impact of pseudo targets. In order to use local information of a target while remaining the holistic structure under the situation of partial occlusion, local tracking that integrates holistic patch and local ones is utilized in [14], in which a tracked object is departed into seven patches, including a global one. The contribution score of each patch is calculated after it is tracked in a frame; afterwards, patches with larger score will be selected for model retraining. To make use of available features of unoccluded parts, in [15], part-based tracking that is similar to the idea in [14] is employed in the state of partial occlusion. Key feature points are extracted to construct the target Gaussian map to obtain the number of patches, thus the correlation filter of each patch is defined. Note that mere global pattern is still utilized when the target is not occluded. For the recovery of a target after full occlusion, owing to the fact that important target information has been preserved by the ICSK model in [4] at the moment before the period of full occlusion, it is usually essential to use the information of the target in the frames before full occlusion, after all of this period belongs to partial occlusion. To preserve the important target information, detected object samples are still selected to update the classifier when the target is partial occluded thanks to the ability to determine scale and position of ICSK, meanwhile ICSK parameters are also preserved. During full occlusion, the parameter set of the optimal classifier is selected according to the energy formulation to identify the reappearance of the target.

The tracked target cannot be identified as being completely responsible, as background pixels may exist together with foreground ones in the bounding box more or less. Even though the background pixel masking process [15] can help to alleviate the interference of background pixels, the existence of noise might not ensure the correct mask of each pixel, thus the background-removed foreground template might not be credible. Up to now, many frameworks, such as dictionary learning (DL) and sparse coding (SC), utilize multiple-class and local-representation structures, i.e., local background and foreground dictionaries are respectively modeled to check out how much background information takes up in the representation of a tracked target so as to correctly track unoccluded parts of a target and enhance the ability to discriminate the background from foreground of some generative models. Owing to the sparsity of image information during partial occlusion, visible parts of the tracking result are used for the encoding of template patches [16], where the corresponding template patches less represent the occluded parts and other parts are regularly updated. Three types of dictionaries are constructed in [17], namely $\mathbf{D}$, $\mathbf{D}_o$, and $\mathbf{D}_b$, which respectively donate the tracking dictionary, target dictionary, and background dictionary to enhance the ability to separate the background from foreground for better target locations. A tracking result is classified into three types of patches, namely stable patches, valid patches, and invalid patches, in which a stable patch is constantly represented by the patch at the same region of the template during some period, while valid ones are the patches that are represented with less error by foreground template patches than background ones and invalid ones are more frequently represented by background template patches. A tracking result is regarded as reliable when the number of valid patches is no less than an extent and the total number of valid patches and stable ones is also no less than a certain threshold; therefore, $\mathbf{D}$ and $\mathbf{D}_o$ are respectively updated, in which $\mathbf{D}_o$ is updated while using valid patches.

2.2. Update Using Response Maps

To judge the responsibility of a tracked object, in the past two years, response maps have been widely utilized in the field of visual tracking. A response map shows the probability of each pixel belonging to the target, whose maximum value point is near to the center of the Gaussian map of the target when the target is normally tracked, and when it is projected to a three-dimensional coordinate, it appears to have only one sharp peak around which the values sharply decrease with farther distance to it. When occlusion or background clutter comes across, more than one peak value can appear in the same response map, or even there is only one peak, the peak appears not so high enough or it is not sharp enough. Processed forms of the response map i.e., *PSR*, *PAR*, *APCE*, etc., are widely adopted in some researches to identify the presence of occlusion or background clutter, which are the derived parameters that measure the responsibility of a tracking result.

A tracking result is only judged to be reliable when the three-dimensional (3-D) response map of the frame image has only one sharp peak. *PAR* [18] is defined to represent the fluctuation of a response map to reflect the reliability of a tracked target, whose formulation is

$$PAR = \frac{R_{\max}^2}{\text{mean}\left(\sum\limits_{w,h} R_{w,h}^2\right)}, \tag{1}$$

in which $R_{\max}$ represents the maximum response value, $R_{w,h}$ is the value at a specific position, and the mean function calculates the average value of the map. Higher *PAR* indicates a more reliable tracking result. When the *PAR* and $R_{\max}$ are both greater than a predefined threshold, the result is judged as reliable, thus the correlation filter model in [18] is updated. Similarly, *APCE* is defined in [19], as

$$APCE = \frac{(F_{\max} - F_{\min})^2}{\text{mean}\left(\sum\limits_{w,h} (F_{w,h} - F_{\min})^2\right)}, \tag{2}$$

where $F_{\max}$ and $F_{\min}$, respectively, denote the maximum and minimum value of the response map, and this parameter also reflects the fluctuation of the map. The context correlation filter in [12] is updated when *APCE* and $F_{\max}$ are both higher than the threshold.

The parameter *PSR* is also similarly defined, except that the sharpness of the peak is not put into consideration, which is calculated by firstly subtracting the mean value and then dividing by the standard deviation, as (3) in [20]

$$PSR = \frac{R_{\max} - \mu}{\sigma}, \tag{3}$$

where μ and σ, respectively, represent the mean value and the standard deviation of the response map. A tracking result is regarded as responsible when the *PSR* is above 10 [20], and thus the long-term and short-term filter memory models are updated; otherwise, the target is occluded and then further face recognition is started.

However, most researches merely take the response map of the target in the frame justly tracked into account, in other words, the influence of the maps in the previous frames are neglected. To be specific, parameters, like *PAR* and *APCE*, etc., vary with different trends during different periods'—usually the variation goes faster when the target is being gradually occluded or it moves away from the occluding background object during the period of partial occlusion. So as to capture the process of the variation of the response map under partial occlusion, the parameter *FCDS* is proposed in [21] to learn the variation feature of the *APCE* in all past frames for the identification of occlusion or background drift, which is formulated as in (4)

$$FDCS = \frac{\text{mean}(\text{max}_N(APCE[0:n])) - APCE_t}{\text{mean}(\text{max}_N(APCE[0:n]))},$$ (4)

where $\text{max}_N(APCE[0:n])$ is the largest N values of *APCE* in all previous frames and $APCE_t$ is the value in frame t. The correlation filter is regarded as not so reliable when its *FDCS*, namely $FDCS_{cf}$, reaches a threshold, thus an update of the filter tracker and the color tracker is stopped. Otherwise, the two trackers are respectively updated according to their discrimination scores.

2.3. Update Using Similarity Measurement

Multiple-template models are usually adopted in generative models, i.e., sparse coding, in which template sets are updated along with the appearance variation of targets-in usual cases, a target appearance model is updated when the appearance of the tracking result is similar enough to the templates, while it needs to be updated when the similarity is not too low but relatively lower than the normal value, which indicates an apparent appearance change. Commonly adopted similarity measurements are cosine similarity, L1 norm, Euclidean distance, etc.

A template set can well represent a tracking result if the similarity values between it and the majority of candidates are high enough; therefore, it needs to not be updated for calculation reduction, while drift might occur when the similarity falls below a degree. Cosine similarity [22] is used for measurement of the similarity between the tracking result and the templates, where the template with a low similarity value is replaced by the tracked object when the similarity value is between 0.65 and 0.85 to avoid excessive mixture of background pixels. Similar update mechanism is utilized by the extreme learning machine (ELM) framework in [23]; however, the ELM model need not be updated only when the similarity is above the threshold, since the semi-supervised learning mode of ELM model and its strong discrimination ability guarantees the quality of the tracked targets. Soft cosine similarity [24] is defined for the measurement rather than conventional cosine similarity to cope with combined challenging factors, i.e., out-of-plane rotation and apparent scale change simultaneously occur during a period. In [24], a tracking result is departed into several parts, anyone of which does not contain too many background pixels when the soft cosine similarity between it and the corresponding template is no less than a predefined value, therefore that template part is updated in a linear interpolation way, otherwise the update is prohibited. Of the multiple-feature pattern, the absolute error gets lower as the similarity between the specific feature template and the corresponding feature of the tracked target goes larger. The sum value of L1 norm of the subtraction matrix of all the template features and the tracking result is used to reflect the total difference, which is greater than a certain threshold when some of the features have undergone evident variance to measure the difference between the result and the templates. The feature template with the smallest weight is then updated to adjust to the change of this feature of the target.

2.4. A Conservative Updating Strategy

Usually, the reliability of the tracked object needs to be estimated no matter how frequently the model is updated in the regular cases. However, drifts may occur when the surrounding patches that are similar to the object are mistakenly identified as foreground, incurring fatal impacts in the consequent frames if the errors are not erased in time. Under this situation, it is sometimes hard to discriminate true appearance change and occlusion when the difference between the tracked target and the template gets bigger.

A conservative update strategy is proposed in [25], in which the reliability of the tracked object is not considered, to reduce impacts of drifts under background interference. During tracking, a whole sequence is departed into several long time periods, each of which is further divided into smaller ones, and several rather than one trackers are established, of which the amount is equal to the amount of small time periods within each larger ones. Each tracker is distributed with a specific update policy, but the public update must be performed frame-by-frame in the first small period of each big ones, thus each tracker stops updating after a certain amount of small periods and then restarts.

The beginning of the next big period is shown in Figure 1. The tracking framework in [25] is named MT.

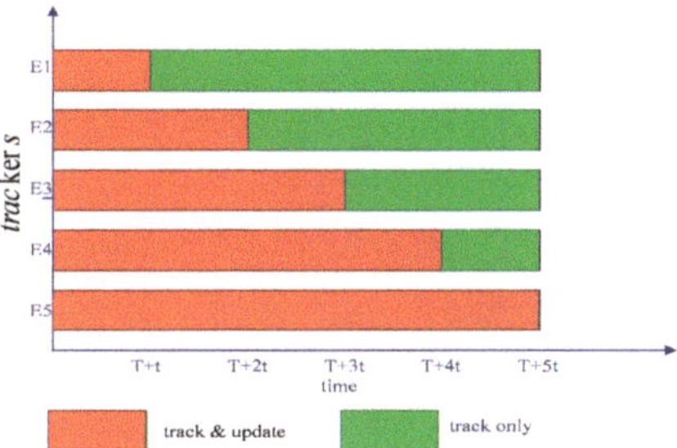

Figure 1. Update Policy of MT.

At the end of a big period, each tracker might track to a position different with which tracked by other ones-some trackers are able to correctly capture the target, while others might fail; therefore, how to select an optimal tracker needs to be further considered. So as to measure the trackers' performances, each one of them tracks the object backwards from the terminal position for a long period equaling to which in the forward tracking, of which the update policy is also the same as in the forward tracking stage. Trajectories of the forward and backward tracking of each tracker are both recorded after the entire process. For a tracker with better performance, the distance between the trajectories of the two different directions is usually comparably lower than others, thus the tracker with the least distance in a big period is selected as optimal.

The tracker that is composed of feature-specific ones named MTM is designed on the basis of the single-feature tracker model named MTS when considering that different features can also bring about different influences to the tracking effect, thus the total amount of trackers equals the product of the amount of features and that of the small periods. The optimal tracker is chosen from all those ones after a round of forward and backward tracking process.

2.5. Combination of Long and Short Term Update

For trackers in many researches, the target model is also updated when a sudden appearance variation or occlusion occurs in addition to when the scheduled update time is up in order to resist drifts that are caused by abrupt target appearance changes or partial occlusion brought about from fixed-time-interval update. It is called update in combination of long and short terms (LST).

For the resistance of impacts of scale variation, deformation, and some other sudden factors, "semantic segmentation" mechanism is introduced in [26], where the correlative parameters of HoG and RGB feature maps between target-based "segmentation map" and position-based "tracking map" are respectively calculated. As long as the target state suddenly changes, the correlation parameter between the "segmentation map" and the hybrid feature map goes higher than that between the "tracking map" and the hybrid feature map, thus the "segmentation signal" is comparably more reliable than the "tracking signal". An immediate target model update is needed to satisfy the real-time changes in this case. Unlike conventional fixed-time-interval update, in this research a frame is regarded as a key frame when the tracking result is judged to be reliable, hence the tracking network is updated when the number of key frames reaches a certain amount rather than frame of a specific index is reached. To avoid erroneous update aroused from occlusion or background clutter, a kind of drift and occlusion detection method is proposed in [27], in which the target model and dual network model are short-term updated while using the best latest tracking results; In addition, the long-term update is performed every ten frames. For adequate use of earlier target information, the score of a tracked target is calculated in [28], which is above 0.5 if the result is regarded as responsible, thus the frame number is added into both the long-term frame number queue (contains 100 frames for most) and the short-term frame number queue (contains 20 frames for most). Appearance variation is detected when the positive classification score is less than 0.5; hence, positive samples from the frames in the short-term queue are used for the network update to meet the demand of pacing with the instant variation. The long-term update is also performed every constant ten frames, when the positive samples from the long-term queue that are rich of previous target information are selected to update the network.

2.6. Module Summary

This module discusses commonly utilized model update occasions. Basic update occasions are listed at the beginning and limitations about time-scale-based update method are briefed next. Recently adopted update occasion determination methods are then illustrated in detail that three kinds of tools for measurement of target's responsibility—occlusion detection, response map, and similarity measurement—and two kinds of newly-proposed hybrid updated occasions—the so-called MT with a conservative update mechanism and LST are respectively illustrated. A reliability check of the tracked object ahead of track can well prohibit erroneous update of the target and tracker model. Additionally, the mixture of long and short term update that fuses the advantages of different update occasions further enhances the adaptability of the trackers. Further solutions to disturbance of similar objects in the target's surrounding area are required in future researches. According to this problem, response check on surrounding background regions should be utilized for the recognition of the true target-the real position can be obtained by comparison of the similarity between the characters of the surrounding background and which of the surroundings templates or utilizing the response maps of the surroundings patches, which might help to alleviate background drifts.

3. Review on Target Model Update Strategies

The design of the model update strategy is a hard project in the work of target tracking. The strong abilities to discriminate the foreground and the background and recover the target after temporary disappearance are not the only requirements for a robust tracker, lower time, and memory consumption as well as an excellent data structure are also essential demands of a good update strategy. In recent years, increasing researches on object tracking have focused on how to balance the robustness of a tracker and low expense of time and memory space. Updated strategies that are based on four commonly-used tracking frameworks—correlation filter (CF), sparse coding (SC), bag-of-words (BoW), and neural network are respectively illustrated below.

3.1. Update Strategy Based on Correlation Filter

Correlation filter (CF) has become one of the most popular utilized models for moving target tracking, especially since Kernelized Correlation Filter (KCF) was first proposed in 2015, and nowadays a large number of researchers have paid attention to the design of filter models with much higher speed, owing to the character of fastness, preciseness, and low expense of time and memory space. Improvement measures of CF model update are also proposed in recent years, having created great breakthroughs over the traditional CF model update method.

Traditional CF target and parameter model update is the linear interpolation of the previous model and the model just trained by the samples from the current frame, as in (5) and (6), which respectively formulates the update of the target model and the parameter model

$$x_t^* = (1 - \alpha)x_{t-1} + \alpha x_t, \tag{5}$$

$$A_t^* = (1 - \mu)A_{t-1} + \mu A_t, \tag{6}$$

where x_t and A_t respectively represents the tracking result in the current frame and the tracker parameters, α and μ respectively means the learning rate of the appearance and parameter model. Constant learning rate is widely used in early models [15,29–33]; however, fixed learning rate cannot properly reflect the real variation of the target appearance, owing to the uncertainty of target variation. If the rate remains high when the target is occluded, some background characters will unavoidably mix into the appearance model; otherwise, if it remains a lower value, the target model will not be able to catch up with faster variations of the target [8,28]. Most recent researches have adopted adaptive learning rates that are adjusted to the extent of target appearance variation and the reliability of the tracking result, which increases the robustness of the tracker model to a great extent, in order to avoid drawbacks of the constant learning rate.

In the last two years, response maps are widely utilized to measure the reliability of the tracking results, of which the simplest method is to use the maximum value. A parameter in [34] is defined to adjust the learning rate according to the response of the tracked target, which is equal to the ratio of the maximum value of the response map in current frame to the maximum of all the response values in previous frames in order to avoid impacts aroused from drastic target appearance variations led by background drift, as formulated in (7)

$$\mu = \frac{F(t)}{\max\{F(i)\}_{i=1}^{t-1}},\tag{7}$$

in which $F(t)$ denotes the maximum value of the response map in frame t; μ gets smaller when improper background drift or heavy occlusion happens, so as to prevent the template model from being contaminated by the tracking result in current frame. The target appearance model is updated as (8), where γ_{init} is the initial learning rate.

$$\hat{x}_t = \mu\gamma_{init}\mathbf{x}_t + (1 - \mu\gamma_{init})\hat{\mathbf{x}}_{t-1},\tag{8}$$

Owing to the fact that target appearance varies in a continuous form, the variation remains stable as time goes on in normal situations; hence, response maps in each frame of a sequence are not independent, especially relevant between two adjacent frames. The reliability parameter (denoted as S_t in (9)) is defined in [35], which is the product of negative exponent of the distance between the target center in the adjacent frames and the PSR value in the current frame, to more effectively represent the stability of the appearance variation of a tracking result. Additionally, to put the temporal stability into consideration, previous movement information is further assembled and an increasing sequence $W = \{\theta^0, \theta^1, \ldots, \theta^{\Delta t - 1}\}$, $(\theta > 1)$ is introduced for providing the latest scores with more weights. The learning rate keeps unchanged when the value in the current frame is above μ (is set to 0.7 in the experiment) time of the weighted average of it in the last Δt (=5) frames; otherwise, it decays to the ratio of the reliability value in the current frame to i time of the weighted average of it in the last five frames, as in ((9), (10) and (13))

$$S_t = \exp(-\frac{1}{\sigma^2}\|C(b^t) - C(b^{t-1})\|^2) \times PSR^t,\tag{9}$$

$$\bar{S} = \frac{1}{\Delta t}\sum_i \omega^i S^i,\tag{10}$$

$$A_t = (1 - \eta)A_{t-1} + \eta A_t^*,\tag{11}$$

$$x_t = (1 - \eta)x_{t-1} + \eta x_t^*,\tag{12}$$

where $C(b^t)$ denotes the center of the tracked target in frame t PSR^t is the PSR value that is introduced in the second module of Section 2; ω^i is the weight in frame i, where the index $i \in [t - \Delta t + 1, t]$, and $\omega^i = \theta^i / (\sum_i \theta^i)$, θ^i is the $(i - t + \Delta t)$-th element in the sequence W; $\bar{S}$ is the weighted average reliability of the last Δt frames. (11) and (12) are, respectively, the formulation of parameter and target appearance model update, η is the adaptive learning rate, which can be formulated as in (13)

$$\eta = \begin{cases} \eta_{init} & S^t > \mu \bar{S} \\ \eta_{init}[S^t/(\mu\bar{S})]^\beta & other \end{cases}, \tag{13}$$

where μ is the fixed parameter that equals to 0.7 and β is the decay factor. This update strategy works well during the process of partial occlusion—when the target is being gradually occluded, the size of its visible part is getting smaller. The shape of the response map become increasingly irregular and the target response value goes lower correspondingly; therefore, the reliability value S_t also drops, and the learning rate is adapted lower to avoid improper update (as the lower formulation in (13) when $S^t \leq \mu\bar{S}$). For the other case, when the target is leaving off the occluding background, the size of the visible part continuously grows, and the response map gradually recovers to the normal shape, thus the reliability value S_t increases. However, the learning rate remains unchanged in this period to inhibit the excessive integration of new target characters that cuts down the universal usage of the model (as the upper formulation in (13) when $S^t > \mu\bar{S}$).

The decrease of response parameters might not be only related to the interference of pseudo targets, self-variation of the target appearance can also bring about the temporary drop in the current and last few frames. The target model is badly in need of an instant update at the moment but it might be disabled if this decrease is mistakenly regarded as the consequence of unreliable variation. The authors in [36] believe that the variation of the target is proportional to its instant speed. Hence, dynamic update of the target model should also be paced with the variation of the speed of the target in addition to the changes of its characters. The learning rate in [36] is determined by two aspects—target moving speed and its feature variation. To get over the problem of partial occlusion that makes it difficult to update, as well as avoid the defect of the speed measurement by distance description, it is believed that the variation of target speed and appearance features are complementary; therefore, the learning rates that are relevant to them ought to be respectively defined, i.e., θ_1 and θ_2 respectively in (14) and (15), which increases with the speed and similarity between the template and the tracked target, respectively. The final learning rate is formulated, as in (16)

$$\theta_1 = \frac{1}{1 + \left(\frac{6}{1+v}\right)^5}, \tag{14}$$

$$\theta_2 = \frac{1}{2} \frac{e^{5c-\frac{5}{2}} - e^{\frac{5}{2}-5c}}{e^{5c-\frac{5}{2}} + e^{\frac{5}{2}-5c}}, \tag{15}$$

$$\theta = \alpha\theta_1 + \beta\theta_2, \tag{16}$$

in which v and c denotes the speed that is measured by the distance between target centers in two adjacent frames and the similarity between the tracking result and the template, respectively; e is the natural exponent base; α and β respectively denotes the adaptive coefficients of θ_1 and θ_2. To learn more about the derivation of θ_1 and θ_2, please refer to [36] for more detail.

The linear interpolation update calculation makes the model sustain the old target appearance as well as introduce new appearance features. The single template model is not able to adequately reflect historical target appearances, although the learning rate can be real-time adjusted according to the response of the target. To overcome this limitation, multiple-template structure, which is being more commonly adopted in generative models, is utilized in some CF trackers, as in [8], to get over the difficulty of calculating the learning rate. Two sets of templates $\mathbf{H}_f^* = \{H_i^*\}_{i=1}^n$ and $\mathbf{H}_s^* = \{H_i^*\}_{i=1}^n$ are established respectively for the first and second tracking in [8], the former of which is asserted by the tracking result X_t in each frame, i.e., $\mathbf{H}_f^* = \mathbf{H}_f^* \cup \{H_t\}$, $H_t = G/X_t$, where G denotes the trained filter parameter image. In the meantime, a template with a relatively larger difference from the result and lower confidence value is removed from the set. Similar to the representation form of sparse coding, the tracking result is linearly represented by the target template set $\mathbf{F} = \{f_i\}_{i=1}^{n}$

$$f_t \approx \mathbf{F}a = \sum_i f_i a_i, \tag{17}$$

in which the coefficient vector a can be solved through sparse coding and it is used for the generation of candidate regions for the first track in the next frame. The second track template is acquired by the combination of the first track template and the original template in the next frame, which is used for the selection of the optimal candidate as the tracking result. The formulation of the second track template is as in (18)

$$\mathbf{H}_{si}^* = (1-p)H_t^* + p\mathbf{H}_{fi}^*, \tag{18}$$

where p is the proportion parameter.

A multiple-filter template structure is adopted in [37] to form a strong CF classifier based on the CFs from current and previous frames in order to utilize historical parameter models. To reduce calculation complexity and memory consuption led by storing similar CFs from adjacent frames, CFs are clustered. After the target in frame n is tracked, the CFs in the last r frames, including #n, are firstly added into the CF set while those in other n-r frames are clustered into K classes; afterwards, the CF with the lowest classification error in each cluster is added into the CF set. The $K + r$ CFs are combined with different weights to form the final strong CF, which can be formulated as in (19), and ρ_n^i is the weight of the i-th filter in frame #n calculated as in (20), where e_i denotes the training error of the filter calculated as in (21), in which (x_t, y_t) is the new training sample of the t-th frame, whose spatial size is $M \times N$ and each sample $x_t^{(k)}$ of x_t is a d-dimensional vector $[x_t^{(k)(i)}]_{i=1}^{l}$ $\hat{x}_t$ and $\hat{f}_t$ are Discrete Fourier Transforms (DFT) of x_t and f_t and w_k denotes the weights of all samples $x_t^{(k)}$, which is defined as in (22). However, the CF set is updated every certain frames rather than in each frame to cut down calculation burden and prevent useless operations.

$$f_n^{strong} = \sum_{i=1}^{K+r} \rho_n^i f_n^i, \tag{19}$$

$$\rho_n^i = \frac{1}{2} \ln\left(\frac{1-e_i}{e_i}\right), \tag{20}$$

$$e_i = \sum_{k=1}^{M \times N} w_k \left(F^{-1}\left\{\sum_{l=1}^{d} \hat{x}_n^{(k)(l)} \bullet \hat{f}_n^{i(k)(l)}\right\} - y_n^{(k)}\right)^2, \tag{21}$$

$$w_k = \begin{cases} Y_k / \sum_{m,n} \exp\left(-\sigma\left((m-M/2)^2 + (n-N/2)^2\right)\right) & \text{at the beginning} \\ \frac{w_k}{\sum_k w_k} \exp\left[\rho_n^i\left(F^{-1}\left\{\sum_{l=1}^{d} \hat{x}_n^{(k)(l)} \bullet \hat{f}_n^{i(k)(l)}\right\} - y_t^{(k)}\right)^2\right] & \text{others,} \end{cases} \tag{22}$$

3.2. Update Strategy Based on Dictionary Learning and Sparse Coding

Dictionary learning (DL) and sparse coding (SC) are common generative frameworks of visual tracking. The template set is usually made up of the tracking results from each frame, while at the beginning stage of tracking it consists of the positive and negative samples drawn in the first frame. Two common ways generates the dictionary [38], one of which is through learning methods, i.e., principal component analysis (PCA), where the dictionary is acquired by the form of iterative training of samples in specific frames, the other is to directly insert the tracking result into the template set and then select a subset. The latter method is more popularly adopted in the recent year.

The dictionary model needs to be constantly constructed with the appearance variation of the target and background. When considering that there are some slight differences between two adjacent frame images, from each frame positive and negative samples should be added into the sample set. However, there are at least two variables must be iteratively solved in the normal dictionary learning framework—the dictionary and the sparse coefficients, obviously calculation burden will increase if the dictionary is updated every frame that unnecessary updates may have consumed a lot of time. For the balance of tracking accuracy and efficiency, in [39], foreground and background samples are preserved after tracking in each frame, but the dictionary is updated every T (=15) frames, which is mainly trained while using the target and background samples in the last 15 frames and is emptied whenever the dictionary update is finished. Target samples in the first frame and the sample that is calculated as the mean image of all the best tracked results are also used for training and never deleted after updates to overcome the impacts of bad positive samples arouse from occlusion or background drifts. Similar dictionary learning way is utilized in [40], whereas background samples are not used for dictionary training, and the method in [40] is the improvement of the space sparse learning (SSL), which fixes too much attention to positive samples in the latest frames while ignoring the contributions of distant tracked frames, which might unavoidably make the template integrate with too many newest characters that makes the tracker hard to re-detect the target after full occlusion or out-of-view.

Currently, the latter dictionary construction method that selects a set of reliable tracking results as the dictionary has been more popularly utilized, which is termed as sparse coding, in order to cut down the calculation burden brought about by dictionary training and alleviate the impact arouse from irregular sample distribution generated from fixed-time-interval update. The simplest way is to directly use the tracking result in the current frame as the new template and insert it into the set or replace one with the least similarity in the set with it. However, owing to the reality that the image of a tracked object is often interfered by pseudo target pixels or noises aroused from irregular illumination, the target model might get distorted if the raw tracking result is directly added to the set. To eliminate the influence of noises, trivial templates [41,42] are usually used for target image representation, which is expressed as in (23)

$$\min_{z} \|g - \mathbf{B}z\|_2^2 + k\|z\|_1, \text{ s.t. } \mathbf{B} = [\mathbf{E}, \mathbf{I}], z = [a', h'], \tag{23}$$

in which g denotes the raw target image, $\mathbf{B}$ is the template set that is composed of a denoised template set $\mathbf{E}$ and a trivial template $\mathbf{I}$, and a' and h' are their coefficients correspondingly. The denoised target image $\mathbf{T} = \mathbf{E}a'$ rather than the raw image is used to update the template model. So as to overcome the defect of less enough contribution of the denoised templates due to the excessive sparsity effect on them, the sparsity constraint is only imposed on the trivial template set in [42], which is formulated as in (24)

$$\min_{q,e} \|p - \mathbf{U}q - e\|_2^2 + k\|e\|_1, \tag{24}$$

where p is the image of the raw tracking result, q and e are respectively the coefficients of the denoised template set and the trivial template set, and $\mathbf{U}$ is the eigenbasis of p. The final image $\tilde{p} = \mathbf{U}q$ is inserted into the template set. Although the template set is also updated every a few (=5) frames, to make it more representative that it should not contain too much newest characters or too old ones, the set composed of 10 templates is established, where the target in the first frame is permanently preserved in the first room, while tracking results are stored in other nine rooms in time order. The templates in room 2, 5, and 8 are removed and the denoised results in three editions are added at the rear.

A global update of templates makes the model less complicated and the calculation burden is thus alleviated. However, the representation of each target part should not be the same due to the truth that different features are contained in different regions of a target image. Besides, the sparsity constraint does not work well if a template set that can only globally represent image is used. When considering different variation form of each part and the effect of partial occlusion, target dictionaries are not only the subset of a template set according to the theories in newest researches, patch dictionaries are usually established instead of holistic ones that a specific region of the templates are used to construct the local dictionary of that target region [16,17,41,42]; therefore, different update policies are utilized on different local patches. For more robust representation of visible object parts when other parts of the object are occluded, the object is represented in a different form from the situation of no occlusion in [16]. During partial occlusion, the contribution of each template patch is calculated while using the tracking result—occluded patches contribute much less to the representation, therefore template patches with a higher contribution value can be effectively updated while the update of other patches is temporarily prohibited. To eliminate the impact of background pixels in a target image and make the tracker model more robust to deformation and rotation, object patches are classified into three types: stable patches, valid ones, and invalid ones, and three types of dictionaries, called total dictionary, object dictionary, and background dictionary are constructed in [17], which has been illustrated in the first subsection of Module 2. The target dictionary $\mathbf{D}_o$ is updated while using valid patches.

3.3. Update Strategy Based on Bag-of-Words

Objects in each frame of a tracking sequence can be only classified into two classes—object and background. In terms of animal's vision mechanism, the classification of two different types of objects is usually according to the characters that are not the same among them, which gives the inspiration of bag-of-words (BoW) model in the domain of visual tracking, for the fact that in general characters contained in the foreground are distinguished from that in the background, thus there should be plenty of symbolic features to assist in object classification. However, there have not been too many tracking algorithms that are based on this framework when compared to other ones up to now, and less robustness has been shown in the tracking performances, for the reason that most of them neglect the consideration of the holistic structure of the target and background.

Visual "words" are the visual characters from the area of the target and background in a tracking frame that are used as training samples in discriminative frameworks. For instance, in [42], the "words" are classified in a supervised way while using SVM. During the update process, new visual foreground "words" and background ones are extracted from the region of the object and a random background patch, respectively.

However, the background and foreground in one frame might share some "words" with similar features, therefore a background character might be mistakenly classified as a target if it is much too similar to some features in the target feature bag. Hence, the target "words" like these cannot be used for discrimination. In [7], the authors believe that target occlusion might well happen when there exist features in the bounding box that are similar to or even the same as those in the surrounding area. If the number of these features is larger than usual, occlusion can be surely regarded to have occurred. In usual condition where no occlusion happens, foreground and background features in the bounding box are respectively merged into the target feature set and the background one; afterwards, other background features are searched from the surrounding background in the past few frames and then merged into the new background feature set, which has made the background more distinguishable that false targets have lower probability to be misidentified as the true one. A similar unsupervised way is utilized in [43], in which if the distance between a word v_i in the context bag M_B^t and its nearest neighbor word v_n from the bag of the last frame is lower than the threshold τ_B, a new word v_{new} in combination of the two words is added into the word bag in the current frame, as in (25), where C denotes the flag of background or object and $á$ is the proportion parameter; otherwise, when the background word bag M_B^t is updated, v_i is directly merged into the bag: $M_t^B = M_t^B \cup v_i$. If it is time to update the object word bag M_t^O, there is a need to check whether the current word v_i is reliable, which is measured by the distance between it and its neighbor word v_m from the newly updated background word bag M_B^t and that between it and the neighbor v_n from object bag M_{t-1}^O of the last frame. The word is regarded as reliable if the latter distance $d(v_i, v_n)$ is smaller than the former, named $d(v_i, v_m)$; thus, it is merged into the object word bag in the current frame: $M_t^O = M_t^O \cup v_i$; otherwise, no bag is expanded. In addition, when any of the two bags is full, some words are randomly removed from the bag.

$$v_{new} = (1.0 - \alpha)v_n + \alpha v_i,$$
$$M_t^C = M_t^C \cup v_{new}, \qquad (25)$$

3.4. Update Strategy of Neural Network Models

A series of neural network framework have been widely adopted in researches of visual tracking because of its strong capability of feature extraction and image classification, of which researches on the improvement of accuracy, speed, as well as the structural layouts are gaining rapid progress. Quantities of labeled images are used for iterative training and during training features of different depths that describe the trained samples from different aspects are extracted, thus a set of parameters with high validity are finally determined thanks to the neural structure of it, which greatly alleviates the tedious process of handcrafted feature extraction in traditional machine learning models. A huge challenge of visual tracking under neural network framework today lies in the shortage of training samples as well as in the sensitivity to irregular sample distribution and noisy samples [44], of which the sample distribution and quality of training samples decides the capacity of a network to a large extent. So as to further boost the capacity of tracking networks, the hot topic of tracking under deep neural network has recently transferred to the further procession of training samples, which is a credible mark of progress in the research of deep learning.

The distribution of foreground and background stays stable during tracking in a short period. The samples used for model update should possess two characteristic to make the network adjust to the appearance change of the target: firstly, the frames that the positive samples are selected from should be as close as possible to current frame to ensure the real-time requirement; secondly, it must contain a correctly tracked object that is without the influence of occlusion or drift. In other words, it must be responsible enough. Based on these two characteristics, during the stochastic (short term) update reliable samples are picked out for model retraining in [27]. To make the target model less dependent on newest appearances and cope with the lack of positive samples when temporary target loss occurs during periodically (long term) update, positive samples from the first frame are also used for the update as supplement in addition to from the best tracked frames. The similar method that takes the samples in the first frame into account is also utilized in [45], where Gaussian maps of each frame image also take part in the update training.

The initial appearance is preserved in a network model if the target samples drawn from the first frame are put into consideration when updated, which is helpful for target re-detection after its reappearance after temporary disappearance. Pessimistically believed in [46], from the author's point of view, only the positive sample from the first frame is completely reliable, whereas contamination and decision mistake must exist in other frames to some extent, which is also deemed to be true in [47] that there must exist error a bit or too much in each frame, except in the first one. However, optimistically speaking, thanks to the close appearances from the two adjacent frames, a trend of the variation can be foreseen within a small period (no above than three frames); therefore, there exists a high confidence of making sure whether the tracking result is responsible. As a matter of fact, the target appearance might have undergone variances plenty of times after hundreds of frames of tracking, it is not sufficient to achieve re-detection only through the target appearance in the first frame; since, in usual cases, the real appearances of the target in the last few frames are much closer to that in the re-detection frame, as the assumption that target samples that satisfy the two conditions listed in last paragraph should be more important. Target reliability detection is utilized in some researches so as to use more reliable samples, whereas the best-fitted positive samples are selected for retraining. A read-and-write memory structure is established in [46], to which the tracked object is inserted and the sample with the lowest confidence is removed from it unless it is full. During the update, scores of importance are given to the selected samples from the memory for calculation of the gradient descent parameters. For adequate use of the reliable samples in the past frames, the self-paced selection model is adopted in [48] to control the selection of positive samples, those with the lowest loss value based on the current loss function are selected for network retraining, and the criterion of the samples to be selected for retraining in a frame is based on the overall reliability of the samples in the previous frame.

3.5. Module Summary

In this module, the update strategies under four mainstream tracking frameworks—correlation filter, sparse coding (dictionary learning), bag-of-words (features), and deep neural network—are discussed, and the progresses are illustrated according to the specific examples in recent researches. Questions regarding the challenges that remain in the existed update strategies are summarized as below: (1) How to build a template set structure that includes more abundant information about the target but consumes the least amount of memory as possible; (2) How to more effectively choose training samples that contain various kind of target appearances and control the distribution of the sample set for deep neural network update; (3) How to deal with visible parts of the occluded target and make good use of them for update to boost the network's adaptation to newest appearances; and, (4) How to separately use different features of the target and utilize feature-specific update methods to make the tracker more robustly adjust to the variation of each feature. Contributions of each feature or convolution layer should also be considered for the update at the global level.

4. Background Model Update

The environment of the target existence is background. With the movement of the target, the background also varies its appearance, so the correct estimation of targets' surrounding background is the premise of correct location of the target. Characters of the background regions that surround the target are especially essential to prevent drift to similar objects in the background, which should be distinguished from the characters of the target [49]. Compared with the target, the background occupies much larger area in the view of a frame, whose appearance features appears more complicated, hence there ought to be plenty of available negative sample sources, therefore how to more credibly select background samples is also a key part in the work of update. Background model update occasion and strategies are discussed below.

4.1. Background Sampling Methods

Sampling of background samples is the key part of the update work, owing to the fact that the number of background patches is far larger than that of foreground patches. Random selection is adopted in some researches, for instance, background "words" are extracted from random regions outside the target area in [31]. Yet, an object must exist in a specific environment—it must possess an exact position in the background area. Based on this truth, the authors in [2] hold the view that all non-overlapped background patches are not equal, and background regions with different features affect apparently differently on the sample classification. In this research, sampled background patches are clustered into multiple groups; afterwards, the specific SVMs are trained using each group of the background and target samples. Negative samples distant from the target area are drawn for update to make the foreground samples more distinguishable, where the sampling method is often utilized in extreme learning machine (ELM) [3,50] frameworks. Some SC based trackers also use background patches faraway from the object, i.e., [41,51].

Nevertheless, not all of the background characters are of valuable use. On the basis of animal's visual tracking mechanism, the background regions near to the area of the moving object contain the most valuable information that can help with target location; hence, they ought to be the most available parts through the entire background, while the influence of the information of background far from the position of target are of far less importance. The examples of background sample selection policies in last paragraph overlook the relationships between the target and its context, which violates the mechanism of animal's selective attention, despite the fact that the ELM frameworks are robust enough to fight against the diversity of sample appearances. Luckily, there are an increasing number of researchers who have realized that mechanism that background characters close to the target area ought to be given the highest importance. For instance, background samples that are drawn near to the target region are used for the dictionary model update in [52], which is the spatial constraint of the data sampling in the article, in which the temporal constraint is that the samples selected for training should be from the latest few frames. This distance constraint is also satisfied in [46] by the update of the network model. Of the bag-of-words (features) based tracking frameworks, as in [7] and [43], words or features in the surroundings near to the target are used to update the context (background) bags when the foreground bags are usually updated in parallel, which has been illustrated in detail in the third subsection of Module 3. The parameter of intersection over union (IoU) is usually used to identify whether the patch that is selected around the target is foreground, the patch is regarded as a positive sample when which is above a higher threshold, or a negative one if below a lower threshold. In [28] and [53], IoU is used to help draw positive and negative samples for network update. Samples whose IoU are between the two thresholds are also picked out for network retraining in order to increase the robustness of target position and make abundant use of visible parts of a tracked target when partial occlusions occur.

Dense sampling is commonly utilized as for the density of sampling, like some particle filter based sampling methods, i.e., [51]. Dense sampling means that positive and negative samples are drawn within a length of radius according to a given distribution, i.e., Gaussian Distribution, in which there is a large overlap between any two of the samples of the same class. The advantage of this kind of sampling approach lies in that it not only makes abundant use of the background information around the target thus strengthen the discrimination capability of the tracker, but it also helps to provide more sufficient source of samples, which boosts the robustness of deep networks.

4.2. A Kind of Background Unity Estimation Approach: TBE

Up to now, most tracking algorithms have concentrated a lot on the construction and update of target models, while those of background models have been rarely researched. The distribution of the feature of the target's surrounding area is usually irregular, owing to the complexity of the background. Therefore, the requirement of accurate target location cannot only be satisfied through simple target matching methods. When the target is occluded, its appearance has gotten incomplete that available target characters have become less, which makes it hard to distinguish from the background. An original method, named Tracking by Background Estimation (TBE), is proposed in [12], which includes the approach of background modeling and update strategy by which foreground pixels are extracted out for target detection and location, to achieve more accurate target location especially in the state of partial occlusion.

TBE is based on the principle of background subtraction, through which the preserved area of foreground pixels is used for target detection and location; afterwards, the appearance model of the target is learned. Suppose that the entire image f_i is composed of a target t_i and a background b_i i.e., $f_i = \{t_i, b_i\}$, where i is the frame index, and the mask of the background b_i in frame i is identified as m_i. All the pixels in the image domain of f_i compose the set P_i. Given a pixel $x \in P_i$, if x belongs to the background, there is $b_i(x) = f_i(x)$ and $m_i(x) = 1$; otherwise, $b_i(x) = 0$ and $m_i(x) = 0$. To eliminate the influence of background illumination, "mean-background" is defined and suppose $\tilde{b}_i$ is the mean-background in frame i, the corresponding mask of which is $\tilde{m}_i$. All of the pixels in the image domain of $\tilde{b}_i$ compose the set $\tilde{P}_i$.

Assume that the camera is stationary, the background in two adjacent frames is completely the same, thus $t_i = f_i - b_{i-1}$, and the target can be recognized by means of the subtraction of the frame images. Yet, in almost all cases, the camera is in movement sometimes, which brings about the deformation and scale variation of the background. Based on this factor, the warped image in frame i is identified as $\hat{b}_i$, which is transformed from the mean-background in the last frame, as in (26)

$$\hat{b}_i = H_i * \tilde{b}_{i-1}, \tag{26}$$

where $*$ is the transform operator and H_i is the calculated homography matrix. The warped $\hat{b}_i$ from the mean-background in frame $i - 1$ suits to the background in the current frame i, making the background subtraction applicable. Thus, the mean-background in frame i is calculated by the weighted sum of $\hat{b}_i$ and $\hat{b}_i$, as formulated in (27)

$$\tilde{b}_i(x) = w_i^T \bullet \left(\hat{b}_i(x), b_i(x) \right), \text{ s.t. } x \in \tilde{P}_{i-1} \cup P_i, \tag{27}$$

Some background regions in the previous frame do not appear in the current frame and new background regions may appear due to the movement of the background. Besides, the target must exist in the shared parts of the background regions, i.e., x if $x \in t_i$; hence, the weight w_i is defined as in (28)

$$w_i = \begin{cases} (1,0)^T & x \in \tilde{P}_{i-1} \wedge x \notin P_i \\ (\hat{m}_i(x), m_i(x)^T)/(\hat{m}_i(x) + m_i(x)) & x \in \tilde{P}_{i-1} \cap P_i \\ (0,1)^T & x \notin \tilde{P}_{i-1} \wedge x \in P_i, \end{cases} \tag{28}$$

in which $\tilde{m}_i$ is the warped mask. Subsequently, $\tilde{m}_i$ is calculated as in (29)

$$\tilde{m}_i(x) = \begin{cases} \hat{m}_i(x), & x \in \tilde{P}_{i-1} \wedge x \notin P_i \\ \min\{T, m_{i-1}(x) + \hat{m}_i(x)\}, & x \in \tilde{P}_{i-1} \cap P_i \\ m_i(x), & x \notin \tilde{P}_{i-1} \wedge x \in P_i, \end{cases} \tag{29}$$

where T is the predefined threshold that upper bounds the maximum of $m_i(x)$, ensuring the contribution of the latest frame, without which the weight of the mean-background might rise to a high value and the weight of the input frame will be negligible.

An update of the background model is performed after the target is tracked in every frame that the parameters $\tilde{b}_i$ and $\tilde{m}_i$ are obtained and the warping operation is done before target detection in the next frame. Afterwards, background subtraction is conducted for the detection and location of the target.

4.3. Occasions of Background Update

Because variation of the background mainly relies on its movement, though some of its features may passively vary with the environment, it must exist in every frame, the reliability of it should not be given too much consideration, therefore sophisticated discussion regarding the background update occasions is not necessary. The background appearance is temporarily stable thanks to the variety of background patches and the movement along with the target. Usually, fixed-time-interval background update is adopted in SC based and deep neural network based models, and unsupervised models, like BoW (or BoF), update the background model along with the target model frame-by-frame. Negative samples drawn from the latest frames are used for model retraining, which is the guarantee of the requirement of the adaptation of the tracker to the newest background features.

4.4. Module Summary

This module discusses background update strategies and occasions, including a new background model update strategy named TBE. Although the update of background model seems to be much simpler than that of target model, there are still needs of improvements in many respects. The questions remaining about the background update are as below. (1) How to utilize the background information that is useful for discriminating the target and the surroundings for the extraction of key background characters that can help with target location; (2) How to determine the density of background patch sampling. Background regions containing much more valuable information ought to be more densely sampled to boost the efficiency of the tracker; (3) How to build the holistic structure of the background.

Algorithms about the background update in the global level should better be designed in future tracking researches, as patches or visual words drawn from background are placed in order in the original image.

5. Analysis on Experimental Results

Challenging factors in visual tracking include occlusion, in-plane and out-of-plane rotation, illumination variation, background clutter, fast motion, abrupt deformation, and scale variation, etc. The robustness of a tracker is measured by its performances under these situations on specific sequences. A successful track means that a tracker is able to track the target without drift through the whole sequence in spite of any of those factors in the video. Whether a tracker can successfully track the target in a sequence depends on the quality of the model update to a large extent. This Section will discuss the tracking experiments from recent researches, where performances under those challenging factors are talked about in detail. The advantages in contrast to the benchmark trackers as well as some failure cases are listed and analysis on the merits and drawbacks with respect to the update strategies are then illustrated. Improvement measures are proposed among the analysis.

5.1. Update Strategies from Recent Researches

Some typical tracker models are listed in this subsection to illustrate the merits and drawbacks of recent trackers, as in Table 1. Table 2 lists abbreviations for the names of the listed frameworks.

Table 1. Model update strategies in recent years and their merits and drawbacks.

Tracker	Framework	Update Strategy	Performances
L3SCM [24]	PF	A local region of the template is updated when the similarity between it and the same region in the target image is no less than a threshold.	Targets can be correctly tracked no matter any challenging factor happens.
MSRBTP [54]	PF	For each feature, the weight is cut down by multiplying a positive value smaller than 1.0 when there is a classification mistake.	Targets can be stably tracked when there are illumination changes. In sequences of *Skiing* and *MotorRolling*, the targets are failed to be detected after presence of scale variation, out-of-plane rotation and out-of-view.
TBE [12]	Background Similarity	Background image model is updated every frame. The target is relocated with the help of the new background model when there is no full occlusion and the target appearance model is updated when no occlusion happens.	Occlusion cases can be correctly identified and the tracker can re-detect the objects after long-time full occlusion. The tracker remains its robustness even if the target constantly changes its appearance, especially when background clutter or out-of-plane rotation happens. However, it is not able to recapture small-sized targets.
ALIEN [7]	BoF	Occlusion does not happen when the number of background features in the bounding box is small, thus target features in the box are transformed and then merged into the foreground feature set while background features in the search regions of the last few frames are merged into the background feature set.	The target can be re-detected in a short time after full occlusion. The tracker is not sensitive to similar objects in the surroundings and is able to accurately measure the size of the target.
ELMAE [3]	ELM	Target template is updated when the distance between the template of the newest result and the template of the target in the first frame is lower than the threshold. Negative samples faraway from the targets are used to update the background model.	Performances on typical sequences that include mixed challenging factors and severe occlusions are much better than benchmark trackers, especially able to deal with the problem of constant rotation in *Freeman1*.
PML [50]	ELM	Positive samples and negative samples far from the target area are selected to update the ELM model, which is performed every certain frames.	The tracker is able to tracker 12 challenging sequences. It is able to accurately detect the target when there is severe in-plane or out-of-plane rotation or scale variation.
SPDCT [48]	DNN	Positive samples with lowest loss values are chosen for network model update every five frames.	The tracker can handle severe problems such as deformation, occlusion and background clutter compared to the benchmarks.
DNCT [45]	DNN	Tracking results in the last six frames and positive samples from the first frame are used for model update when the maximum value of the response map is higher than the threshold.	Targets can be re-detected after full occlusion even if they are much smaller than the normal size.
HCF [29]	CNN+CF	Regular linear interpolation method is adopted during the update of the filters of each CNN layer.	The targets are failed to be tracked in the sequences of *Girl2* and *Lemming*; For *Singer2* sequence, the darker foreground is extremely hard to be distinguished from the brighter background for the reason that combined information by multiple layers are used.
DNT [27]	CNN	Short term update is performed using the best tracking results in the latest frames when occlusion or background drift happens while long term update is done using recent results and the target samples in the first frame every certain frames.	Targets' scale and position can be accurately determined even though in the situations of fast motion or background interference.

Table 1. *Cont.*

Tracker	Framework	Update Strategy	Performances
WALSA [22]	SC	The tracking result is added into the template set and an old template is randomly removed when the similarity value between them is within the range of 0.65 and 0.85.	Targets can be stably tracked under any challenging situations.
TPS [14]	LR	Training approaches of SVMs are applied for the vector regression model SVR. Contribution values of each target part is gotten for the local update.	Strong robustness is displayed in the situations of partial occlusion and deformation.
ODLR [39]	DL	The dictionary which includes background samples is updated every 15 frames, during which positive samples from the first frame and a set of tracking results in the last few frames as well as the mean sample of historical best tracked targets are used, afterwards the set consisting of recently tracked objects is emptied.	Targets can be correctly located whichever any challenging factor comes across. However, for *Pedestrain2* during the reappearance after the disappearance of the walker, false samples are used to construct the object dictionary. For *Skiing*, the dictionary also failed to be constructed as the size of the target becomes too small.
SALSC [41]	SC	Denoised tracking results are used for update. Three templates in the set are replaced by different forms of the result.	Under various kinds of mixed challenging situations like occlusion + background interference, illumination change + rotation, scale variation + background clutter + rotation + illumination change, etc., the targets are still able to be stably tracked.
approach in [40]	DL	The dictionaries are updated using latest tracking results every 15 frames.	The tracker performs excellently at dealing with newly varied appearances.
CRSRCF [55]	CF+SD	The correlation filter and the weight map of the saliency map are updated in each frame.	Objects with irregular shape and heavily deformed targets can be correctly tracked.
LSHR [33]	CF+CNN	The model is updated each frame. When the distance between the target's exact position and the estimated position is bigger than the threshold the state of the target is recalculated using features extracted by shallower layers.	Targets in more challenging videos can be well accurately tracked, especially for the sequence of *Ironman*, only the proposed tracker is able to track whole of it.
DSARCF [56]	CF+SD	Target feature maps from the first frame to the current are used to update the CF in the next frame. The CF and the spatial weight map are updated every two frames.	Under occlusion or heavy appearance changes, targets can be successfully detected. In *Girl2* sequence, when the girl's face reappears after occluded by a man's face, it can be correctly tracked for quite a long time. Yet the saliency map does not work well in the situation of fast motion. In sequences of *Matric* and *Drangonbaby*, the targets failed to be detected using the saliency maps when low resolution or background clutter occurs.
CLIP [34]	CF+SVM	The learning rate is adjusted according to the ratio of the maximum response value to the sum of which in all previous frames. Image patches with highest SVM classification scores are used for the update of SVM as long as the maximum response value of the synthetic features is above the threshold.	Compared with the benchmarks, the tracker performs with much more robustness no matter any challenging situation occurs.

Table 1. *Cont.*

Tracker	Framework	Update Strategy	Performances
SRKCF [35]	KCF	The credibility value is calculated by the distance parameter between the tracking results in adjacent frames and the PSR value. The learning rate is sustained if the credibility value in current frame is above the average of which in the past few frames, otherwise it is reduced by the ratio of them.	The proposed tracker has better performances than other compared KCF trackers, typically it outperforms others at occlusion handling. In the sequences with background clutter and deformation like *Basketball* and *Bolt2*, only the proposed tracker is able to accurately track the targets.
AECF [37]	multiple CF	The final strong CF is updated every 5 frames. The CFs in last several frames are firstly preserved while others are clustered into many groups, thus one CF is picked out from each cluster. These CFs are combined to generate the strong CF.	For *Skating* where the target reappears, it can be retracked. The tracker is also robust in coping with the background clutter problem in the sequences of *Shaking*, *Panda* and *Dragonbaby*.
OSAMCF [57]	CF	Position CF and scale CF are separately updated, whereas the target model from the first frame is also used to update the former one.	Targets can be stably detected no matter any challenging problem comes across.
HDT [58]	CNN	The regret value of each frame that is the cumulative value of the loss values in all past frames is updated by the distance and the appearance difference to further calculate the weight of each feature. The network is updated incrementally using samples in current frame.	Drifts can be well avoided in the sequence of *Basketball* where there exist objects similar to the tracked player in the surroundings.
MLFF [18]	CNN	The model is updated only when the maximum value of the response map and the PAR value are both above the average of the historical values.	The proposed tracker performs superior to FCNT, SiamFC and CF2 under most of challenging situations.
PMC [59]	KCF	The basic (first) classifier is never updated. The first and second classifiers are updated when the scores of them are no less than that of the third classifier and the predefined threshold, while the third classifier is updated when its classification score is above than that of the other two classifiers.	The proposed tracker performs extraordinarily well under the mixed challenging factors of partial occlusion and rotation, i.e., *Girl*.
RDLT [51]	KCF+SC	The CF model is updated unless the HoG and color score are both above the average, meanwhile histograms of the foreground and background as well as the sample templates used for re-detection are also updated.	Targets can be correctly recovered after drift loss. However, for *Face-ce*, due to the high similarity between the character of the occluding object and the tracked object, recovery of the target is failed. Also the algorithm does poorly in handling the fast motion problems in *MotorRolling* and *Bike-ce2*.
LSA [17]	SC+PF	The total dictionary is updated when there are enough stable patches and valid ones. The object dictionary is updated using valid patches while background dictionary is updated using local background patches around the target.	Problems of out-of-plane rotation and illumination variation can be greatly handled and partial occluded targets can be accurately tracked. But the proposed tracker is not able to cope with severe scale changes.
NMC [42]	SC	The background template is updated 5 frames; When the number of the background patches that take part in the representation of the tracking result is no more than one, there is no severe occlusion, thus the target template set is updated.	The proposed tracker performs excellently on many sequences with challenging factors.
CBOD [9]	KCF	The kernelized correlation filter is regularly updated unless no occlusion happens.	The proposed tracker performs excellently under various occlusion situations, i.e., sequences of *Tiger1*, *Coke*, *Basketball*, *Football*, *FaceOCC1*, *CarScale*.

Table 2. Abbreviations for name of the frameworks.

Abbreviations	Full Name
PF	particle filter
DNN/CNN	deep/convolutional neural network
(K)CF	(kernelized) correlation filter
SC	sparse coding
DL	dictionary learning
BoW/BoF	bag-of-words/bag-of-features
ELM	extreme learning machine
LR	linear regression
SD	saliency detection

5.2. Qualitative Advantage Analysis of Some Trackers' Performance on Typical Sequences

To evaluate the quality of a tracker, its performances under those challenging situations, such as occlusion, in-plane or out-of-plane rotation, etc., are usually the accordance, which essentially depends on the quality of the model update strategy. This subsection gives analysis on specific cases where the performances as well as advantage analysis of the recent trackers under the factors of occlusion, background interference, rotation, scale variation, and deformation are respectively illustrated.

(1) *Occlusion*: Occlusion is a hard problem that almost occurs in all sequences, the update strategy under which situation measures the robustness of a tracker to a maximum degree. In sequences of *Jogging-1* and *Subway*, the walkers are respectively occluded by the telephone pole and other passers-by, only the tracker in TPS [14] and the benchmarks of TGP, SCM, and KCF are able to stably track them, which explains that updating in local patterns has provided assistance in tracking partial occluded objects via visible parts. Local feature representation is adopted in [54], where the global feature pattern is fused with local ones to represent the tracked object. In the sequence of *Walking* when the walking woman reappears after occluded by the man, the tracker in [54] can successfully recapture the woman, while the compared benchmarks, like OAB, MIL, and COM, fail to retrack it. The local-patterned update is also adopted by L3SCM [24], which has gained better performances than the compared benchmarks. SC based LSA [17] shows strong robustness in handling occlusion thanks to the use of stable patches and valid ones for update. In the sequence of *Jogging2*, after the occlusion of the walker by the telephone pole, the compared KCF and DSST fail to cope with the drift problem. Different template patches are used to represent the tracked object by NMC [42], whereas the distribution of foreground and background templates is used for the detection of occlusion, which shows its superiority in occlusion handling in *Suv* and *Jogging2*.

Utilization of background models is the key of correct target localization. The target is completely occluded in the frames #27 to #36 of *Uav*, thanks to the constant utilization of background model in TBE [12], the appearance of the target is preserved before the start of its full occlusion; therefore, it is able to be retracked after it reappears, while other compared trackers fail to re-detect it. In the sequence of *Thuyx*, the characters of the surrounding is similar to that of the target, still only the proposed TBE can correctly track it while drifts to the surroundings occur when using other compared trackers. These cases have given us the inspiration that the background model is typically essential in dealing with occlusions. Bag-of-feature based ALIEN [7] effectively prohibits the drift problem in the sequence of *FaceOCC1* due to the use of the background characters. The background information in the tracking bounding box are used to describe the reliability of the tracking result in [40], thus the target model update is prevented if there is too much background information, so for the sequences where there are partial occlusions, i.e., *Coke*, *Girl*, *Lemming* and *Tiger1*, the tracker performs well.

Valuable use of positive training samples plays an essential role in dealing with target re-detection. In frame #131 of *Girl2*, where the man's head moves away and the girl's head return visible, DSARCF [56] is able to perfectly retrack the girl's head while other trackers fail, due to the reliability check of target training samples that are used for the spatial weight update; PMC [59] also performs well on this sequence, even though the face slightly rotates in the process of being occluded, which can be attributed to the complementary update policy of the three classifiers. For *Human3* after the entire occlusion of the target, CLIP [34] is able to recover the correct track, while drift occurs when using the compared trackers, like MUSTer and LCT, which, thanks to the preserved long-term target appearance information that can help to re-detect the targets after recovery. Due to the target samples from the first frame that are used for model retraining, ELMAE [3] shows excellent results in the sequences of *Jogging* and *Suv*. Positive samples in the first frame are also used for the network update by DNCT [45], which is able to recapture the recovered targets, even if they are much smaller than normal, i.e., *Skiing*. The targets in *Lemming* and *Jogging2* simultaneously rotate and change their appearance, in the meantime both of them are in the state of occlusion. Owing to the dynamic reliability parameter that is used for occlusion detection, SRKCF [35] can more effectively handle those more complicated occlusion problems, the center location error (CLE) of which is relatively lower than its compared benchmark trackers.

(2) *Background interference*: Background clutter is also one of the most challenging factors, performances of a tracker under which situation is a key point of the measurement of its robustness. For *Basketball* and *Bolt2*, where there exist objects sharing too many characters with the true target in the surroundings, SRKCF [6,60] is able to track the true target while other compared trackers, like SRDCF, LCT, and SAMF, drift to the false ones. This is because of the fact that SRKCF has made use of the distribution of foreground and background pixels that is useful to feature update, which is combined with the parameters of the target location distance and the *PAR* to prevent similar but unrelated background pixels from contaminating the target template. MLFF [18] adopts integrated features extracted by multiple network layers to distinguish the true objects when considering that the true target is not completely the same as the false one in the background, which performs well on the challenging sequencesm such as *FaceOCC2*, *Football*, *Sylvester*, *CarDark*, and *Singer2*. Cluttered backgrounds in some frames, like frame #51 of *Davidoutdoor*, frame #146 of *Bicycle*, frame #53 of *Thusy*, and frame #105 of *Gymnastics* may impact the feature extraction of the targets therefore drifts probably appear when the sequences are tracked while using some benchmark trackers. Thanks to the approach utilized in TBE [12] that separates the target from the background and regards the background as the Gaussian model, which is able to resist many kinds of background interference, hence the appearance information of the targets can be correctly used for more concise target location. For instance, for the sequences of *Bicycle* and *Uav*, owing to the fact that initial location of the target might be incorrect because of background noises, the separated target appearance model can be used to obtain the more accurate location. For the framework of dictionary learning based ODLR [39], target dictionary and background dictionary are independently constructed while using positive and negative samples respectively describing the target and the background, which is helpful in the detection of complicated backgrounds. The tracker performs well on *Deer* sequence, while the benchmarks, such as ALSA, IVT, SCM, and VTD, do comparably poorly. The distance of the estimated target locations in two adjacent frames might be larger than normal due to the interference of the false target in the background, based on which problem, the relocation mechanism in LSHR [33] makes good use of target features that are extracted by different layers, hence it is able to accurately track some videos with background clutters i.e., *Ironman* entirely, while the benchmark trackers cannot perfectly handle the problems.

(3) *Illumination variation*: Illumination variation of a target is a kind of passive appearance change that the illumination of the target is influenced by the environment it exists in. For example, some noises, such as too light or too dark spots, caused by unusual environment illumination may appear in the target area. Due of the samples from the past latest frames used for the target dictionary update in [40], the dictionary can well encode the latest appearance of the target, especially when there are intensive changes on some features. In frame #127 of *Davidindoor*, when the tracked man suffers intense illumination variation, the proposed tracker in [40] is able to more perfectly capture his immediate appearance change as compared to SSL. Strong performances are shown by LSA [17] on the sequences of *Sylvester* and *Shaking*, in both of which there are tense illumination changes, which can be attributed

to the stable patches and valid ones that are of excellent use for the representation of deformed objects. MSRBT [54] gives more concern about the features that are more apparent for target and background discrimination, while repressing the ones not so available. It makes use of those distinguishable features for the detection of the targets in the frames with illumination variation, i.e., frame #156 of *Singer2*, frame #22 of *Crowds*, and frame #408 of *David*, while the compared benchmarks SCM, L1APG, and ALSA perform worse owing to the use of illumination-sensitive gray features. By fusing the features of color names (CN), color histograms (CH), and HoG in appropriate proportions, CLIP [34] is able to encode the appearance of a target from diverse aspects, which performs apparently better than HCF and SiamFC on *Singer2*, in which drastic illumination variation comes across, for the reason that the benchmarks have made excessive use of semantic features that are not of good use for the discrimination in that situation. Illumination-insensitive HoG feature is emphasized by TPS [14], which shows strong robustness on the frames of #528, #615, and #703 of *Sylvester*. The use of the "mean-background" that eliminates the influence of illumination variations makes TBE [12] more robust, which shows excellent results on frame #177 of *Bicycle* and frame #202 of *Woman*.

(4) *Rotation*: Rotation of a target can be regarded as a type of target appearance variation; however, other challenging factors, such as occlusion or scale variation, may occur in the meantime during target's rotation, thus in-plane and out-of-plane rotations are also hard problems to tackle with. In the sequence of *Skating*, the target athlete is rotating in-plane and out-of-plane alternatively, in addition in frame #304, it suffers intensive illumination variation; SALSC [41] is still able to capture the athlete after frame #304, while its compared benchmarks have lost the target. Excellent performance is also shown on *Car4* by SALSC. These good performances are thanks to the template update mechanism that gives new target appearances and old ones with equal importance. When considering that rotation is the appearance change of a target that its local parts are rearranged within the target area of an image, LSA [17] makes use of valid target patches for the representation of newest target appearances that have undergone in-plane or out-of-plane rotation. It tracks the targets in the sequences of *Basketball* and *Bolt*, in which out-of-plane rotation happens much more favorably correctly as compared to L1APG and Struck, which do not have the capability of rotation handling. Among the network of LSHR [33], the midst layer does the best in coping with the rotation problems, in addition features that are extracted by the shallowest layer are also adopted, thus the network is able to deal with challenging situations in mixture of low resolution and rotation, whereas excellent tracking results are performed on the sequences of *David* and *FleetFace*. In-plane rotated targets can be spontaneously separated from the surroundings, owing to the background subtraction mechanism of TBE [12]. Though new characters of the target can appear if it has undergone out-of-plane rotation, because of the principle that TBE has acquired abundant background information that is of valuable use of background discrimination, newly appeared target characters are detected as background correctly; therefore, TBE also performs much better under this situation. Robust performances are shown on frames #309 and #353 of *Polarbear* and frames of #101 and #961 of *Lemming*, whereas other benchmark trackers can hardly achieve such correct tracking.

(5) *Scale variation* and *deformation*: The scale of a target varies continuously with indeterminacy in frames, the shape of which might also vary along with its initiative scale change, because of the movement of targets and the camera. In the sequence *Sylvester*, the tracked doll severely deforms in frames #676 and #1078; DSARCF [56] can well capture the doll and precisely estimate its scale and shape, while the compared trackers lose the target or wrongly calculate the size during tracking. This is owed to the saliency information DSARCF adopts when updating the spatial weight map, after all the saliency map of a target can naturally reflect the size and shape of it. For the sequence of *Bolt* where the player deforms his body, MSRBT [54] does excellent in tracking him, owing to the local multiple feature pattern. In *Singer1* the size of the target singer constantly varies due to frequent camera distance variation between him LSA [17] tracks the singer much more correctly than the benchmarks due to the state search mechanism based on PF. However, it lacks the capability to deal with more drastic scale variations.

5.3. Analysis of Failure Cases

Target re-detection is an essential part in the whole process of tracking, into which consideration has to be put by all model update strategies. There are some failure tracks that the reappeared targets that have ever been out of the scene are retracked in vain due to the lack of re-detection mechanism by the tracker or improper update methods serving for re-detection. For instance, the walker reappears in the scene after long time of out-of-view in *Pedrstrian2*. ODLR [39] fails to capture it again due to the use of false positive samples for the dictionary construction. This is attributed to the lack of the re-detection process that ODLR has poor ability in relocating targets after heavy background drifts or target losses. Additionally, for *Suv*, where the target reappears after occlusion, MSRBT [54] does poorly in recognizing it. HCF [29] fails to recapture the targets in *Girl2* and *Lemming* when they return visible.

There remains a question of making use of valuable features of small targets in visual tracking. ODLR performs not so well on *Skiing*, owing to the fact that there is not sufficient target information for the construction of the target dictionary because of much too small size of the target, thus the target cannot be properly described by the model. For TBE [12], which puts important attention to the background, the background occupies nearly the entire image when the target is excessively small, hence appearance features of the target are hard to learn by the tracker, thus drift problems exist in some snatches of the sequences where the targets are comparably much smaller.

There are also some failure cases in some videos where situations of fast motion, rotation, background clutter, irregular illumination distribution, etc. exist. Although DSARCF [56] can handle scale variation and deformation problems perfectly due to the use of saliency information, it fails in utilizing the feature information of targets with faster moving speed, especially when the background moves together with the target, where the saliency map loses its function. For instance, the background moves upwards or downwards along with the diving athlete in *Jump*, in which the backgrounds in the adjacent frames have more differences than in the normal conditions, which prohibits the filter in the previous frame from valid detection in the current frame. The saliency map cannot work well either on the sequence of *Matrix*, where there are influences of low resolution and background clutter, bringing about target loss in tracking the later part of the sequence. For *Dragonbaby*, where the face of the baby disappears and its arm becomes distinct, the bounding box permanently drifts to the region of the arm. CLIP [34] is not to able to cope with the rotation problems in which handcrafted target features are used, leading to the drift in *MotorRolling* where the target rotates and translates rapidly simultaneously, which is also failed to track by MSRBT that also utilizes handcrafted features with limited robustness only, owing to the fact that target rotation implies the transformation of its spatial orientation. Despite the capability that PF can well calculate the states of targets in silent videos, it does not work well in estimating the states of moving objects, thus performs much poorer in the videos where targets move much more drasticly. For instance, LSA [17] wrongly estimates the target's states in *CarScale*. Besides, LSA has poor ability in distinguishing responsible patches from occluded ones owing to the mechanism of linear regression, hence it has poor performances on the sequences with mixed challenging factors, such as *Ironman* and *MotorRolling*. Due to the fact that deeper layers of a network extract more semantic features, HCF can not discriminate the dark target singer and the bright background, since the features that are extracted by the first layer are reliable enough to complete the classification.

5.4. Module Summary

In this module, approaches of model construction and update strategies that are based on paper researches in recent years are listed and some typical performances of the trackers are briefed in the first subsection. In the second subsection, excellent performances of the recent trackers under challenging factors, which are occlusion, background interference, illumination variation, rotation, scale variation, and deformation are illustrated in order in detail. Advantages of the update strategies of each tracker are illustrated based on specific tracking cases and analyses regarding model construction and update mechanism are given to account for those merits. In the third subsection, some failure tracking cases are listed and the remaining problems of the listed trackers are explained.

Through the performance results, we can draw a conclusion that the framework of a tracker lays the foundation of its basic quality, while the update approach reflects its robustness and adaptability. Detailed conclusions are drawn by the analysis of diverse update mechanisms, which are illustrated below.

(1) Local representation of a target makes the tracker much easier to detect the local parts of the target. An independent update of each local patch guarantees that the tracking model can well capture more reliable local appearances of target local parts. On the basis of animal's selective attention mechanism, it is not necessary to fix attention on the whole object when tracking it, whereas only the typical characters of the target rather than others are sufficient for use as the attention for visual detection. In addition, it is better to design a multiple-tracker framework that each part of a target is independently tracked to reduce the complexity of training samples and the irregularity of sample distribution. Lots of researches have proved that frameworks with local patterns perform stronger robustness under many challenging situations as compared to that with mere global patterns, especially under the state of partial occlusion, although of which state the occluded parts of the target template cannot be updated, the remaining visible parts can still be used for detection and location and their corresponding parts of the template can be dynamically updated to make the model adapt to the newest target appearance. Examples of L3SCM [24], TPS [14], MSRBT [54], etc., have verified the robustness of tracking under partial occlusions. In addition, rotation problems can also be dealt with by local patterns. When in-plane rotation happens, all of the target parts remain visible and there is just the rearrangement of places order of the parts, while during out-of-plane rotation, some of target parts remain visible. Under these cases, the old visible parts contain valuable information regarding the location of target parts, thus independent update of each target part makes sure that symbolic target parts provide the most assistance for the location of the whole target. Some researches also have shown the effectiveness of local patterns in dealing with rotations, as for the instance of LSA [17], where the valid patches provide a lot of contributions to the target detection under the states of in-plane and out-of-plane rotations.

(2) The utilization of multiple features makes the tracker much more excellent in figuring out the target under some special situations. Independent update of different feature parameters can make the tracker avoid the disturbances of environmental changes, which is also an approach that disassembles the complexity of initial training samples. As different features describe a target from diverse respects, the contribution of each feature is not constant in different periods of tracking [58]; hence, it is better to adopt feature-specific update methods. As for the instance of illumination variation, the target passively changes its appearance along with the illumination change of the environment, during which period some features have undergone apparent changes, i.e., gray feature [54], while some do not change so much, i.e., HoG [14]. In this case, illumination-insensitive features, like HoG, are of better use for target detection and larger weights should be given on it, while features sensitive to illumination, such as gray feature, should be given smaller weights to reduce the impact of noise.

(3) Features that are extracted by layers of different depths in a deep neural network also do different performances on tracking, and the highest-level features are not always the most effective. In the process of visual tracking, the only work is to separate the target from the background and then locate it, rather than obtaining the semantic features of the target and its surroundings, thus sometimes features that are extracted by deeper layers are less important than shallower ones. The failure case of tracking the actor in *Singer2* by HCF [29] has indicated the drawback of the high-level feature extracted by deep layers, whereas the features that are extracted by the first layer do the best performance on the contrary under this situation. The example of LSHR [33] has also explained that each layer has its own excellence in the discrimination, where features that are extracted by the first layer are best at dealing with low resolution problems, while features by the third layer do the best in handling with rotations and features by the fifth (deepest) layer performs best in occlusion cases. The update strategy that

takes the contribution of each layer into consideration and gives full play to the advantages of each layer improves the robustness of the tracking network to a great extent.

(4) Target re-detection is an especially essential part in tracking, and the positive samples from the first frame reused for update help with recovery of the target. Tracking failures of MSRBT [54], HCF [29], and ODLR [39] have explained this significance. The positive and negative samples are used for the dictionary construction in [51], in which the dictionaries are adopted for the target re-detection. The RDLT tracker does perfectly under the situations of out-of-view and full occlusion thanks to the re-detection mechanism and update policy. Due to the fact that a target might also vary its appearance in the process of temporary leaving off from the scene or being fully occluded, it is not responsible enough to merely use the latest appearance models in the moment before its disappearance. Like the training approach in object detection, theoretically target images, including all of the target appearances ever appeared, should be used as the retraining sample set. However, though positive samples from some best tracked frames can also help with re-detection, target samples from the first frame are believed as the most credible and share some characters with the recovered target, even if the appearance of them may not be so close due to the impact of noise and other environmental disturbances, thus the update methods that take the positive samples from the first frame into account make better performances on target re-detection. Instances of ELMAE [29] and DNT [27] have also verified this effectiveness.

(5) An excellent update strategy of the background models helps with more precise target location. Owing to the continuity of target movement in the background, positions of the target in the surroundings in adjacent frames are very close; hence, characters of the surroundings of the target contain valuable information for target location in the next frame. Examples of BoW (BoF) based models are supportive of this conclusion. It is good to adopt dense sampling to make each surrounding region more representative since the background has more diversity appearances.

(6) The saliency feature of the target provides the tool for the estimation of target scale and shape. The scale and shape of the target in the saliency map highly reflects those in the original image due to the characteristic of the target that it should be a salient object, thus the saliency feature does well in handling scale variation and deformation. The precise estimation of the target scale and shape by DSARCF [56] shows its function.

6. Summary and Outlooks

This review has given detailed analysis of the visual tracking model update approaches in recent years, where discussions about target model update occasions and strategies as well as approaches of background model update are illustrated in order, and specific performances of sequence tracking are then exampled. The merits and drawbacks of the listed trackers in recent researches are illustrated afterwards and conclusions regarding the performances with respect to model construction and update are briefed. In light of the problems remaining in the latest tracking performances posing challenges to future researches, to make future tracker frameworks more applicable, focuses with respect to the model update training of visual tracking should be fixed on the following aspects.

(1) Adoption of the background information should be further enhanced and algorithms for dynamic background appearance model update need to be designed. In view of the truth that a target must exist in a specific environment, information regarding the background that surrounds the target provides sufficient information for target location, which can wonderfully help with discriminating the true target and the similar objects in the surroundings. Encoding and updating the background model should well be respectively conceived from the angles of the global pattern and local ones, of which the former gives the requirement that relationships among each background parts should be encoded for the holistic description, which provides useful information for the rough location of the target. When the tracked target becomes much too smaller than that in the first frame, the holistic character of the surrounding background rather than the target itself is better to be tracked, hence the problem of the model construction of small targets can be greatly alleviated. Besides, the hard problem of rapidly

moving background can also be well settled. Therefore, a network that has the ability to recognize the position and distance relationships among different background parts should better be designed, which should be an application in object position recognition. The latter namely local patterns requires that the set of characters in the regions of surroundings which includes the most evident symbols for target location should be searched out as the auxiliary feature, which is close to the mechanism of animal's selective visual attention that symbolic background areas ought to be given more visual attentions, thus the moving speed of a target can be well estimated with the help of these background auxiliaries. Based on the tool of the target response map, the response map of background regions should be made use of to decide whether the holistic model or the local model needs to be updated.

(2) Saliency information should be adopted as an important feature. As random variations of the target scale and shape also constitute the challenging factors, though some target state estimation models, like PF, scale CF, etc., perform well in calculating the size and shape of targets, they are not always credible due to the extent of target movement and restriction of datum complexity, whereas the saliency map of the target naturally represents the shape and size of it in the original image; therefore, it can be regarded as a responsible feature for the state estimation of the target, which gives the guidance of target detection under partial occlusions that lays the foundation of target appearance update in this situation. Although there are failures that the saliency map does not work well on some sequences, it is better to be used after the raw detection of the target for further state estimation instead of using as an appearance template before target search, and it should be preserved as a state template for the reference of target state calculation in the next frame.

(3) It makes a tracker more robust to make adequate adoption of features extracted by different layers to achieve more responsible target detection and location, and a multiple-feature based template model should be utilized. Like handcrafted features, neural network features that are extracted by different layers express the target and background from different aspects, making different contributions to the tracking performance, hence the tracking performances by different layers ought to be considered for update. To make deep neural network frameworks give more adequate play in many computer vision applications, it ought to be an excellent idea to make use of layers of any depth for specific usages. In light of the fact that the feature that was extracted by one deeper layer is further processed and extracted from that by the previous shallower layer, the complexity of the feature information increases with the depth of the layer. Yet, the feature information by the higher level sometimes provides more contribution, while in other cases lower level features contribute the most on the contrary. Inspired by the cascade template update approach in [61], cascade adoption of the features by different depth of the layers should be taken into consideration to assess the tracking performance using the performance score of each layer in depth order. Meanwhile, the template corresponding to each layer should be set up. If the performance score of the shallower layer is higher than a predefined threshold, the template corresponding to this layer needs to be updated meanwhile update of the templates corresponding to deeper layers should be temporarily disabled. Target matching should also be done in the cascade mode. if the template corresponding to a layer matches to the target candidates with too high confidence, those to the matching of the templates corresponding to the higher levels should also be stopped in this turn. The proposed cascade method is able to prohibit the interference from irrelevant features, thus time expense aroused from feature selection can be reduced.

Author Contributions: Conceptualization, D.W. and W.C.; methodology, W.C. and W.F.; validation, W.F. and T.S.; formal analysis, D.W. and W.C.; investigation, D.W. and T.C.; resources, W.C.; writing—original draft preparation, D.W.; writing—review and editing, D.W. and W.F.; visualization, D.W.; supervision, W.C.; project administration, W.C.; funding acquisition, W.C.

Funding: This word was funded by National Natural Science Foundation of China, grant number 51874300, National Natural Science Foundation of China and Shanxi Provincial People's Government Jointly Funded Project of China for Coal Base and Low Carbon, grant number U1510115, and the Open Research Fund of Key Laboratory of Wireless Sensor Network & Communication, Shanghai Institute of Microsystem and Information Technology, Chinese Academy of Sciences, grant numbers 20190902 and20190913.The APC was funded by51874300, U1510115 and 20190902.

Conflicts of Interest: The authors declare no conflict of interest.

References

1. Yang, X.; Meng, L. A Survey of Object Tracking Algorithms. *Acta Autom. Sin.* **2019**, *20*, 1–15.
2. Zhu, G.; Porikli, F.; Li, H. Not All Negatives Are Equal: Learning to Track With Multiple Background Clusters. *IEEE Trans. Circuits Syst. Video Technol.* **2018**, *28*, 314–326. [CrossRef]
3. Han, Y.; Deng, C.; Zhao, B. High-Performance Visual Tracking with Extreme Learning Machine Framework. *IEEE Trans. Cybern.* **2018**, 1–12. [CrossRef] [PubMed]
4. Dong, X.; Shen, J.; Yu, D.; Wang, W.; Liu, J.; Huang, H. Occlusion-Aware Real-Time Object Tracking. *IEEE Trans. Multimed.* **2017**, *19*, 763–771. [CrossRef]
5. Ilic, S.; Holzer, S.; Navab, N.; Tan, D.; Pollefeys, M. Efficient Learning of Linear Predictors for Template Tracking. *Int. J. Comput. Vis.* **2015**, *111*, 12–28.
6. Ilic, S.; Holzer, S.; Navab, N. Multipayer Adaptive Linear Predictors for Real-Time Tracking. *IEEE Trans. Pattern Anal. Mach. Intell.* **2013**, *35*, 105–117.
7. Del Bimbo, A.; Pernici, F. Object Tracking by Oversampling Local Features. *IEEE Trans. Pattern Anal. Mach. Intell.* **2014**, *36*, 2538–2551.
8. Ren, Y.; Huang, R. Visual Tracking Using Spatio-Temporal Context Template Set Learning. In Proceedings of the 2017 9th IEEE International Conference on Communication Software and Networks (ICCSN), Guangzhou, China, 6–8 May 2017; pp. 1496–1500.
9. Qiao, Y.; Niu, X. Context-Based Occlusion Detection for Robust Visual Tracking. In Proceedings of the Internation Conference on Image Processing (ICIP), Beijing, China, 17–20 September 2017; pp. 3655–3658.
10. Gu, Y.; Niu, X.; Qiao, Y. Robust Visual Tracking via Adaptive Occlusion Detection. In Proceedings of the ICASSP 2019—2019 IEEE International Conference on Acoustics, Speech and Signal Processing (ICASSP), Brighton, UK, 12–17 May 2019; pp. 2242–2246.
11. Niu, X.; Fang, X.; Qiao, Y. Robust visual tracking via occlusion detection based on staple algorithm. In Proceedings of the 2017 11th Asian Control Conference (ASCC), Gold Coast, QLD, Australia, 17–20 December 2017; pp. 1051–1056.
12. Zhang, S.; Zhao, S.; Zhang, L. Towards Occlusion Handling: Object Tracking with Background Estimation. *IEEE Trans. Cybern.* **2018**, *48*, 2086–2099.
13. Wang, H.; Zhang, X.; Yu, L.; Wang, X. Research on Mean Shift Tracking Algorithm Based on Significant Features and Template Updates. In Proceedings of the 2018 IEEE International Conference on Mechatronics and Automation (ICMA), Changchun, China, 5–8 August 2018; pp. 1199–1203.
14. Huang, L.; Shao, L.; Ma, B.; Shen, J.; He, H.; Porikli, F. Visual Tracking by Sampling in Part Space. *IEEE Trans. Image Process.* **2017**, *26*, 5800–5810. [CrossRef]
15. Lauer, M.; Tian, W. Tracking Objects with Severe Occlusion by Adaptive Part Filter Modeling–in Traffic Scenes and Beyond. *IEEE Intell. Trans. Syst. Mag.* **2018**, *10*, 60–73.
16. Vipin Krishnam, C.V.; Ramya, K.V. Object Tracking Via Boosted Cascade of Simple Features and Coarse and Fine Structural Local Sparse Appearance Models. In Proceedings of the 2018 IEEE International Conference on Communication and Signal Processing (ICCSP), Chennai, India, 3–5 April 2018; pp. 693–697.
17. Nai, K.; Li, Z.; Li, G.; Wang, S. Robust Object Tracking via Local Sparse Appearance Model. *IEEE Trans. Image Process.* **2018**, *27*, 4958–4970. [CrossRef] [PubMed]
18. Kuai, Y.; Wen, G.; Li, D. Learning Fully Convolutional Network for Visual Tracking With Multi-Layer Feature Fusion. *IEEE Access* **2019**, *7*, 25915–25923. [CrossRef]
19. Zou, Q.; Lin, S.; Du, Y. High Confidence Updating Strategy on Staple Trackers. In Proceedings of the 2018 IEEE International Conference of Intelligent Robotic and Control Engineering (IRCE), Lanzhou, China, 24–27 August 2018; pp. 238–241.
20. Soldic, M.; Marcetic, D.; Maracic, M.; Mihalic, D.; Ribaric, S. Real-time face tracking under long-term full occlusions. In Proceedings of the 10th International Symposium on Image and Signal Processing and Analysis, Ljubljana, Slovenia, 18–20 September 2017; Volume 1, pp. 147–152.
21. Zhu, Y.; Wen, J.; Zhang, L.; Wang, Y. Visual Tracking with Dynamic Model Update and Results Fusion. In Proceedings of the 2018 25th IEEE International Conference on Image Processing (ICIP), Athens, Greece, 7–10 October 2018; pp. 2685–2689.

22. Li, Z.; Zhang, J.; Zhang, K.; Li, Z. Visual Tracking With Weighted Adaptive Local Sparse Appearance Model via Spatio-Temporal Context Learning. *IEEE Trans. Image Process.* **2018**, *27*, 4478–4489. [CrossRef] [PubMed]

23. Qiu, S.; Zhang, J.; Qing, S.; Dong, J.; Guo, W. Object Tracking Method Based on Semi Supervised Extreme Learning. In Proceedings of the 2018 International Conference on Information Systems and Computer Aided Education (ICISCAE), Changchun, China, 6–8 July 2018; pp. 308–312.

24. Elharrouss, O.; Moujahid, D.; Tairi, H. Visual Object Tracking Via the Local Soft Cosine Similarity. *Pattern Recognit. Lett.* **2018**, *110*, 79–85.

25. 25. Gao, Y.; Hu, Z.; Yeung, H.W.F.; Chung, Y.Y.; Tian, X.; Lin, L. Unifying Temporal Context and Multi-feature with Update-pacing Framework for Visual Tracking. *IEEE Trans. Circuits Syst. Video Technol.* **2019**. [CrossRef]

26. Ma, D.; Bu, W.; Xie, Y.; Cui, Y.; Wu, X. Segmentation-Guided Tracking with Prior Map Decision. In Proceedings of the 2018 24th International Conference on Pattern Recognition (ICPR), Beijing, China, 20–24 August 2018; pp. 2014–2019.

27. Chi, Z.; Li, H.; Lu, H.; Yang, M.-H. Dual Deep Network for Visual Tracking. *IEEE Trans. Image Process.* **2017**, *26*, 2005–2015. [CrossRef] [PubMed]

28. Han, B.; Nam, H. Learning Multi-Domain Convolutional Neural Networks for Visual Tracking. In Proceedings of the 2016 IEEE Conference on Computer Vision and Pattern Recognition, Seattle, WA, USA, 27–30 June 2016; pp. 4293–4302.

29. Ma, C.; Huang, J.-B.; Yang, X.; Yang, M.-H. Hierarchical Convolutional Features for Visual Tracking. In Proceedings of the 2015 IEEE International Conference on Computer Vision (ICCV), Santiago, Chile, 7–13 December 2015; pp. 3074–3082.

30. Alatan, A.A.; Gundogdu, E. Good Features to Correlate for Visual Tracking. *IEEE Trans. Image Process.* **2018**, *27*, 2526–2540.

31. Dai, K.; Wang, Y.; Yan, X.; Huo, Y. Fusion of Template Matching and Foreground Detection for Robust Visual Tracking. In Proceedings of the 2018 25th IEEE International Conference on Image Processing (ICIP), Athens, Greece, 7–10 October 2018; pp. 2720–2724.

32. Luo, L.; Huang, D.; Chen, Z.; Wen, M.; Zhang, C. Applying Detection Proposals to Visual Tracking for Scale and Aspect Ratio Adaptability. *Int. J. Comput. Vis.* **2017**, *122*, 524–541.

33. Tang, F.; Lu, X.; Zhang, X.; Luo, L.; Hu, S.; Zhang, H. Adaptive convolutional layer selection based on historical retrospect for visual tracking. *IET Comput. Vis.* **2019**, *13*, 345–353. [CrossRef]

34. Hu, Q.; Liu, H.; Li, B.; Guo, Y. Robust Long-Term Tracking Via Instance Specific Proposals. *IEEE Trans. Instrum. Meas.* **2018**, *20*, 1–13.

35. Hu, G.; Liu, Q.; Islam, M.M. Robust Visual Tracking with Spatial Regularization Kernelized Correlation Filter Constrained by a Learning Spatial Reliability Map. *IEEE Access* **2019**, *7*, 27339–37351.

36. Guo, J.; Liu, J.; Hi, S. Correlation Filter Tracking Based on Adaptive Learning Rate and Location Refiner. *Opt. Prec. Eng.* **2018**, *26*, 2100–2111.

37. Zhang, K.; Wang, W.; Lv, M. Robust Visual Tracking Based on Adaptive Extraction and Enhancement of Correlation Filter. *IEEE Access* **2019**, *7*, 3534–3546.

38. Qin, X.; Yang, M.-H.; Wang, G.; Zhong, F.; Liu, Y.; Li, H.; Peng, Q. Visual Tracking via Sparse and Local Linear Coding. *IEEE Trans. Image Process.* **2015**, *24*, 3796–3809.

39. Zhou, T.; Liu, F.; Bhaskar, H.; Yang, J. Robust Visual Tracking via Online Discriminative and Low-Rank Dictionary Learning. *IEEE Trans. Cybern.* **2018**, *48*, 2643–2655. [CrossRef] [PubMed]

40. Lu, X.; Yi, S.; He, Z.; Wang, H.; Chen, W.-S. A New Template Update Scheme for Visual Tracking. In Proceedings of the 2016 7th International Conference on Cloud Computing and Big Data (CCBD), Macau, China, 16–18 November 2016; pp. 243–247.

41. Qi, Y.; Qin, L.; Zhang, J.; Zhang, S.; Huang, Q.; Yang, M.-H. Structure-Aware Local Sparse Coding for Visual Tracking. *IEEE Trans. Image Process.* **2018**, *27*, 3857–3869. [CrossRef]

42. Gong, C.; Liu, F.; Zhou, T.; Fu, K.; He, X.; Yang, J. Visual Tracking Via Nonnegative Multiple Coding. *IEEE Trans. Multimed.* **2017**, *19*, 2680–2691.

43. Zeng, F.; Ji, Y.; Levine, M.D. Contextual Bag-of-Words for Robust Visual Tracking. *IEEE Trans. Image Process.* **2018**, *27*, 1433–1447. [CrossRef]

44. Kristan, M.; Leonardis, A.; Matas, J.; Felsberg, M.; Pflugfelder, R.; Zajc, L.C.; Vojir, T.; Hager, G.; Lukezic, A.; Eldesokey, A.; et al. The Visual Object Tracking VOT2017 Challenge Results. In Proceedings of the 2017 IEEE International Conference on Computer Vision Workshops (ICCVW), Venice, Italy, 22–29 October 2017; pp. 1949–1972.

45. Huo, H.; Lu, X.; Fang, T.; Zhang, H. Learning Deconvolutional Network for Object Tracking. *IEEE Access* **2018**, *6*, 18032–18041.

46. Wang, L.; Zhang, L.; Wang, J.; Yi, Z. Memory Mechanisms for Discriminative Visual Tracking Algorithms with Deep Neural Networks. *IEEE Trans. Cogn. Dev. Syst.* **2019**, 1. [CrossRef]

47. Zhang, S.; Lan, X.; Yao, H.; Zhou, H.; Tao, D.; Li, X. A Biologically Inspired Appearance Model for Robust Visual Tracking. *IEEE Trans. Neural Networks Learn. Syst.* **2017**, *28*, 2357–2370. [CrossRef] [PubMed]

48. Ge, D.; Song, J.; Qi, Y.; Wang, C.; Miao, Q. Self-Paced Dense Connectivity Learning for Visual Tracking. *IEEE Access* **2019**, *7*, 37181–37191. [CrossRef]

49. Zhang, K.; Liu, Q.; Wu, Y.; Yang, M.-H. Robust Visual Tracking via Convolutional Networks without Training. *IEEE Trans. Image Process.* **2016**, *25*, 1. [CrossRef] [PubMed]

50. Deng, C.; Wang, B.; Lin, W.; Huang, G.-B.; Zhao, B. Effective visual tracking by pairwise metric learning. *Neurocomputing* **2017**, *261*, 266–275. [CrossRef]

51. Zhou, W.; Wang, N.; Li, H. Reliable Re-Detection for Long-Term Tracking. *IEEE Trans. Circuits Syst. Video Technol.* **2019**, *29*, 730–743.

52. Sui, Y.; Wang, G.; Zhang, L.; Yang, M.-H. Exploiting Spatial-Temporal Locality of Tracking via Structured Dictionary Learning. *IEEE Trans. Image Process.* **2018**, *27*, 1282–1296. [CrossRef]

53. Jin, X.; Zhang, J.; Sun, J.; Wang, J.; Li, K. Dual Model Learning Combined with Multiple Feature Selection for Accurate Visual Tracking. *IEEE Access* **2017**, *20*, 1–9.

54. Zhang, S.; Lan, X.; Yuen, P.C.; Chellappa, R. Learning Common and Feature-Specific Patterns: A Novel Multiple-Sparse-Representation-Based Tracker. *IEEE Trans. Image Process.* **2018**, *27*, 2022–2037.

55. Han, R.; Guo, Q.; Feng, W. Content-Related Spatial Regularization for Visual Object Tracking. In Proceedings of the 2018 IEEE International Conference on Multimedia and Expo (ICME), San Diego, CA, USA, 23–27 July 2018.

56. Feng, W.; Han, R.; Guo, Q.; Zhu, J.; Wang, S. Dynamic Saliency-Aware Regularization for Correlation Filter-Based Object Tracking. *IEEE Trans. Image Process.* **2019**, *28*, 3232–3245. [CrossRef]

57. Hou, Z.; Wang, X.; Yu, W.; Jin, Z.; Zha, Y.; Qin, X. Online Scale Adaptive Visual Tracking Based on Multilayer Convolutional Features. *IEEE Trans. Cybern.* **2019**, *49*, 146–157.

58. Zhang, S.; Qi, Y.; Qin, L.; Huang, Q.; Yao, H.; Lim, J.; Yang, M.-H. Hedging Deep Features for Visual Tracking. *IEEE Trans. Pattern Anal. Mach. Intell.* **2019**, *41*, 1116–1129.

59. Feng, D.; Gao, F.; Wang, X.; Wang, G.; Dai, H. Cascaded Iterative Training Model and Parallel Multi-Classifiers for Visual Object Tracking. *IEEE Access* **2019**, *7*, 63099–63112. [CrossRef]

60. Wang, T.; Ling, H.; Lang, C.; Feng, S.; Jin, Y.; Li, Y. Constrained Confidence Matching for Planar Object Tracking. In Proceedings of the 2018 IEEE International Conference on Robotics and Automation (ICRA), Brisbane, QLD, Australia, 21–25 May 2018; pp. 659–666.

61. Wang, Y.; Dai, K.; Yan, X. Long-Term Object Tracking Based on Siamese Network. In Proceedings of the 2017 IEEE International Conference on Image Processing (ICIP), Beijing, China, 17–20 September 2017; pp. 3640–3643.

Article

Characterization and Correction of the Geometric Errors in Using Confocal Microscope for Extended Topography Measurement. Part I: Models, Algorithms Development and Validation

Chen Wang [1], Emilio Gómez [1] and Yingjie Yu [2,*]

[1] Department of Mechanical Engineering, Chemical and Industrial Design, ETS of Engineering and Industrial Design, Technical University of Madrid, 28012 Madrid, Spain
[2] Department of Precision Mechanical Engineering, Shanghai University. No.333, Nanchen Rd., Shanghai 200444, China
*   Correspondence: yingjieyu@staff.shu.edu.cn

Received: 2 June 2019; Accepted: 26 June 2019; Published: 27 June 2019

Abstract: This work presents a method for characterizing and correcting the geometric errors of the movement of the lateral stage of Imaging Confocal Microscope (CM) in extended topography measurement. For an extended topography measurement, a defined number of 2D images are taken and stitched by correlation methods. Inaccuracies due to linear displacement, vertical and horizontal straightness errors, angular errors, and squareness errors based on the assumption of the rigid body kinematics are described. A mathematical model for the scale calibration of the X- and Y- coordinates is derived according to the system kinematics, the axis chain vector of CM, and the geometric error functions and their approximations by Legendre polynomials. The correction coefficients of the kinematic modelling are determined by the measured and certified data of a dot grid target standard artefact. To process the measurement data, algorithms for data partitions, fittings of cylinder centers, and determinations of coefficients are developed and validated. During which methods such as form removal, K-means clustering, linear and non-linear Least Squares are implemented. Results of the correction coefficients are presented in Part II based on the experimental studies. The mean residual reduces 29.6% after the correction of the lateral stage errors.

Keywords: geometric errors correction; kinematic modelling; lateral stage errors; Imaging Confocal Microscope; K-means clustering; data partition; Least Squares method

1. Introduction

The calibration of measuring machines is both important for test acceptance and error compensation [1]. Use of software techniques started from the very beginning for the correction of the systematic errors of measurement instruments, as the mechanical accuracy is expensive while repeatability cost little [2]. Over the last several decades, measurement accuracy and error compensation have been an area of intensive investigation [3–5].

Lateral calibration of the X- and Y- coordinates serves as a calibration of the magnification of the X- and Y-axis scale [6]. Lateral calibration/correction of confocal microscope (CM) can be classified mainly into two groups: imaging system calibration and machine system calibration [7]. Calibrations of the imaging system studies aberrations of refractive systems of objective lenses [8–10], axial distortion by point spread function or refractive-index mismatches [11,12], etc. The machine system calibrations investigate *geometric errors* generated by the movement of probes [2,13,14]. The first study of using kinematic geometric errors for error compensation of coordinate measurement traces back to the work of G Zhang et al. [15]. From then on, characterization and correction of kinematic geometric errors has mainly focused on coordinate measuring machines (CMM) [16,17], and has seldom been applied for optical measurement instruments.

However, areal measurement by CM is realized by both imaging systems and machine systems. CM scan in the lateral and vertical directions using different physical mechanisms [18]. To obtain a surface topography, CM captures a series of two-dimensional images by stepping either the specimen holding base or the objectives along the Z-stage [19]. For a single topography measurement, the 2D images are acquired in lateral directions either all at once by a CCD array (Imaging CM) or by a raster scan (Laser Scanning CM) [20,21]. For an extended topography measurement, a defined number of 2D images are taken and stitched by correlation methods. Accordingly, calibrating the imaging system only is not sufficient for a CM, as its measurement is realized by both imaging and machine systems. Lateral calibration or machine system calibration is necessary and important. It is worth mentioning that some works have been carried out on characterization of metrological characteristics defined in ISO 25178-600 [22], such as amplification coefficient, linearity deviation, x-y perpendicularity deviation, etc. [23–26]. These metrological characteristics imply the whole instrumentation system's characteristics to some extent.

In this work, we present a method to characterize and correct the geometric errors of the lateral stage of the CM, with implementation of a standard artefact of dot grid target. Extended topography measurements are influenced by kinematic errors of the measurand holding base, which moves horizontally in the X- and Y-directions. This work first introduces the theory and mathematics of the 21 rigid body geometric errors applied generally in the study of error compensations for CMM and machine tools in Section 2 Afterwards, we develop our own kinematic modelling based on the theory of kinematic and geometric errors for the lateral stage movement of CM in the same Section. Section 3 introduces the methodology of the study. Mathematical models and algorithms for measurement data processing are presented in Section 4. Section 5 validates our developed models and algorithms using synthetic data. Section 6 draws the conclusions. Experimental studies of the calibration and uncertainty evaluation are presented in Part II, published as another paper in the same journal.

2. Mathematical Model for the X- and Y-Scale Calibration

For each of the three axes there exist three translational deviations, i.e. the linearity deviation, the straightness deviation and the ortoganality deviation. The straightness deviation is defined as the displacement orthogonal to the axis of motion, with two straightness deviation functions for each of the orthogonal directions. Rotational movements of stages are roll, yaw and pitch, with roll denoting a screw like movement, and pitch and yaw denoting angular deviation functions that describe movements within a plane [1,15,27]. Some parameters of mechanical systems, such as bearing spacing and guideway geometric errors, determine axis motion errors of machine axes [28]. T. O. Ekinci et al. categorized machine errors into three levels, i.e., geometric errors, joint kinematic errors, and volumetric errors [29]. This work clarifies the terminologies and highlights the usefulness of a machine error modeling approach.

For the measurement of a specimen using the 3D Imaging CM, the workpiece is placed on the X- and Y-stage above the supporting base, which is usually made of granite. As shown in Figure 1a, to focus the laser beam onto the measurand surface, two movements are carried ou, i.e., movement along the X- and Y-axis, controlled by the lateral stage, and movement along the Z-axis, controlled by the vertical stage. For example, to measure the point P on the measurand surface, its position can be denoted as $\vec{\mathbf{P}}(P_x, P_y, P_z)$ with respect to its coordinate origin $\vec{O}(0,0,0)$. For the focus of the probe and the point of workpiece, the first movement include translation of X carriage followed by the translation of Y carriage; the second movement is the translation of Z carriage. It can be suggested that vectors $\vec{\mathbf{X}}, \vec{\mathbf{Y}}$ and $\vec{\mathbf{Z}}$ represent the translations of X, Y and Z carriages, the rotation matrices $\mathbf{R}(\mathbf{X})$, $\mathbf{R}(\mathbf{Y})$ and $\mathbf{R}(\mathbf{Z})$ denote the angular motion errors caused by the translations of X, Y and Z carriages, the vector $\vec{\mathbf{T}}$ represents the X, Y and Z abbe offsets of the probe with respect to the carriage to which the probe is attached. For our Imaging CM, the probe is attached to the Z-stage. Therefore, it is obvious that the translation of the X and Y carriage, with an offset of the position $\vec{P}$, stops at the same point reached by the translation of the Z carriage with a probe offset $\vec{T}$, which can be understood as the focus of the probe with the measurand. According to this conclusion, a diagram of the axis chain vector of 3D imaging microscopy is plotted and shown in Figure 1b.

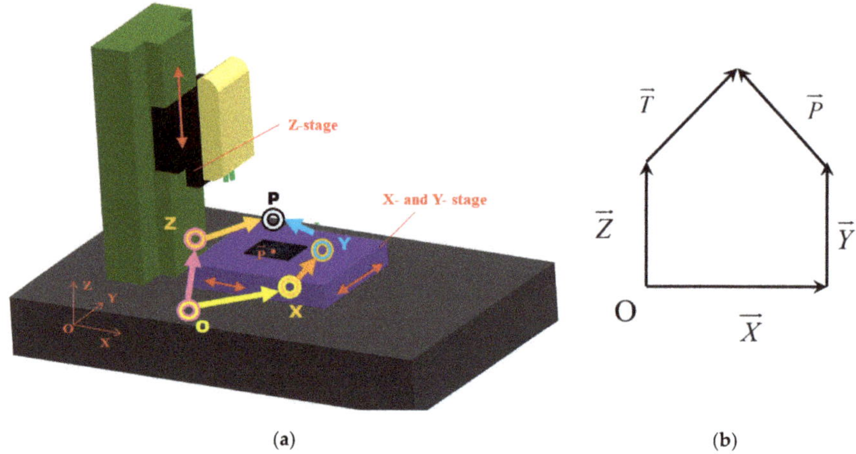

(a) (b)

Figure 1. Manifestation of the vectors in 3D Imaging CM: (**a**) 3D view; (**b**) diagram of the axis chain vector.

As shown in Figure 1b, there are two kinematical paths from the coordinate origin $\vec{O}(0,0,0)$ to the laser beam focus, i.e., $\vec{Z}\rightarrow\vec{T}$ and $\vec{X}\rightarrow\vec{Y}\rightarrow\vec{P}$. In each kinematical path, the actual movement of that axis is affected by the rotational error of its predecessors. The movement of $\vec{T}$ is affected by the rotational error of the movement of $\vec{Z}$. $\vec{Y}$ is affected by the rotational error of $\vec{X}$. $\vec{P}$ is affected by the rotational errors of $\vec{X}$ and $\vec{Y}$. Therefore, the chain vector shown in Figure 1b for the two kinematical paths, from the same origin reaching the same laser beam focus, can be expressed by Equation (1).

$$\vec{Z} + \mathbf{R}^{-1}(Z)\vec{T} = \vec{X} + \mathbf{R}^{-1}(X)\vec{Y} + \mathbf{R}^{-1}(X)\mathbf{R}^{-1}(Y)\vec{P} \tag{1}$$

After rearrangement of the above equation, vector $\vec{P}$ can be expressed by all the other vectors and matrices, as the same result presented by G. Zhang et al. [1].

The rotation can be expressed by the infinitesimal rotation matrix. The vectors $\vec{X}$, $\vec{Y}$, $\vec{Z}$ and $\vec{T}$ can be substituted by their position matrix. This has been widely accepted and implemented by many former investigators [1,15,27,30].

In the 3D Imaging CM, there are no real X- and Y-carriages, the workpiece is carried by a lateral stage mounted on a fixed base [13,27], the Z-stage is mounted on another fixed base. In addition to that, the beam focus is considered to be the probe. As the focus is always on the workpiece surface and the information is acquired exactly at that point, there is no Abbe offset on x, y and z, nor is there an angular term. This means $\vec{T} = 0$ and $\mathbf{R}(Z) = 0$.

In this work, only geometric errors along the X- and Y-directions are studied. After dropping out the Z-stage variables, the geometric error functions for X- and Y-stage are:

$$P_x = x + \delta_x(x) + \delta_x(y) - y \cdot \sigma_z(y) \tag{2}$$

$$P_y = y + \delta_y(x) + x \cdot \sigma_z(x) + \delta_y(y) + x \cdot \sigma_z(y) \tag{3}$$

Those errors in Equations (2) and (3) are functions, which can be simulated by Legendre polynomials. Implementation of Legendre polynomials is computationally simpler and provides a reduction of around 2% higher than using Chebyshev polynomials [31]. After substituting the error functions in Equations (21) and (22) by using the approximation of Legendre polynomials, the final mathematical models for the beam focus point $\overrightarrow{\mathbf{P}}(P_x, P_y, 0)$ with error corrections can be acquired:

$$
\begin{aligned}
P_x = \ & x + a_1 x + \tfrac{3}{2}a_2 x^2 + \tfrac{3}{2}a_3(5x^3 - 3x) + \\
& + b_1 y + \tfrac{3}{2}b_2 y^2 + \tfrac{3}{2}b_3(5y^3 - 3y) - \\
& - y \cdot (c_1 y + \tfrac{3}{2}c_2 y^2 + \tfrac{3}{2}c_3(5y^3 - 3y))
\end{aligned}
\tag{4}
$$

$$
\begin{aligned}
P_y = \ & y + d_1 x + \tfrac{3}{2}d_2 x^2 + \tfrac{3}{2}d_3(5x^3 - 3x) + \\
& + x \cdot (e_1 x + \tfrac{3}{2}e_2 x^2 + \tfrac{3}{2}e_3(5x^3 - 3x)) + \\
& + f_1 y + \tfrac{3}{2}f_2 y^2 + \tfrac{3}{2}f_3(5y^3 - 3y) + \\
& + x \cdot (c_1 y + \tfrac{3}{2}c_2 y^2 + \tfrac{3}{2}c_3(5y^3 - 3y))
\end{aligned}
\tag{5}
$$

3. Methodology and a Brief Introduction of Experimental Design

To calibrate the kinematic geometric errors of the lateral stage of a dual core 3D CM, a standard artefact with dot grid target on glass is measured and a series of algorithms are developed for the process of the raw measurement data, which include separations of cylinders and flat, fitting of cylinder centers, determination of coefficients, etc. Figure 2 illustrates the standard, which was provided by Max Levy Autograph, Inc. Those dots are deposited on the substrate, which is glass with a thickness of 1.5 mm. Table 1 gives an indication of the size, the dot spacing, the diameter, the accuracy, etc., of the dot pattern.

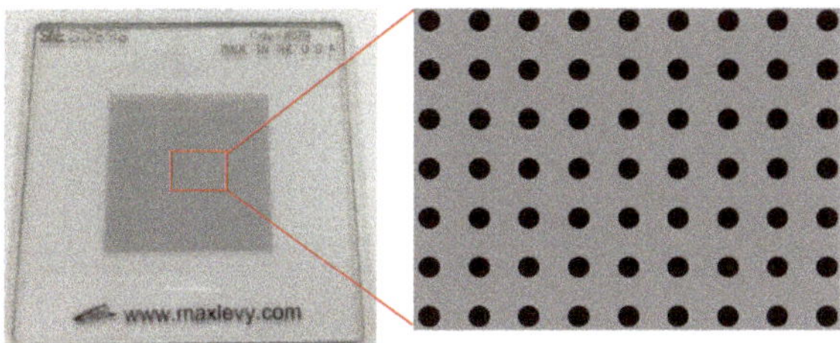

Figure 2. Illustration of the standard artefact of the dot grid target.

Table 1. Detailed information about the dot grid target standard artefact.

Width [mm]	Dot Diameter [mm]	Dot Spacing [mm]	X and Y Axis Accuracy [mm]	Dot Array		
				X	Y	Total
25	0.0625	0.125	±0.001	201	201	40401

This work aims at characterizing and correcting the kinematic geometric errors of the movement of the lateral stage, by calculating and applying the coefficients of Equations (4) and (5). To calculate the coefficients, the introduced dot grid target standard is measured, the measurement data are processed for obtaining the coordinate values (x_i, y_i) of the dots' centers, the coordinate values (x_i, y_i) and their corresponding certified values are put into Equations (4) and (5), the coefficients are obtained by solving the equations using non-linear Least Squares method. The obtained coefficients and Equations (4) and (5) are implemented for correcting the new measurement data. Another different area of the introduced dot grid target standard is measured, and the measurement data is corrected. By comparing the residuals before correction and after correction, the significance and practicality of our model for kinematic geometric error correction can be observed.

The algorithms used for measurement data processing and for calculating the coefficients are introduced in Section 4. Validations of those algorithms are introduced in Section 5. The details of the experimental procedures, parameters, and analysis are presented in Part II, which is published in another paper in the same journal.

4. Algorithms and Procedures for Measurement Data Processing

4.1. Algorithm and Procedure for the Separation of Flats and Cylinders

The raw measurement data usually contains defects, outliers, unmeasured points, etc. [32]. The measured surface often inclines, as it is impossible to locate the measurand surface perfectly perpendicular to the Z-axis. Figure 3 gives an example of the measurement of the dot grid target, implementing an objective of magnification 50×, numerical aperture of 0.90. The acquisition parameter of measurement area was defined as the topography stitching measurement, with 8 × 8 extended topographies, covering an area of 1.78 × 1.33 mm^2. The parameter of the overlapping area is 25%, and the correlation takes the XYZ option. The level of resolution is 2, and the measured extended topographies contain 2673 × 2003 pixels.

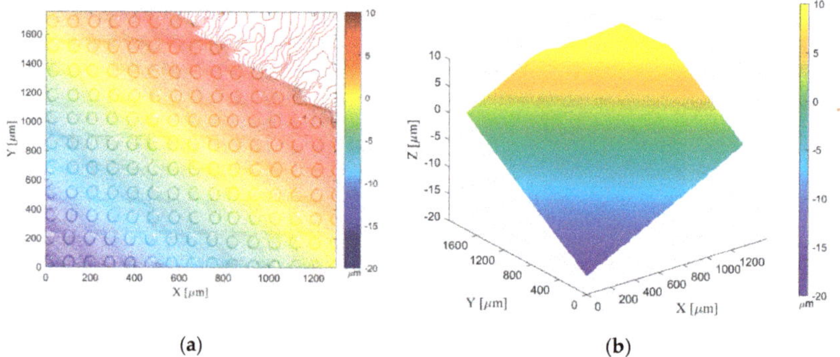

(a)   (b)

Figure 3. Reconstruction of the measured surface with its raw measurement data: (**a**) contour plot; (**b**) surf plot.

This measurement has many unmeasured points in the right-top corner due to the Z-range parameter setting. The first step is to choose an area of this raw surface with appropriate size and location, targeting at minimizing the influence of defects and removing outliers [32] by selecting valid grid positions with loss of the uniformity of the grid. Figure 4 shows a surface reconstruction by trimming the raw data with 10% edges along the X- and Y-directions. As there are too many measurement defects when $y > 800$ μm, this part is also abandoned.

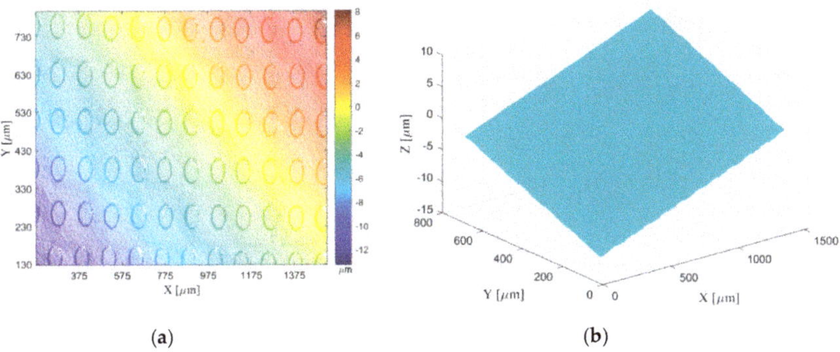

Figure 4. Reconstruction of the surface with trimmed raw data: (**a**) contour plot; (**b**) areal plot.

After obtaining an appropriate area for analysis, it is necessary to eliminate its form. The principal axes of the distribution of data points of the measured point cloud are determined by solving the eigen value problem. The data are rotated by principal axis transformation to align them in the direction of the eigen vectors. Figure 5 gives an example of the rotated surface, which is rotated from the surface shown in Figure 4. Figure 5a was plotted by the Matlab® function 'plot3', while Figure 5b was plotted by the function 'surf'. It can be found that after rotation, the range of the Z-axis is much smaller, and many details of the surface can be observed. It is obvious that this surface is constructed using point clouds, which form a flat plane and many cylinders perpendicular to this flat.

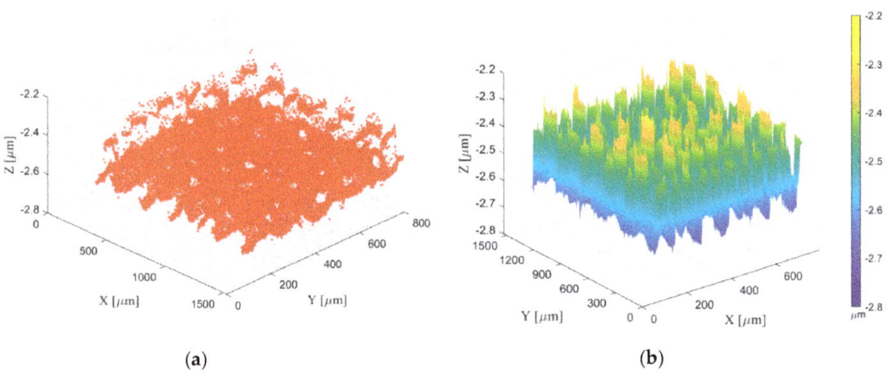

Figure 5. Reconstruction of the rotated surface: (**a**) point plot; (**b**) surfplot using Z for the color data.

This rotated surface is now ready for the separation of flats and cylinders. Figure 6 gives an indication of the distribution of the heights of the surface. Figures 7 and 8 give examples of the separations of the rotated surface shown in Figure 5.

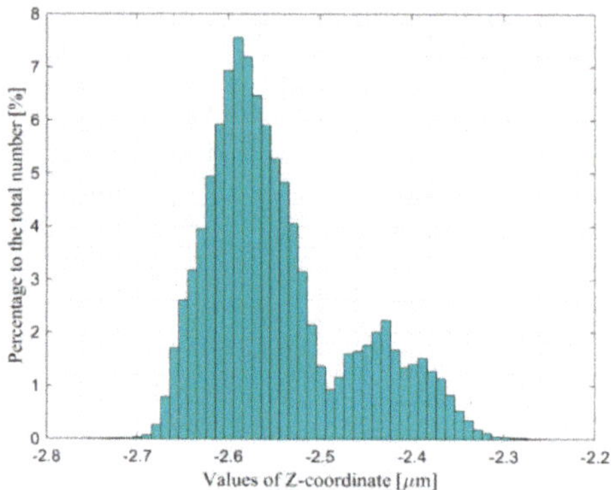

Figure 6. Histogram of the Z values of the rotated surface.

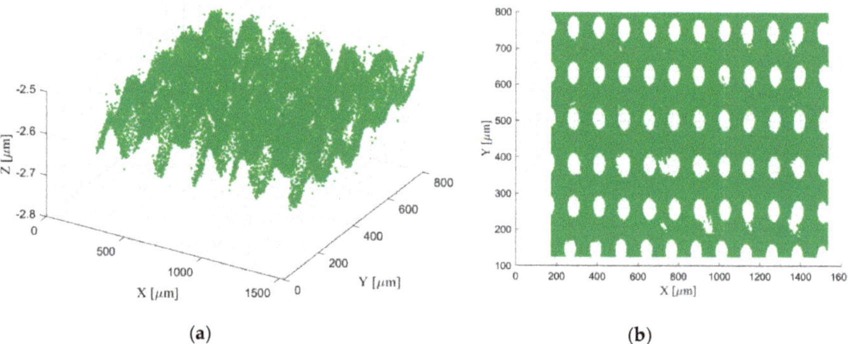

(**a**) (**b**)

Figure 7. The separated flat plane of the rotated surface: (**a**) 3D view; (**b**) view from top down.

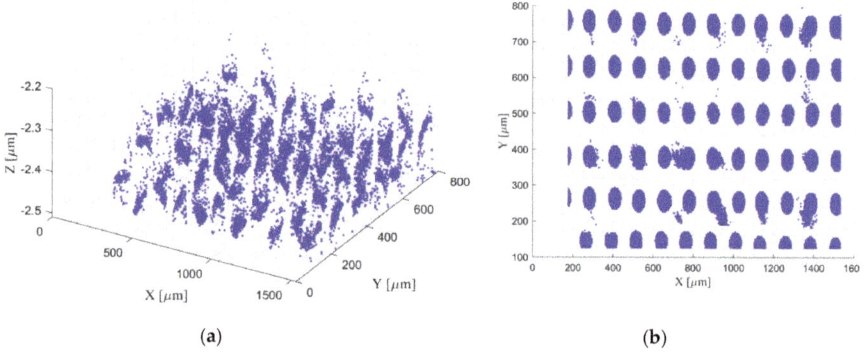

(**a**) (**b**)

Figure 8. The separated cylinders of the rotated surface: (**a**) 3D view; (**b**) view from top down.

After separating the cylinders from the flat, the next step is to obtain the three coordinates of each cylinder point cloud. Figure 9 gives an example of the separation of cylinders into individual clusters, characterizing the cylinders with different colors. Here the squared Euclidean distance metric is applied for distance calculation, as shown in Equation (6). Moreover, each centroid of the point cloud is the mean of the coordinates' values of the points.

$$d(\mathbf{x}, c) = (\mathbf{x} - c)(\mathbf{x} - c)^T \tag{6}$$

where $\mathbf{x}$ is the initial observations, c is the centroid.

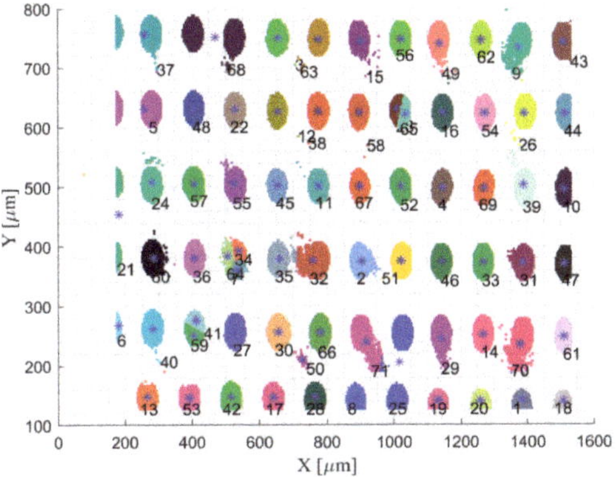

Figure 9. Separation of cylinders into individuals.

According to the previous investigations, there are several efficient algorithms for data partitioning, such as the K-means clustering algorithm [33,34] and Lloyd's algorithm [35]. These two algorithms are iterative data-partitioning algorithms, which assign the initial observations into k clusters according to the squared Euclidean distance metric and the centroids. The number of clusters k is defined before data partition. In our work, the Lloyd's algorithm is used for finding the centroids of k clusters while the K-means clustering algorithm is used for partitioning the initial observations into k clusters according to the squared Euclidean distance metric with respect to the centroid of each cluster. The K-means clustering algorithm requires the input as a matrix of M points in N dimensions as well as a matrix of K initial cluster centers in N dimensions. Denoting the number of points in cluster L is $NC(L)$, and the Euclidean distance between point I and cluster L is $D(I, L)$, the general procedure of the K-means clustering algorithm is to find a K-partition of clusters with locally optimal within-cluster sum of squares by changing the initial observations from one cluster to another [33], as indicated by Equation (7).

$$\min_{D^2} \sum_{i=1}^{K} \sum_{j=1}^{NC(L)} D^2(I, L) \tag{7}$$

The procedures of the algorithm for partitioning the filtered cylinders into individuals are shown in Figure 10, which consists of 6 steps. It is worth mentioning that the data are subjected to dimensionality reduction before data partitioning. These cylinders are a point cloud of three dimensions. The aim of dimensionality reduction is to produce a compact low-dimensional encoding of a given high-dimensional data set [36]. The dimensionality can be reduced in two ways [37]. The first is by only keeping the most relevant variables from the original dataset, which is called feature selection. The second way is to exploit the redundancy of the input data and to find a smaller set of new variables, each being a combination of the input variables containing basically the same information as the input variables. In this work, the dimensionality of the data is reduced to 2D from 3D by projection along Z-axis.

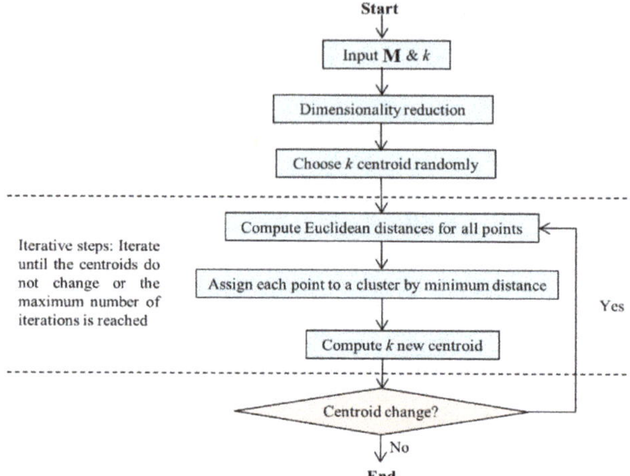

Figure 10. Flowchart of the algorithm for data partition.

Since the equations characterizing lateral linearity deviations, Equations (4) and (5) are independent of the vertical axis z, the dimension of the 3D point cloud is reduced to the two dimensions of the x-y-plane.

4.2. Algorithm for Determination of Coefficients

This section aims at the determination of the coefficients, $a_1, a_2, a_3, b_1 \ldots , f_3$, defined in Equations (4) and (5). The objective is to fit the 18 coefficients to obtain the least squared residuals between the corrected X- and Y- coordinates and the nominal position. That is:

$$\sum_{i=1}^{n} \|P_{nom}{}^{i}(x,y) - P^{i}{}_{corr}(x,y)\|_2{}^2 = \min \tag{8}$$

where, $P_{nom}{}^{i}(x,y)$ represents the i_{th} nominal position defined by the X- and Y- coordinates, $P_{corr}{}^{i}(x,y)$ denotes the i_{th} corrected position obtained from the measured data and Equations (4) and (5).

Therefore, the objective function is

$$\begin{aligned}
\min F(x,y)\big|_{a_1,a_2,\ldots,g_3} &= \min \sum_{i=1}^{n} \|P_{nom}{}^{i}(x,y) - P^{i}{}_{corr}(x,y)\|_2{}^2 \\
&= \min\left(\left(x^i_{nom} - x^i_{corr}\right)^2 + \left(y^i_{nom} - y^i_{corr}\right)^2 \right) \\
&= \min \left[\begin{array}{l}
\left(\begin{array}{l} x^i_{nom} - x^i_{meas} \;\; +a_1 x^i_{meas} + \frac{3}{2}a_2\left(x^i_{meas}\right)^2 + \frac{3}{2}a_3(5\left(x^i_{meas}\right)^3 - 3x^i_{meas})+ \\ +b_1 y^i_{meas} + \frac{3}{2}b_2\left(y^i_{meas}\right)^2 + \frac{3}{2}b_3(5\left(y^i_{meas}\right)^3 - 3y^i_{meas})- \\ -y^i_{meas} \cdot (c_1 y^i_{meas} + \frac{3}{2}c_2\left(y^i_{meas}\right)^2 + \frac{3}{2}c_3(5\left(y^i_{meas}\right)^3 - 3y^i_{meas})) \end{array} \right)^2 \\
+ \\
\left(\begin{array}{l} y^i_{nom} - y^i_{meas} + d_1 x^i_{meas} \;\; + \frac{3}{2}d_2\left(x^i_{meas}\right)^2 + \frac{3}{2}d_3(5\left(x^i_{meas}\right)^3 - 3x^i_{meas})+ \\ +x^i_{meas} \cdot (e_1 x^i_{meas} + \frac{3}{2}e_2\left(x^i_{meas}\right)^2 + \frac{3}{2}e_3(5\left(x^i_{meas}\right)^3 - 3x^i_{meas}))+ \\ +f_1 y^i_{meas} + \frac{3}{2}f_2\left(y^i_{meas}\right)^2 + \frac{3}{2}f_3(5\left(y^i_{meas}\right)^3 - 3y^i_{meas})+ \\ +x^i_{meas} \cdot (c_1 y^i_{meas} + \frac{3}{2}c_2\left(y^i_{meas}\right)^2 + \frac{3}{2}c_3(5\left(y^i_{meas}\right)^3 - 3y^i_{meas})) \end{array} \right)^2
\end{array} \right]
\end{aligned} \tag{9}$$

The minimum value of F arrives when the gradient is zero [38,39], i.e.,:

$$\frac{\partial F}{\partial \beta_i} = 0 \tag{10}$$

where F is the objective function, β_i represents the parameters of coefficients to be determined. There are 18 coefficients in this part of the work, and hence there are 18 partial derivatives:

$$\frac{\partial F}{\partial a_1} = \sum_{i=1}^{n} 2 \left(\begin{array}{l} x^i_{nom} - x^i_{meas} \quad +a_1 x^i_{meas} + \frac{3}{2}a_2\left(x^i_{meas}\right)^2 + \frac{3}{2}a_3\left(5\left(x^i_{meas}\right)^3 - 3x^i_{meas}\right)+ \\ \qquad\qquad +b_1 y^i_{meas} + \frac{3}{2}b_2\left(y^i_{meas}\right)^2 + \frac{3}{2}b_3\left(5\left(y^i_{meas}\right)^3 - 3y^i_{meas}\right)- \\ \qquad\qquad -y^i_{meas} \cdot \left(c_1 y^i_{meas} + \frac{3}{2}c_2\left(y^i_{meas}\right)^2 + \frac{3}{2}c_3\left(5\left(y^i_{meas}\right)^3 - 3y^i_{meas}\right)\right) \end{array} \right) \cdot x^i_{meas} = 0 \quad (11)$$

$$\frac{\partial F}{\partial a_2} = \sum_{i=1}^{n} 2 \left(\begin{array}{l} x^i_{nom} - x^i_{meas} \quad +a_1 x^i_{meas} + \frac{3}{2}a_2\left(x^i_{meas}\right)^2 + \frac{3}{2}a_3\left(5\left(x^i_{meas}\right)^3 - 3x^i_{meas}\right)+ \\ \qquad\qquad +b_1 y^i_{meas} + \frac{3}{2}b_2\left(y^i_{meas}\right)^2 + \frac{3}{2}b_3\left(5\left(y^i_{meas}\right)^3 - 3y^i_{meas}\right)- \\ \qquad\qquad -y^i_{meas} \cdot \left(c_1 y^i_{meas} + \frac{3}{2}c_2\left(y^i_{meas}\right)^2 + \frac{3}{2}c_3\left(5\left(y^i_{meas}\right)^3 - 3y^i_{meas}\right)\right) \end{array} \right) \cdot \frac{3}{2}\left(x^i_{meas}\right)^2 = 0 \quad (12)$$

$$\frac{\partial F}{\partial f_3} = \sum_{i=1}^{n} 2 \left(\begin{array}{l} y^i_{nom} - y^i_{meas} \quad +d_1 x^i_{meas} + \frac{3}{2}d_2\left(x^i_{meas}\right)^2 + \frac{3}{2}d_3\left(5\left(x^i_{meas}\right)^3 - 3x^i_{meas}\right)+ \\ \qquad\qquad +x^i_{meas} \cdot \left(e_1 x^i_{meas} + \frac{3}{2}e_2\left(x^i_{meas}\right)^2 + \frac{3}{2}e_3\left(5\left(x^i_{meas}\right)^3 - 3x^i_{meas}\right)\right)+ \\ \qquad\qquad +f_1 y^i_{meas} + \frac{3}{2}f_2\left(y^i_{meas}\right)^2 + \frac{3}{2}f_3\left(5\left(y^i_{meas}\right)^3 - 3y^i_{meas}\right)+ \\ \qquad\qquad +x^i_{meas} \cdot \left(c_1 y^i_{meas} + \frac{3}{2}c_2\left(y^i_{meas}\right)^2 + \frac{3}{2}c_3\left(5\left(y^i_{meas}\right)^3 - 3y^i_{meas}\right)\right) \end{array} \right)$$
$$\cdot \frac{3}{2}\left(5\left(y^i_{meas}\right)^3 - 3y^i_{meas}\right) = 0 \qquad (13)$$

These equations can first be solved by solving the equation for the initial solutions. The initial solutions might be far from the correct solutions, as there are too many more constraints than unknown coefficients. Those initial solutions can be used as the starting point for iteratively finding the nonlinear least squares solutions. Extensive work has been done on the nonlinear least squares algorithms [40,41]. Let the model for data fitting be [41]:

$$\begin{aligned} E(y) &= f(x_1, x_2, \ldots, x_m; b_1, b_2, \ldots, b_k) \\ &= f(\mathbf{x}, \mathbf{b}) \end{aligned} \qquad (14)$$

where $x_1, x_2, \ldots, x_m$ are independent variables, $b_1, b_2, \ldots, b_k$ are k parameters of coefficients to be determined, $E(y)$ is the expected value of the dependent variable y. Denote the data points as: $(Y_i, X_{1i}, X_{2i}, \cdots X_{mi})$, where, $i = 1, 2, \cdots, n$. The objective is to calculate the k parameters of coefficients which will minimize the squares of the residuals:

$$\min \sum_{i=1}^{n} \left(Y_i - \hat{Y}_i\right)^2 = \min \|\mathbf{Y} \text{-} \hat{\mathbf{Y}}\|^2 \qquad (15)$$

This problem can be written as an objective function which aims at optimizing the coefficients of each function $Y_i - \hat{Y}_i$, denoting the vector of the parameters to be optimized as $\mathbf{t}$, where $\mathbf{t} = [a_1, a_2, \cdots, f_3]$ in this work. The objective function is:

$$\min F(\mathbf{t}) = \sum_{i=1}^{k} f_i^2(\mathbf{t}) \qquad (16)$$

where, $f_i(\mathbf{t})$ are functions of the to-be-optimized vector $\mathbf{t}$. When $f_i(\mathbf{t})$ are nonlinear functions with respect to $\mathbf{t}$, this problem is about nonlinear optimization. The solution is to use Taylor expansion to convert $f_i(\mathbf{t})$ into linear functions. As only the first order of the Taylor series is linear, here we approximate it by the first order:

$$\begin{aligned} \varphi_i(\mathbf{t}) &= f_i(\mathbf{t}^{(k)}) + \nabla f_i(\mathbf{t}^{(k)})^T (\mathbf{t} - \mathbf{t}^{(k)}) \\ &= \nabla f_i(\mathbf{t}^{(k)})^T \mathbf{t} - \nabla f_i(\mathbf{t}^{(k)})^T \mathbf{t}^{(k)} + f_i(\mathbf{t}^{(k)}) \end{aligned} \qquad (17)$$

where $\nabla f_i(\mathbf{t}^{(k)})$ is the value of the first derivative of $f_i(\mathbf{t}^{(k)})$ on vector $\mathbf{t}$ evaluated at the point $\mathbf{t}^{(k)}$.

Substituting $f_i(\mathbf{t})$ in Equation (16) with its Taylor Expansion approximation indicated by Equation (17), the approximation of $F(\mathbf{t})$ is:

$$
\begin{aligned}
\min\phi(\mathbf{t}) &= \sum_{i=1}^{k} \varphi_i^2(\mathbf{t}) \\
&= \sum_{i=1}^{k} \left(\nabla f_i(\mathbf{t}^{(k)})^T \mathbf{t} - \nabla f_i(\mathbf{t}^{(k)})^T \mathbf{t}^{(k)} + f_i(\mathbf{t}^{(k)}) \right)^2
\end{aligned}
\tag{18}
$$

To simplify the above equation, use $\mathbf{A_k}$ to represent $\nabla f_i(\mathbf{t}^{(k)})^T$, and $\mathbf{b}$ to represent $\nabla f_i(\mathbf{t}^{(k)})^T \mathbf{t}^{(k)} - f_i(\mathbf{t}^{(k)})$. Equation (18) is simplified into:

$$
\min\phi(\mathbf{t}) = \sum_{i=1}^{k} (\mathbf{A_k} - \mathbf{b})^2
\tag{19}
$$

where,

$$
\mathbf{A_k} = \begin{bmatrix} \nabla f_1(\mathbf{t}^{(k)})^T \\ \vdots \\ \nabla f_m(\mathbf{t}^{(k)})^T \end{bmatrix} = \begin{bmatrix} \frac{\partial f_1(\mathbf{t}^{(k)})}{\partial t_1} & \frac{\partial f_1(\mathbf{t}^{(k)})}{\partial t_2} & \cdots & \frac{\partial f_1(\mathbf{t}^{(k)})}{\partial t_n} \\ \vdots & \vdots & \vdots & \vdots \\ \frac{\partial f_m(\mathbf{t}^{(k)})}{\partial t_1} & \frac{\partial f_m(\mathbf{t}^{(k)})}{\partial t_2} & \cdots & \frac{\partial f_m(\mathbf{t}^{(k)})}{\partial t_n} \end{bmatrix}
\tag{20}
$$

$$
\mathbf{b} = \begin{bmatrix} \nabla f_1(\mathbf{t}^{(k)})^T \mathbf{t}^{(k)} - f_1(\mathbf{t}^{(k)}) \\ \vdots \\ \nabla f_m(\mathbf{t}^{(k)})^T \mathbf{t}^{(k)} - f_m(\mathbf{t}^{(k)}) \end{bmatrix} = \mathbf{A_k}\mathbf{t}^{(k)} - \mathbf{f}^{(k)}
\tag{21}
$$

$$
\mathbf{f}^{(k)} = \begin{bmatrix} f_1(t^{(k)}) \\ f_2(t^{(k)}) \\ \vdots \\ f_m(t^{(k)}) \end{bmatrix}
\tag{22}
$$

Therefore, it can be obtained:

$$
\phi(\mathbf{t}) = (\mathbf{A_k}\mathbf{t} - \mathbf{b})^T (\mathbf{A_k}\mathbf{t} - \mathbf{b})
\tag{23}
$$

The solution of this equation is:

$$
t^{(k+1)} = t^{(k)} - (\mathbf{A_k}^T \mathbf{A_k})^{-1} \mathbf{A_k}^T f^{(k)}
\tag{24}
$$

This solution can be simplified as:

$$
t^{(k+1)} = t^{(k)} - \mathbf{H}_k^{-1} \nabla F(t^{(k)})
\tag{25}
$$

where, $\mathbf{H}_k$ is the Hessian matrix:

$$
\mathbf{H}_k = 2\mathbf{A}_k^T \mathbf{A}_k
\tag{26}
$$

$$
\nabla \mathbf{F}(x^{(k)}) = 2\mathbf{A_k}^T f^{(k)}
\tag{27}
$$

Equation (25) can be solved iteratively by starting with an initial solution.

5. Validation of the Algorithms with Synthetic Data

5.1. Validation of the Algorithm for Determination of Dots' Centers and Distance

To validate the algorithm for determination of dots' centers, synthetic data with known cylinder centers and point cloud distributions is generated. This generated data contains 35 cylinders of point cloud, distributed as 5 × 7 along the X-axis and Y-axis, as shown in Figure 11.

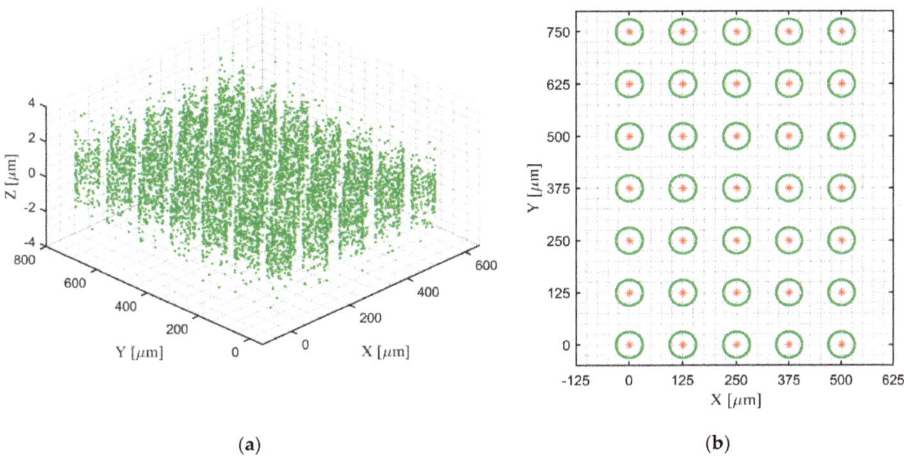

(a)   (b)

Figure 11. Plot of the created data for algorithm validation: (**a**) 3D view of the point cloud; (**b**) view from the Z-axis.

For each point cloud, the radius of the cylinder is 31.25 µm, with a uniform distribution of $r \sim U$ (−0.1, 0.1) µm. The values of the Z-coordinate of each point follows a standard normal distribution $z \sim N(0, 1)$ µm. The centers of each cylinder, as indicated by red asterisks in Figure 11b, have an interval of 125 µm. Their X-coordinates are [0, 125, 250, 375, 500] µm and the Y-coordinates are [0, 125, 250, 375, 500, 625, 750] µm. The generated data contains X-, Y-, and Z-coordinates' values. Those data are saved in a Matlab® file with the suffix name '.mat' for processing.

The generated data are processed using our developed algorithm for the determination of dots' centers. The first step is to separate the data into individual point clouds. The results of the point cloud separation are shown in Figure 12, with each individual being indicated by numbers and different colors. In this step, the algorithm not only separates the point cloud into individuals, but also gives an indication of the initial centroids of each cloud.

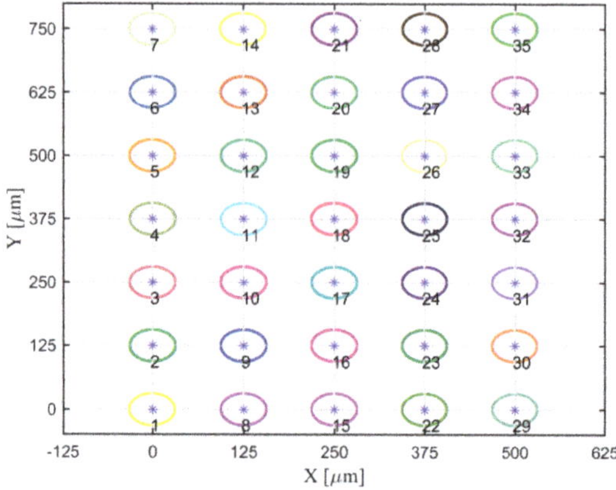

Figure 12. The separation of cylinders into individual point clouds.

After the separation, each individual point cloud is processed by our developed algorithm in order to fit the circle centers. The calculated centers are exactly the same with the generated ones. Figure 13 makes a plot of the calculated centers of the synthetic data generated for algorithm validation.

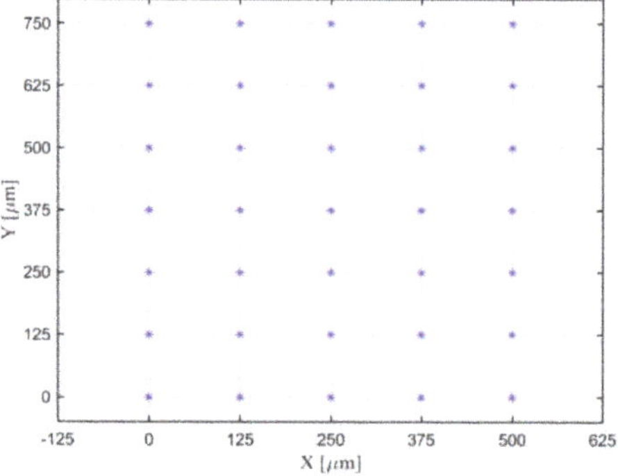

Figure 13. Plots of the calculated centers of the generated data.

The results indicate that the algorithm developed in this work for the determination of the centers of cylinder point cloud is valid.

5.2. Validation of the Algorithm for Determination of Coefficients

Here, we use two methods for the validation of the algorithm for the determination of the coefficients.

The first method is to create a dataset containing points with intervals of 125 µm as certified positions, while making a little displacement with each point. Calculate the correction coefficients and correct the points with the displacement using the algorithm defined in Section 4.2. The points with the displacement with respect to the certified points represent the measured points in the experiments. Therefore, two residuals can be obtained, i.e., the residual between the certified points and the points with displacement as well as the residual between the certified points and the corrected points. Compare those two residuals to check whether the correction is meaningful.

The algorithm first creates the two data described above. Then it rotates the points with displacement to the certified points, aiming at making the axis parallel. This arises from the methodology of this work, which requires locating the standard artefact that is parallel with the axis. As it is impossible to realize it manually, the coordinate axes are adjusted mathematically. After that, the error coefficients are calculated by the algorithm. With those coefficients, the values of the X- and Y-coordinates of the corrected points can be obtained. Figure 14 shows a comparison of the positions of the certified, measured and corrected points. The contour of the errors between certified and measured points is displayed in Figure 15. The contour of the errors between the certified and corrected points is shown in Figure 16. It is obvious that almost all of the errors are smaller after correction.

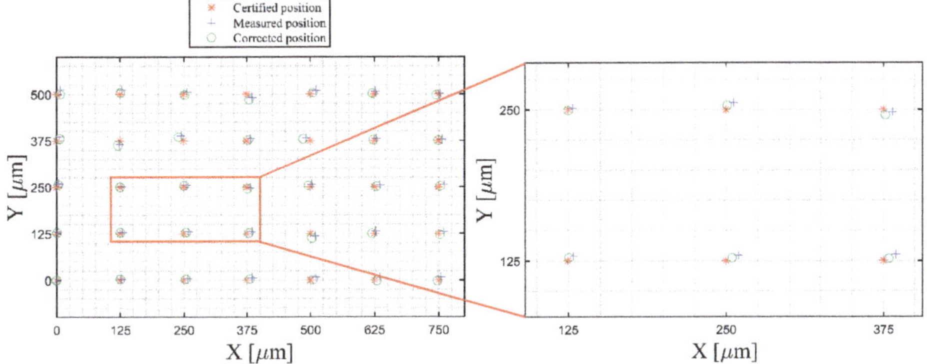

Figure 14. Comparison of the positions of certified, measured and corrected points.

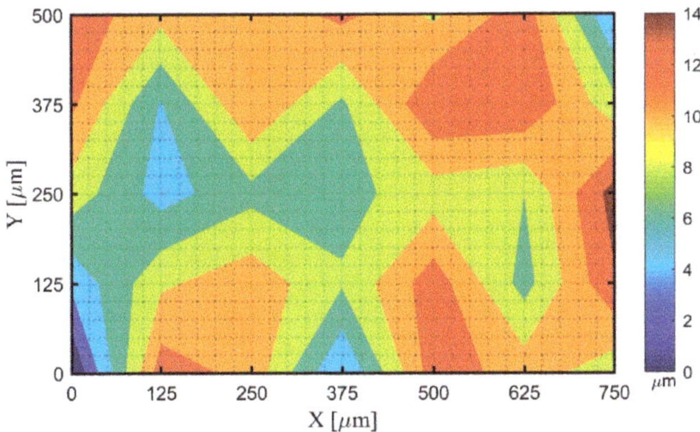

Figure 15. Contour of the errors between the certified and measured points.

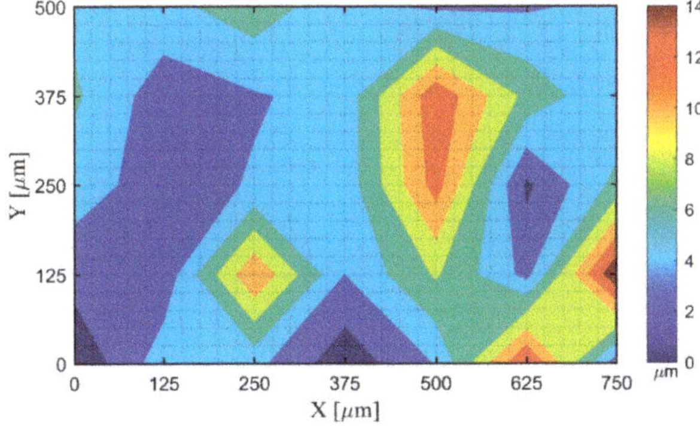

Figure 16. Contour of the errors between the certified and corrected points.

The error is calculated by the Euclidean distance between the certified point and its corresponding point with displacement. The same goes with the error between the certified point and the corrected point. Table 2 makes a comparison of the error before correction and after correction. The difference is calculated as:

$$Diff(i) = Err_c(i) - Err_m(i) \tag{28}$$

where, $Diff(i)$ is the difference between the error after correction and before correction. When $Diff(i)$ is negative, it means the error is smaller after correction. $Err_c(i)$ is the Euclidean distance between the certified point and corrected point. $Err_m(i)$ is the Euclidean distance between the certified point and the measured point. The mean residual after correction is 5.60 μm, while the mean error between the measured and the certified point is 9.65 μm. The square residual of all the 35 corrected points is 1587.99 μm^2, while it was 3684.85 μm^2 before correction. It can be found that some points are more distorted after correction. This is because the fitting method used is the nonlinear least squares, the objective function of which aims at finding the sum of least squares of the residuals of all points, but does not ensure that every residual is smaller after fitting.

Table 2. The values of the error of each individual point.

Point No. i	$Err_c(i)$ [μm]	$Err_m(i)$ [μm]	$Diff(i)$ [μm]	Point No. i	$Err_c(i)$ [μm]	$Err_m(i)$ [μm]	$Diff(i)$ [μm]
1	0	0	0	19	4.66	7.65	−2.99
2	2.92	4.35	−1.43	20	4.72	12.64	−7.93
3	4.87	9.20	−4.31	21	4.16	13.09	−8.94
4	6.36	12.48	−6.12	22	8.26	13.13	−4.87
5	5.55	13.25	−7.70	23	12.87	9.05	3.82
6	5.27	13.13	−7.86	24	13.52	13.92	−0.41
7	2.63	9.64	−7.01	25	3.56	9.58	−6.01
8	3.08	5.16	−2.08	26	14.17	11.12	3.04
9	2.86	5.85	−2.99	27	3.39	7.24	−3.86
10	5.36	10.66	−5.31	28	1.61	8.03	−6.43
11	4.63	11.28	−6.65	29	7.47	13.86	−6.39
12	11.90	11.32	0.59	30	3.73	13.14	−9.41
13	4.17	7.38	−3.20	31	2.88	8.69	−5.81
14	3.82	11.99	−8.17	32	15.93	13.78	2.15
15	7.19	10.38	−3.20	33	6.59	14.88	−8.28
16	0.62	3.85	−3.23	34	4.15	8.68	−4.53
17	3.99	8.23	−4.25	35	4.49	3.82	0.67
18	4.72	7.36	−2.64	-			
mean error [μm]	5.60	9.65					
Sum of Squared error [μm^2]	1587.99	3684.85					

The second method creates another 10 data by varying the displacement from the certified positions based on this data. In these 10 simulations, the certified positions are all the same with those certified positions introduced above. The measured points' positions are assigned with displacements, the values of which are shown in Table 3.

Table 3. Further displacements of the measured points of the 10 simulations.

Simulation	No. 1	No. 2	No. 3	No. 4	No. 5	No. 6	No. 7	No. 8	No. 9	No. 10
X [μm]	−20	−16	−12	−8	−4	0	4	8	12	16
Y [μm]	−15	−12	−9	−6	−3	0	3	6	9	12

The 10 data are processed by this algorithm. The results of each simulation are compared and analyzed. Figure 17 indicates the positions of the certified, measured, and corrected points of those 10 simulations. It is obvious that almost all of the corrected points are closer than the measured points to the certified points.

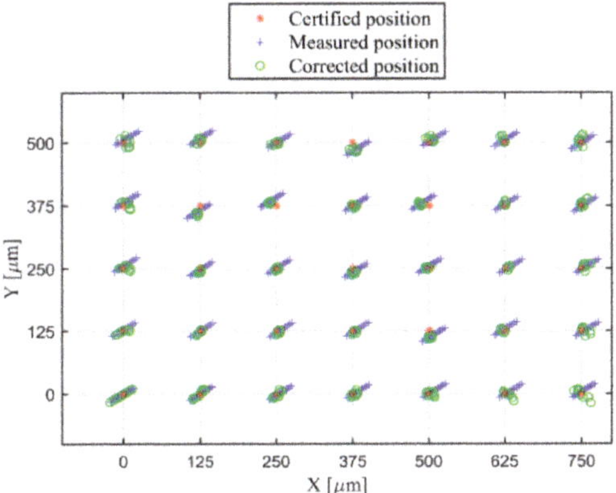

Figure 17. Comparison of the certified, measured, and corrected positions of the 10 simulations.

Table 4 lists two mean residual and two squared residuals. One of the mean residuals arises from the displacements of the measured points from the certified points, while the other mean residual comes from the corrected points from the certified points. The mean residual is calculated by Equations (29) and (30):

$$\text{Res}_{meas}{}^{M} = \frac{1}{n}\sum_{i=1}^{n} \|P_{cert}(i) - P_{meas}(i)\|_2 \tag{29}$$

$$\text{Res}_{corr}{}^{M} = \frac{1}{n}\sum_{i=1}^{n} \|P_{cert}(i) - P_{corr}(i)\|_2 \tag{30}$$

where, $\text{Res}_{meas}{}^{M}$ and $\text{Res}_{corr}{}^{M}$ represent the mean residual of the measured points and the mean residual of the corrected points, respectively, $P_{cert}(i)$ represents the position of the i_{th} point of the certified positions, $P_{corr}(i)$ represents the position of the i_{th} point of the corrected positions.

Table 4. Mean residuals and squared residuals of the measured and corrected points to the certified points.

No. Simulation	Mean Residuals [μm]		Squared Residuals [μm²]	
	Measured	**Corrected**	**Measured**	**Corrected**
1	17.94	8.11	12155.65	3287.68
2	13.13	8.45	6961.49	3269.41
3	8.56	8.13	3517.33	2986.80
4	5.81	5.23	1823.17	1206.33
5	6.42	5.96	1879.01	1288.29
6	9.65	5.60	3684.85	1587.99
7	13.93	6.40	7240.69	1903.75
8	18.53	8.20	12546.53	3101.98
9	23.31	9.66	19602.37	4179.53
10	28.18	11.06	28408.21	5389.67

The two squared residuals include the squared residual of the measured points to the certified points and the squared residual of the corrected points to the certified points. The squared residual is calculated by:

$$\text{Res}_{meas}{}^{S} = \sum_{i=1}^{n} (\|P_{cert}(i) - P_{meas}(i)\|_2)^2 \tag{31}$$

$$\text{Res}_{corr}{}^S = \sum_{i=1}^{n} \left(\|P_{cert}(i) - P_{corr}(i)\|_2 \right)^2 \tag{32}$$

Figures 18 and 19 plot the mean residuals and squared residuals of the 10 simulations. It can be observed that both the mean residual curve and the squared residual curve of the measured points are concave in the middle and convex at the two sides. The two residuals of the corrected points are much smoother. The larger the original residuals, the more the correction can be realized.

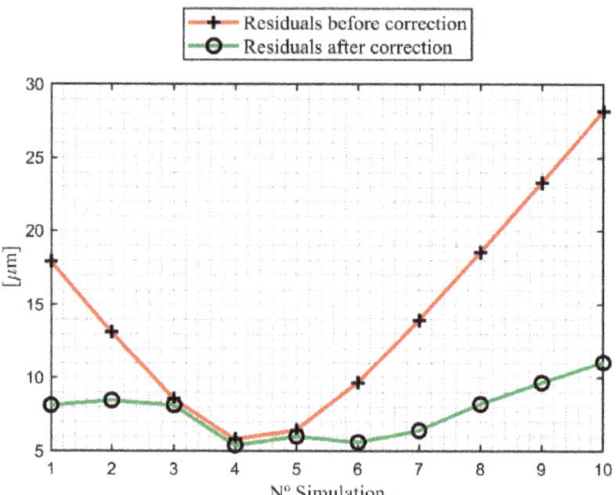

Figure 18. Comparison of the mean residuals before correction and after correction.

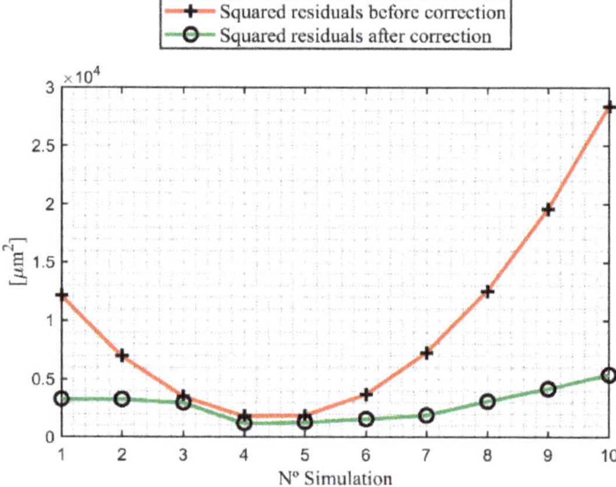

Figure 19. Comparison of the squared residuals before correction and after correction.

The results of the Simulation No. 10 are shown. Figure 20 compares the positions of the certified, measured, and corrected points of Simulation No. 10. Figures 21 and 22 presents contours of the mean residual and the squared residual between certified and measured points of Simulation No. 10. Figure 23 shows the error vectors between the measured and corrected points of Simulation No. 10.

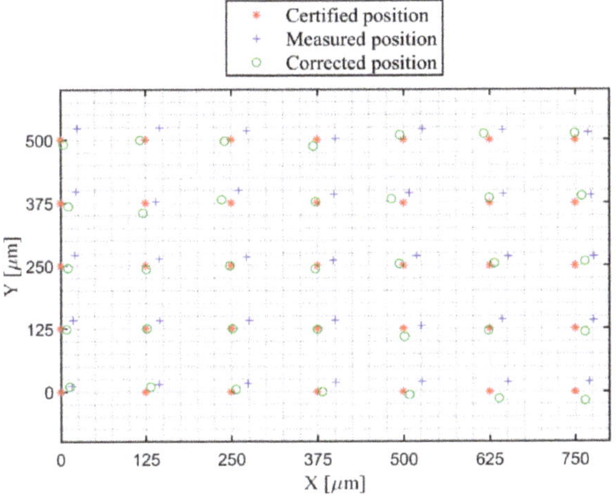

Figure 20. Comparison of the positions of the certified, measured, and corrected points of Simulation No. 10.

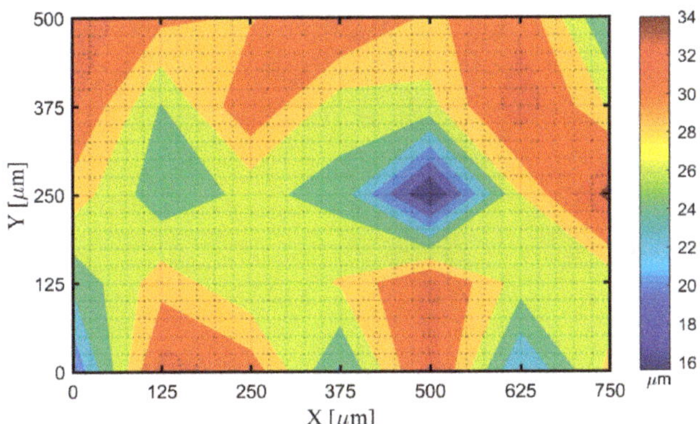

Figure 21. Contour of the mean residual between certified and measured points of Simulation No. 10.

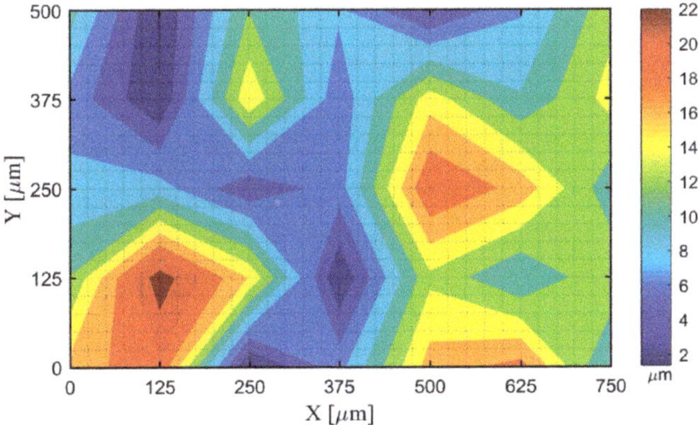

Figure 22. Contour of the mean residual between certified and corrected points of Simulation No. 10.

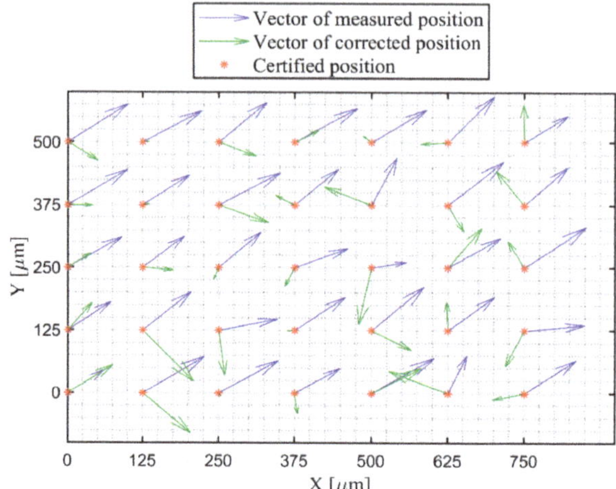

Figure 23. Comparison of the error vectors between the measured and corrected points of Simulation No. 10.

As the results above show, all the measured values are improved after geometric error correction. It can be concluded that our developed algorithm for the determination of correction coefficients is valid.

6. Conclusions

The work presented herein is Part I of the whole work, describing our methods for characterization and correction of lateral stage geometric errors. With implementation of the 21 parametric errors based on the assumption of the rigid body kinematics for calculating the machine volumetric errors of a three-axis machine, a mathematical model for correcting the lateral stage deviations of the CM was developed.

According to the measurement principle of CM, a diagram of the axis chain vector was drawn. Equation (1) was developed according to the kinematic path of the chain vector. After the rearrangement of Equation (1) and its simplification by dropping $\vec{Z}$ and $\vec{T}$, this equation was further simplified due to the study specifications and characteristics of the CM. The function errors were approximated by Legendre polynomials. Finally, Equations (4) and (5) for corrections of kinematic geometric errors were obtained. The methodology for calculation and application of the corrections coefficients defined in Equations (4) and (5) are introduced afterwards. Experiments on a dot grid target standard artefact will be measured and studied in Part II, published in the same journal.

The algorithms that will be implemented for the measurement data processing are presented and validated. These algorithms apply a lot of methods, including form removal, K-means clustering, linear and non-linear Least Squares. Their mathematical models were also introduced. The results of the validation based on synthetic data imply that our mathematical models and algorithms are reliable.

Author Contributions: Conceptualization and Methodology, C.W. and E.G.; Algorithms, Experiments, Data, Manuscript writing, C.W.; Resources, Funding, Supervision, E.G. and Y.Y.; Review and Editing, C.W., E.G. and Y.Y.

Funding: This work is funded by the Spanish State Programme of Promotion of Scientific Research and Technique of Excellence, State Sub-programme of Generation of Knowledge. Project DPI2016-78476-P "Desarrollo Colaborativo de Patrones de Software y Estudios de Trazabilidad e Intercomparación en la Caracterización Metrológica de Superficies", belonging to the 2016 call for R & D Projects. The authors acknowledge the Chinese Scholarship Council (CSC) for funding the first author's doctoral study.

Conflicts of Interest: The authors declare no conflict of interest.

Abbreviations

CM	confocal microscope
CCD	charge-coupled device
CMM	coordinate measuring machine

References

1. Zhang, G.; Ouyang, R.; Lu, B.; Hocken, R.; Veale, R.; Donmez, A. A Displacement Method for Machine Geometry Calibration. *CIRP Ann.* **1988**, *37*, 515–518. [CrossRef]
2. Hermann, G. Geometric Error Correction in Coordinate Measurement. *Acta Polytech. Hung.* **2007**, *4*, 47–62.
3. Krolczyk, G.M.; Maruda, R.W.; Nieslony, P.; Wieczorowski, M. Surface morphology analysis of Duplex Stainless Steel (DSS) in Clean Production using the Power Spectral Density. *Measurement* **2016**, *94*, 464–470. [CrossRef]
4. Krolczyk, G.M.; Krolczyk, J.B.; Maruda, R.W.; Legutko, S.; Tomaszewski, M. Metrological changes in surface morphology of high-strength steels in manufacturing processes. *Measurement* **2016**, *88*, 176–185. [CrossRef]
5. Wang, C.; D'Amato, R.; Gómez, E. Confidence Distance Matrix for outlier identification: A new method to improve the characterizations of surfaces measured by confocal microscopy. *Measurement* **2019**, *137*, 484–500. [CrossRef]
6. Harding, K. (Ed.) *Handbook of Optical Dimensional Metrology*; CRC Press, Taylor & Francis Group: Boca Raton, FL, USA, 2013.
7. Corle, T.R.; Kino, G.S. *Confocal Scanning Optical Microscopy and Related Imaging Systems*; Academic Press: Cambridge, MA, USA, 1996.
8. Thomas, D.; Sugimoto, A. Parametric surface representation with bump image for dense 3d modeling using an rbg-d camera. *Int. J. Comput. Vis.* **2017**, *123*, 206–225. [CrossRef]
9. Bailey, D.G. A new approach to lens distortion correction. *Proc. Image Vis. Comput. N. Z.* **2002**, *2002*, 59–64.
10. Wang, J.; Shi, F.; Zhang, J.; Liu, Y. A new calibration model of camera lens distortion. *Pattern Recognit.* **2007**, *41*, 607–615. [CrossRef]
11. Besseling, T.H.; Jose, J.; Van Blaaderen, A. Methods to calibrate and scale axial distances in confocal microscopy as as function of refractive index. *J. Microsc.* **2015**, *257*, 142–150. [CrossRef]
12. Cole, R.W.; Jinadasa, T.; Brown, C.M. Measuring and interpreting point spread functions to determine confocal microscope resolution and ensure quality control. *Nat. Protoc.* **2011**, *6*, 1929–1941. [CrossRef]
13. Wang, C. Current issues on 3D volumetric positioning accuracy: Measurement, Compensation and Definition. *Proc. SPIE* **2008**, 7128. [CrossRef]
14. Barakat, N.A.; Elbestawi, M.A.; Spence, A.D. Kinematic and geometric error compensation of a coordinate measuring machine. *Int. J. Mach. Tools Manuf.* **2000**, *40*, 833–850. [CrossRef]
15. Zhang, G.; Veale, R.; Charlton, T.; Borchardt, B.; Hocken, R. Error Compensation of Coordinate Measuring Machines. *CLRP Ann.* **1985**, *34*, 445–448. [CrossRef]
16. Schwenke, H.; Knapp, W.; Haitjema, H.; Weckenmann, A.; Schmitte, R.; Delbressine, F. Geometric error measurement and compensation of machines—An update. *CIRP Ann.* **2008**, *57*, 660–675. [CrossRef]
17. Umetsu, K.; Furutnani, R.; Osawa, S.; Takatsuji, T.; Kurosawa, T. Geometric calibration of a coordinate measuring machine using a laser tracking system. *Meas. Sci. Technol.* **2005**, *16*, 2466–2472. [CrossRef]
18. Semwogerere, D.; Weeks, E.R. *Confocal Microscopy*; Taylor & Francis: Abingdon, UK, 2005.
19. Jensen, K.E.; Weitz, D.A.; Spaepen, F. Note: A three-dimensional calibration device for the confocal microscope. *Rev. Sci. Instrum.* **2013**, *84*, 016108. [CrossRef] [PubMed]
20. Senin, N.; Leach, R. Information-rich surface metrology. *Procedia CIRP* **2018**, *75*, 19–26. [CrossRef]
21. Claxton, N.S.; Fellers, T.J.; Davidson, M.W. *Laser Scanning Confocal Microscopy*; Technical Report; Department of Optical Microscopy and Digital Imaging, Florida State University: Tallahassee, FL, USA, 2006.
22. ISO 25178-600. *Geometrical Product Specifications (GPS)—Surface Texture: Areal-Part 600: Metrological Characteristics for Areal-Topography Measuring Methods*; International Organization for Standardization: Geneva, Switzerland, 2019.
23. Leach, R.K.; Giusca, C.L.; Haitjema, H.; Evans, C.; Jiang, X. Calibration and verification of areal surface texture measuring instruments. *CIRP Ann.* **2015**, *64*, 797–813. [CrossRef]

24. Alburayt, A.; Syam, W.P.; Leach, R. Lateral scale calibration for focus variation microscopy. *Meas. Sci. Technol.* **2018**, *29*, 065012. [CrossRef]
25. Leach, R. *Optical Measurement of Surface Topography*; Springer: Berlin, Germany, 2011.
26. Nouira, H.; El-Hayek, N.; Yuan, X.; Anwer, N.; Salgado, J. Metrological characterization of optical confocal sensors measurements (20 and 350 travel ranges). *J. Phys. Conf. Ser.* **2015**, *483*, 012015. [CrossRef]
27. Tong, K.; Lehtihet, E.A.; Joshi, S. Parametric error modeling and software error compensation for rapid prototyping. *Rapid Prototyp. J.* **2003**, *9*, 301–313. [CrossRef]
28. Ekinci, T.O.; Mayer, J.R.R.; Cloutier, G.M. Investigation of accuracy of aerostatic guideways. *Int. J. Mach. Tools Manuf.* **2009**, *49*, 478–487. [CrossRef]
29. Ekinci, T.O.; Mayer, J.R.R. Relationships between straightness and angular kinematic errors in machines. *Int. J. Mach. Tools Manuf.* **2007**, *47*, 1997–2004. [CrossRef]
30. Yang, J.; Ren, Y.; Wang, C.; Liotto, G. Theoretical derivations of 4 body diagonal displacement errors in 4 machine configurations. In Proceedings of the LAMDAMAP Conference, Cransfield, UK, 27–30 June 2005.
31. Aguado, S.; Samper, D.; Santolaria, J.; Aguilar, J.J. Towards an effective identification strategy in volumetric error compensation of machine tools. *Meas. Sci. Technol.* **2012**, *23*, 065003. [CrossRef]
32. Wang, C.; Caja, J.; Gomez, E. Comparison of methods for outlier identification in surface characterization. *Measurement* **2018**, *117*, 312–325. [CrossRef]
33. Hartigan, J.A.; Wong, M.A. A K-means Clustering Algorithm. *J. R. Stat. Soc.* **1979**, *28*, 100–108.
34. Jain, K. Data clustering: 50 years beyond K-means. *Pattern Recognit. Lett.* **2010**, *31*, 651–666. [CrossRef]
35. Lloyd, S.P. Least Squares Quantization in PCM. *IEEE Trans. Inf. Theory* **1982**, *28*, 129–137. [CrossRef]
36. Ghodsi, A. *Dimensionality Reduction—A Short Tutorial*; Technical Report; University of Waterloo: Waterloo, ON, Canada, 2006.
37. Sorzano, S.; Vargas, J.; Montano, A.P. A survey of dimensionality reduction techniques, Technical Report. *arXiv* **2014**, arXiv:1403.2877.
38. Miller, S.J. *The Method of Least Squares*; Brown University: Providence, RI, USA, 2006.
39. Weisstein, E.W. Least Squares Fitting, From MathWorld–A Wolfram Web Resource. Available online: http://mathworld.wolfram.com/LeastSquaresFitting.html (accessed on 1 October 2018).
40. Dennis, J.E.; Gay, D.M.; Welsch, R.E. An adaptive nonlinear least-squares algorithm. *ACM Trans. Math. Softw.* **1981**, *7*, 348–368. [CrossRef]
41. Marquardt, D.W. An algorithm for Least-Squares estimation of nonlinear parameters. *J. Soc. Ind. Appl. Math.* **1963**, *11*, 431–441. [CrossRef]

Article

Characterization and Correction of the Geometric Errors Using a Confocal Microscope for Extended Topography Measurement, Part II: Experimental Study and Uncertainty Evaluation

Chen Wang [1], Emilio Gómez [1] and Yingjie Yu [2,*]

[1] Department of Mechanical Engineering, Chemical and Industrial Design, ETS of Engineering and Industrial Design, Technical University of Madrid, 28012 Madrid, Spain; chen.wang@alumnos.upm.es (C.W.); emilio.gomez@upm.es (E.G.)

[2] Department of precision mechanical engineering, Shanghai University. No.333, Nanchen Rd., Shanghai 200444, China

* Correspondence: yingjieyu@staff.shu.edu.cn

Received: 15 September 2019; Accepted: 15 October 2019; Published: 24 October 2019

Abstract: This paper presents the experimental implementations of the mathematical models and algorithms developed in Part I. Two experiments are carried out. The first experiment determines the correction coefficients of the mathematical model. The dot grid target is measured, and the measurement data are processed by our developed and validated algorithms introduced in Part I. The values of the coefficients are indicated and analyzed. Uncertainties are evaluated using the Monte Carlo method. The second experiment measures a different area of the dot grid target. The measurement results are corrected according to the coefficients determined in the first experiment. The mean residual between the measured points and their corresponding certified values reduced 29.6% after the correction. The sum of squared errors reduced 47.7%. The methods and the algorithms for raw data processing, such as data partition, fittings of dots' centers, K-means clustering, etc., are the same for the two experiments. The experimental results demonstrate that our method for the correction of the errors produced by the movement of the lateral stage of a confocal microscope is meaningful and practicable.

Keywords: geometric errors; rigid body kinematics; lateral stage errors; imaging confocal microscope; MCM uncertainty evaluation; dot grid target

---

1. Introduction

The increasing demands for manufacturing accuracies and quality control due to the rapid development of nanotechnology, ultraprecision machining, micro-, and nanofabrications, etc. [1,2] and the requirements for precision in surface finishing in different technologies such as additive manufacturing [3], mechanical parts with structured surfaces [4], etc., require the use of increasingly sophisticated measurement systems and measurement traceability from a metrological point of view.

Calibration provides a wide range of information about microscope performances. The ever-increasing demand for improved surface quality and tighter geometric tolerances has led to augmentations in the investigations of manufacturing technologies [5]. Measurements using optical microscopes are often affected by common path noise, disturbance in light source and ambient lighting, etc., which cause measurement defects and outliers [6,7], as well as attract investigations on noisy data processing [8]. The need for standardization is becoming ever greater as the range of capturing three-dimensional (3D) information of microscope techniques continues to increase [5,9]. For optical confocal microscopes, the Z-calibrations at nm levels are typically good, while the X- and

Y-accuracies are often left, without further notice than resolution limits of the optics [10,11]. Among the investigations of lateral calibrations, many studies focus on the optical system [12]. For example, H. Ni et al. proposed a new method to achieve structured detection using a spatial light modulator, which modulates the Airy disk amplitude distribution according to the detection function in the collection arm [13] and B. Wang et al. presented confocal microscopy with structured detection in a coherent imaging process to achieve a higher resolution with a comparably large pinhole [14], however, the systematic geometric errors which adversely affect the relative position and orientation between measuring probes and measurands are usually neglected [15,16]. B. Daemi et al. designed a comprehensive verification test by using a high precision metrology method based on subpixel resolution image analysis [17]. The calibration of confocal microscopes usually relies on traceable standard artefacts, which are commonly made up of regular patterns [10].

This paper describes the experimental studies based on the kinematic modeling and algorithms for the correction of the geometric errors developed in Part I [18]. Sections 2 and 3 introduce the methodologies for experiments and uncertainty evaluation individually. Section 4 presents the experiment on a dot grid target for correction coefficients determination and their corresponding uncertainty evaluation. Section 5 implements an experiment and corrects the measured data with determined coefficients, comparing the residuals with respect to certified values before and after corrections. Section 6 presents the conclusions. Following Section 6, acknowledgements and references are included.

2. Methodology for the Experimental Study

The purposes of this experimental study were to, first, determine the error correction coefficients, i.e., defined in the kinematic geometric error correction model developed in Part I [18] and, secondly, apply the determined parameters of coefficients and the correction mathematical model for new measurement data calibration. The dot grid target standard artefact was implemented as the measurand of the experiments. By comparing the residuals of measured points and corrected points with respect to the certified values, the practicality and significance of our developed models and algorithms for lateral stage error calibration were observed.

Two experiments were carried out with our imaging confocal microscope, which is a Leica Confocal Dual Core 3D Measuring Microscope (Leica DCM-3D), at the "Laboratorio de Investigación de Materiales de Interés Tecnológico" (LIMIT) of the Technical University of Madrid. The first experiment measured the dot grid target standard for the determination of the correction coefficients. The second experiment measured another area of the dot grid standard, processing the measurement data with the same developed and validated algorithms implemented in the first experiment. The purpose of this experiment was to observe whether the corrected data improved as compared with the raw measurement data. Because the second experiment used the same measurement parameters, data processing algorithms, and procedures, and measured a different area of the same dot grid standard, this comparison is important as other factors, which might influence the results, could be excluded majorly, such as the uncertainties or inaccuracies generated by algorithms of cylinder separation, center fitting, and movement scope of the lateral stage, etc.

3. Methodology and Procedures for Uncertainty Estimation

A statement of measurement is complete only if it provides an estimate of the quantity concerned, as well as a quantitative evaluation of the estimate's reliability, i.e., the associated uncertainty [18]. Accompany measurement results by quantitative statements about their accuracy is very important particularly when the result are part of a measurement chain tracking back to national standards or when decisions about product specifications are taken [19].

The document issued by BIPM, *Guide to the Expression of Uncertainty in Measurement* (GUM) [20] provides a method and procedure for the evaluation and expression of measurement uncertainties [21]. This method is termed the GUM uncertainty framework in supplement 1 and supplement 2 (GUM-S1 and GUM-S2) [22,23] and other bibliographies [19,24]. The GUM uncertainty framework has two main limitations [2,25]. The first limitation is the lack of generality of the procedure to obtain an interval to contain the values of the measurand with a stipulated coverage probability [25]. In the GUM uncertainty framework, the way a coverage interval is constructed to contain values of the measurand with a stipulated coverage probability is approximate [22]. The second limitation is that insufficient guidance is given for the multivariate case in which there is more than one measurand, namely, more than one output quantity [22,25]. In order to address these limitations, Working Group 1 of the Joint Committee for Guides in Metrology (JCGM) has produced two specific guidance documents, namely GUM-S1 and GUM-S2 [23,24], on the Monte Carlo method (MCM) for uncertainty evaluation and extensions to any number of measurands (output quantity), respectively [22].

The MCM provides a general approach to obtain a numerical representation **G** of the distribution function $G_Y(\eta)$ for **Y**. The heart of the approach is making repeated draws from the probability density functions (PDFs) for the input variables X_i (or joint PDF for **X**) and the evaluation of the output quantity. Assignment of the PDFs for the input variables is dependent on each experiment. The same case applies to the evaluation of the output quantity. The distribution function $G_Y(\eta)$ encodes all information known about the output quantity **Y**. Properties of **Y** can be approximated using $G_Y(\eta)$. The quality of **G** depends on the number of draws made. The symbol **y** represent the output measurement results. It is determined by the input measurement results x_i:

$$\mathbf{y} = f(\mathbf{x}_1, \ldots, \mathbf{x}_N) \tag{1}$$

The relationship between the PDF of output measurement results and input measurement results is:

$$g_Y(\eta) = \int g_{X_1,\ldots,X_N}(\xi_1, \ldots, \xi_N) \times \delta[\eta - f(\xi_1, \ldots, \xi_N)]d\xi_1 \ldots d\xi_N \tag{2}$$

where η denotes possible values that can be distributed to **Y**, $\delta[\cdots]$ denotes the Dirac delta function.

Figure 1 provides an illustration of the propagation of distributions for input and output quantities. The expectation of the output quantities can be obtained by its PDF $g_Y(\eta)$ as:

$$E(Y_i) = \int_{-\infty}^{\infty} \eta_i g_{Y_i}(\eta_i)d\eta_i \tag{3}$$

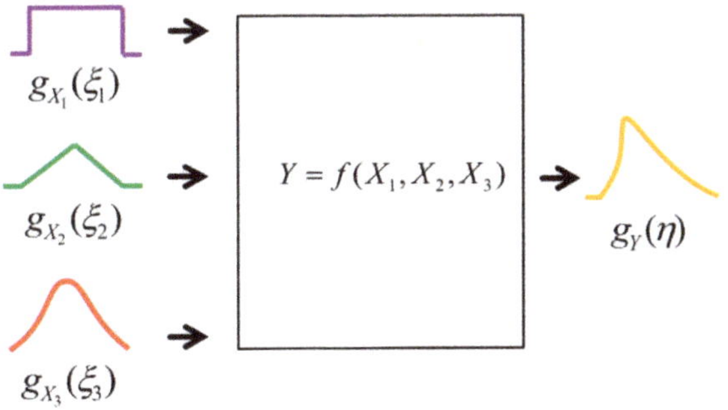

Figure 1. Illustration of the propagation of distributions for input quantities and the obtained output quantities [19].

The variance of the output quantities can be obtained by its PDF $g_Y(\eta)$ as:

$$V(Y_i) = \int_{-\infty}^{\infty} [\eta_i - E(Y_i)]^2 g_{Y_i}(\eta_i)d\eta_i \tag{4}$$

The covariance of the output quantities can be obtained by its PDF $\mathbf{g_Y(\eta)}$ as:

$$Cov(Y_i, Y_j) = Cov(Y_j, Y_i) = \int_{-\infty}^{\infty} \int_{-\infty}^{\infty} [\eta_i - E(Y_i)]\left[\eta_j - E(Y_j)\right] g_{Y_i,Y_j}(\eta_i, \eta_j)d\eta_i d\eta_j \tag{5}$$

where $g_{Y_i,Y_j}(\eta_i, \eta_j)$ is the joint PDF for the two random variables Y_i, Y_j.

The correlation of the output quantities can be obtained by its PDF $\mathbf{g_Y(\eta)}$ as:

$$Corr(Y_i, Y_j) = Corr(Y_j, Y_i) = \frac{Cov(Y_i, Y_j)}{\sqrt{V(Y_i)V(Y_j)}} \tag{6}$$

4. Experiment for Determination and Uncertainty Evaluation of the Error Correction Coefficients

The first experiment is the measurement of the standard artefact of the dot grid target. The purpose of this experiment is to obtain the parameters of the coefficients of the kinematic rigid body errors. Uncertainties of the obtained correction coefficients are also evaluated.

4.1. Determination of Error Coefficients

The standard artefact of the dot grid target is introduced in Part I of [18]. This artefact was measured in an environment under a controlled temperature of 20 ± 1 °C, using the Leica DCM-3D Confocal Microscopy at the Metrology Laboratory of the Technical University of Madrid. A magnification objective of 50x was used, with a numerical aperture of 0.9. The acquisition parameter of measurement area was defined as the topography stitching measurement, with a 4 × 4 extended topographies, covering an area of 0.828 × 0.621 mm². The parameter of overlapping area was 25% and the correlation takes XYZ option. The level of resolution was 1, and the measured extended topographies contained 2496 × 1872 pixels.

After the measurement, the confocal system generated a file with suffix name ".dat", containing three vectors, which are values of the X, Y, and Z coordinates. Data of this file was imported and analyzed by our developed algorithms. The raw measurement data is shown in Figure 2, which is an inclined surface with some outliers. This surface was aligned to be parallel with the X and Y coordinate plane using our developed surface rotation methods introduced in Part I [18]. The aligned surface is shown in Figure 3. The distribution of the values of the X, Y, and Z coordinates of the surface after rotation is shown in Figure 4.

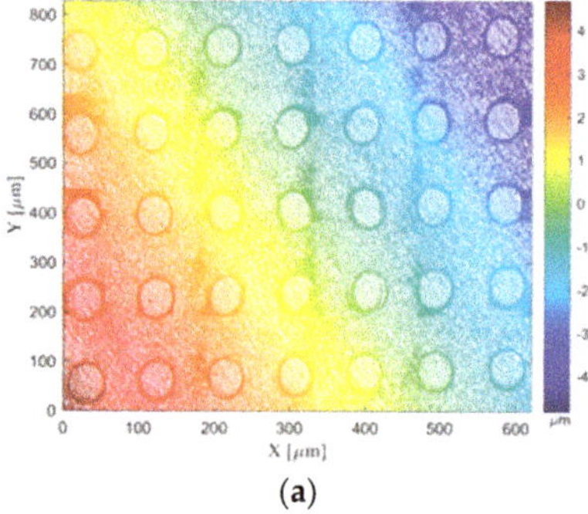

(a)

Figure 2. *Cont.*

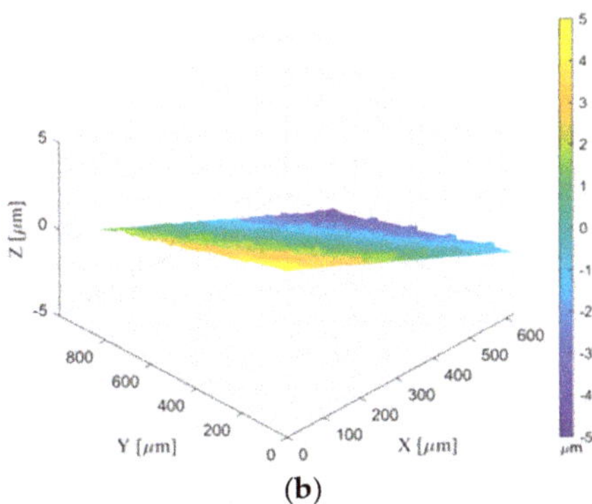

Figure 2. Surface reconstruction by the raw measurement data: (**a**) two-dimensional (2D) contour and (**b**) three-dimensional (3D) surf.

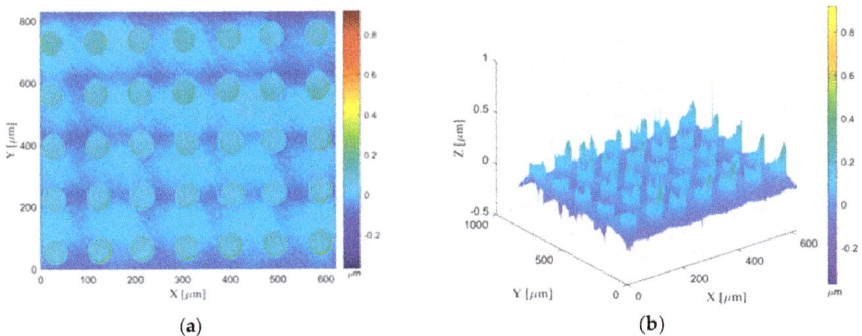

Figure 3. Surface reconstruction after rotation of the raw measured surface: (**a**) 2D contour and (**b**) 3D surf.

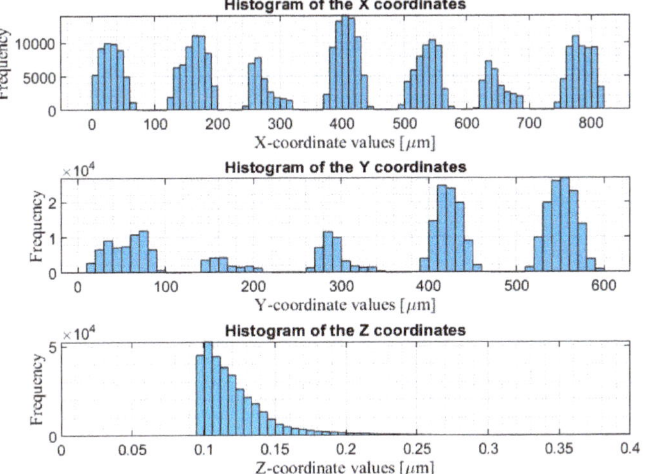

Figure 4. Histograms of the X, Y, and Z coordinates of the aligned surface.

After rotation, the data were separated into surface plane and cylinders. The surface reconstruction of the data of cylinders is shown in Figure 5. It is obvious that this data has many outliers. Those outliers are detected and deleted by the method introduced in our previous work [8]. The distribution of the measurement values of the X and Y coordinates, as well as the threshold for outlier detections are shown in Figures 6 and 7 individually. The surface reconstruction by the data of cylinders with outliers removed is shown in Figure 8.

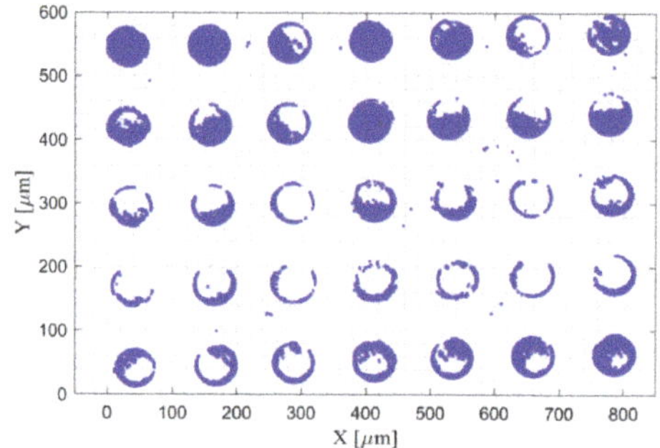

Figure 5. Surface reconstruction of the data of cylinders.

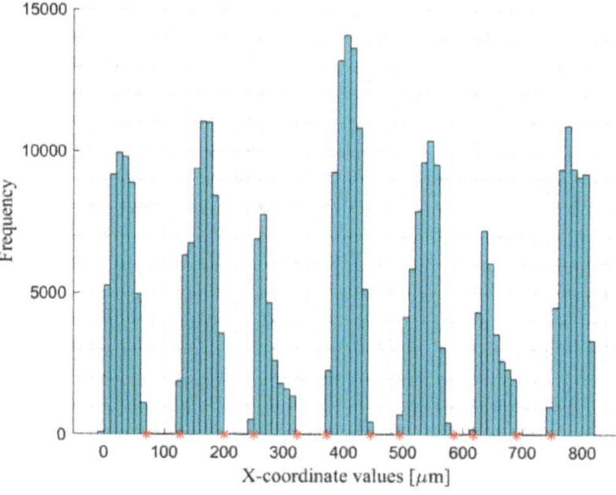

Figure 6. Distribution of the X coordinate values and the thresholds for outlier detection.

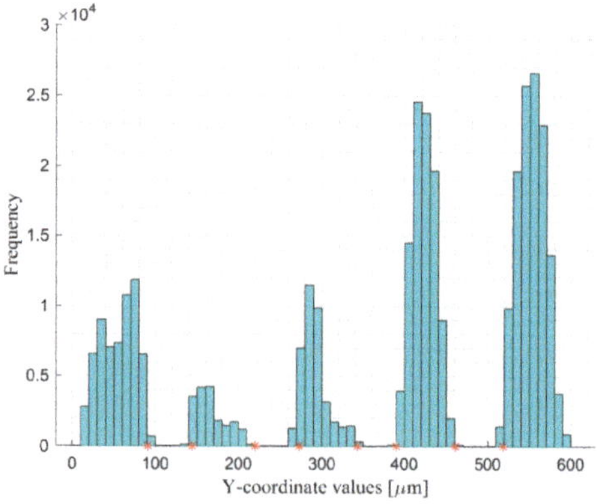

Figure 7. Distribution of the Y coordinates values and the thresholds for outlier detection.

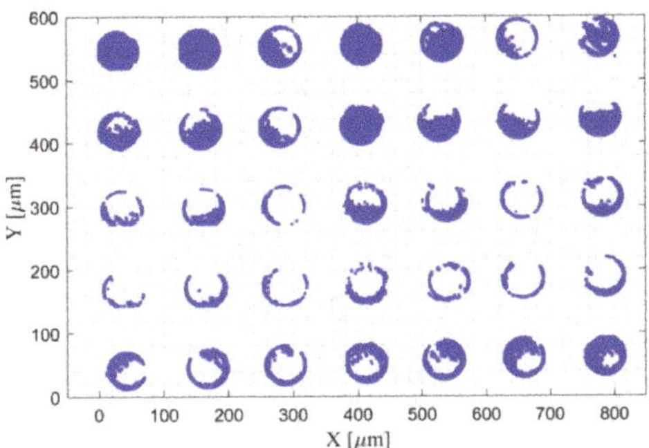

Figure 8. Surface reconstruction of the data of cylinders with outliers removed.

The separated cylinders are shown in Figure 9. They are denoted by different colors and numbers.

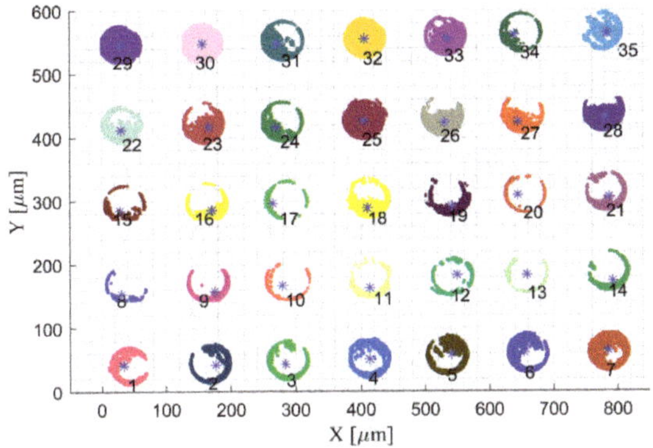

Figure 9. Separation of the cylinders.

Those individual point clouds of cylinders are processed by our introduced algorithms. Their centers are fitted and shown in Figure 10, in comparison to their corresponding certified positions.

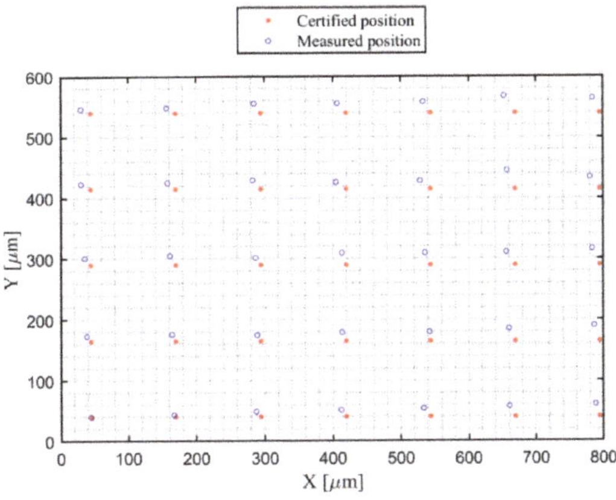

Figure 10. Plot of the cylinders' centers and their corresponding certified positions.

The measured cylinder centers, shown in Figure 10, are all distorted in one direction. This might be caused by the location of the measurand, as it is impossible to locate the measurand parallel with the x-axis. Therefore, those measured centers are first aligned to be parallel with the x-axis. Then, they are calibrated to the certified points using our developed mathematical models and algorithms. From this, the results of the coefficients defined in the mathematical model, i.e., Equations (4) and (5) in Part I [18], can be obtained. The results are indicated in Table 1. Moreover, rotation of the calibrated values of the cylinder centers to align them with the measured ones are also carried out for the validation of the rotation capability.

Table 1. Parameters of the coefficients of the mathematical models for lateral geometric error correction.

Parameter	Value	Parameter	Value	Parameter	Value
a_1	7.38×10^{-3}	c_1	4.36×10^{-6}	e_1	6.00×10^{-6}
a_2	-1.46×10^{-6}	c_2	1.05×10^{-8}	e_2	-6.24×10^{-9}
a_3	3.19×10^{-11}	c_3	-1.20×10^{-12}	e_3	9.69×10^{-14}
b_1	1.05×10^{-3}	d_1	-7.22×10^{-3}	f_1	-2.01×10^{-2}
b_2	-1.11×10^{-5}	d_2	4.00×10^{-6}	f_2	1.25×10^{-5}
b_3	-3.15×10^{-11}	d_3	-1.01×10^{-10}	f_3	-3.10×10^{-9}

With those obtained parameters of the coefficients, the corrected points are calculated according to Equations (4) and (5) in Part I [18]. Figure 11 shows the corresponding positions of certified, measured, and corrected points. The results of the mean errors, maximum errors, sum of squared errors, and standard deviations of the errors are indicated in Table 2.

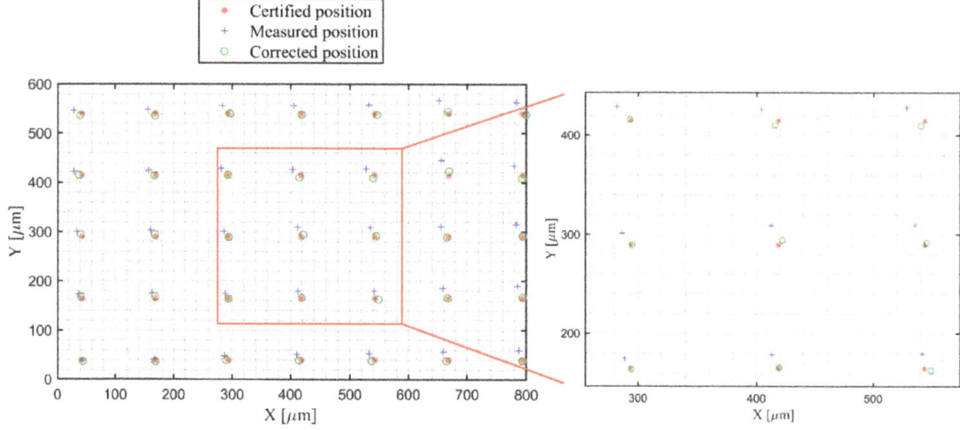

Figure 11. Comparison of the positions of certified, measured, and corrected points.

Table 2. Errors with respect to the certified positions before correction and after correction.

Data Types	Mean Error [μm]	Maximum Error [μm]	Sum of Squared Errors [μm²]	Standard Deviations of the Errors [μm]
Measured points	18.3	33.1	1.4×10^4	7.0
Corrected points	3.8	8.9	628.1	1.9

The Euclidean residuals of each point are plotted by contours, as shown in Figures 12 and 13, which is the contour of the Euclidean residuals of the measured and corrected points with respect to the certified values.

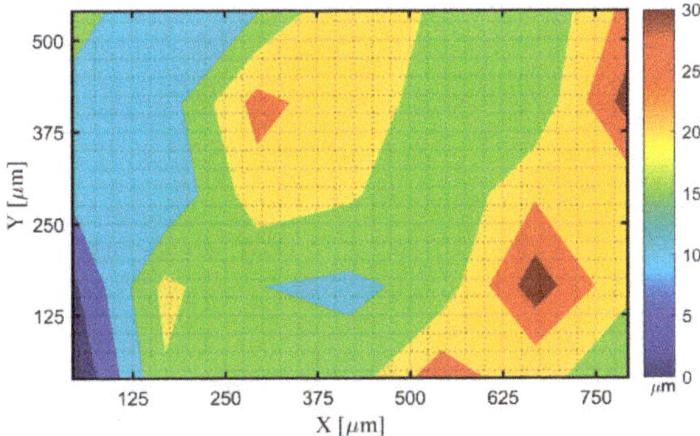

Figure 12. Contour of the Euclidean residuals between each certified and measured points.

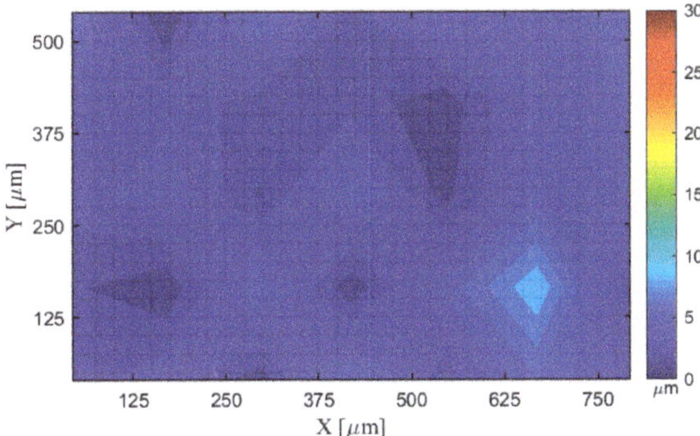

Figure 13. Contour of the Euclidean residuals between each certified and corrected points.

It can be concluded that the determined coefficients for kinematic geometric error correction works very well in this measurement.

4.2. Uncertainty Evaluation

Uncertainty evaluation of the geometric error coefficients is based on the algorithms introduced in Section 3. The heart of the approach is making repeated draws from the PDFs for the input variables X_i (or joint PDF for $\mathbf{X}$) and the evaluation of the output quantity. Here, we define the determinations of the number of repeated draws, namely, the number of simulation trials, the PDFs for the input variables, and the evaluation of the output quantity.

According to GUM-S2 [23] the main stages of uncertainty evaluation constitute formulation, propagation, and summarizing describes as follows:

1. The first stage of formulation includes:

    (a) Define the output quantity, namely, the geometric error correction coefficients $\mathbf{Cc} = (a_1, \ldots g_3)$;

    (b) Determine the input quantity upon which $\mathbf{Cc}$ depends, namely, the measurement results $(\mathbf{x}, \mathbf{y})$ and their corresponded certified values $(\mathbf{P_x}, \mathbf{P_y})$;

    (c) Develop a measurement function f or measurement model relating the input and output quantities, namely, Equations (4) and (5);

    (d) On the basis of available knowledge, assign PDFs to the components of the input quantities. As indicated by Table 2 in Part I [18], the certified values $(\mathbf{P_x}, \mathbf{P_y})$ follow a rectangular distribution $U(-1, 1)$ µm. As there is no more information about the sources of uncertainties for $(\mathbf{P_x}, \mathbf{P_y})$ or information for the measured values of X and Y coordinates $(\mathbf{x}, \mathbf{y})$, here does not assign more uncertainties to the input quantities, for not introducing unnecessary uncertainties.

2. The second stage, propagation, includes: Propagate the PDFs for the components of input quantities through the model to obtain the (joint) PDF for the output quantity.

3. The final step, summarizing, includes: Use the PDF for the output quantity to obtain the expectation of the output quantity, the uncertainty matrix, also named covariance matrix, associated with the expectation of the output quantity, and a coverage region containing the output quantity with a specified probability p(the coverage probability).

The simulation was repeated 1×10^4 times. Mean values, expanded uncertainties ($k = 2$), as well as lower and upper boundaries for a 95% coverage interval are listed in Table 3. Distributions for the output quantities are shown in Figure 14 and we observe that non-symmetric distributions and not assimilable to normal distributions are obtained.

Table 3. Parameters of the coefficients of the mathematical models for lateral geometric error correction.

Parameter	Mean Value	Expanded Uncertainty	95% Coverage Interval	
			Lower Boundary	Upper Boundary
a_1	7.50×10^{-3}	4.74×10^{-3}	5.07×10^{-3}	1.61×10^{-2}
a_2	-2.13×10^{-6}	5.41×10^{-6}	-1.28×10^{-5}	-6.12×10^{-8}
a_3	2.00×10^{-10}	7.14×10^{-10}	-1.35×10^{-10}	1.21×10^{-9}
b_1	-5.91×10^{-6}	8.85×10^{-3}	-1.62×10^{-2}	2.99×10^{-3}
b_2	-9.91×10^{-6}	9.19×10^{-6}	-1.50×10^{-5}	2.08×10^{-6}
b_3	-6.86×10^{-12}	2.36×10^{-9}	-8.94×10^{-10}	2.67×10^{-9}
c_1	3.41×10^{-6}	7.89×10^{-6}	-6.22×10^{-6}	7.41×10^{-6}
c_2	1.26×10^{-8}	1.86×10^{-8}	2.21×10^{-10}	3.86×10^{-8}
c_3	-8.16×10^{-13}	3.44×10^{-12}	-5.74×10^{-12}	1.56×10^{-13}
d_1	-6.57×10^{-3}	4.97×10^{-3}	-8.53×10^{-3}	3.52×10^{-3}
d_2	3.93×10^{-6}	1.94×10^{-6}	1.39×10^{-6}	5.37×10^{-6}
d_3	-5.88×10^{-11}	3.78×10^{-10}	-4.59×10^{-10}	1.53×10^{-10}
e_1	5.89×10^{-6}	2.95×10^{-6}	2.00×10^{-6}	8.04×10^{-6}
e_2	-6.57×10^{-9}	6.71×10^{-9}	-1.44×10^{-8}	-4.88×10^{-11}
e_3	-2.34×10^{-15}	3.80×10^{-13}	-2.36×10^{-13}	3.30×10^{-13}
f_1	-2.21×10^{-2}	1.42×10^{-2}	-5.15×10^{-2}	-1.43×10^{-2}
f_2	1.40×10^{-5}	1.73×10^{-5}	6.71×10^{-6}	4.98×10^{-5}
f_3	-2.94×10^{-9}	2.92×10^{-9}	-5.37×10^{-9}	-1.14×10^{-11}

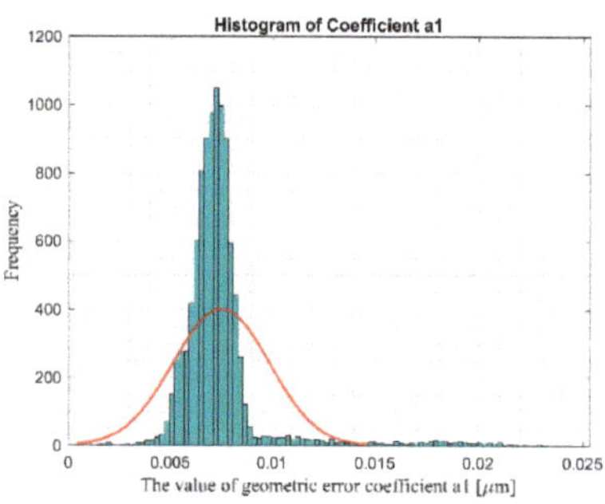

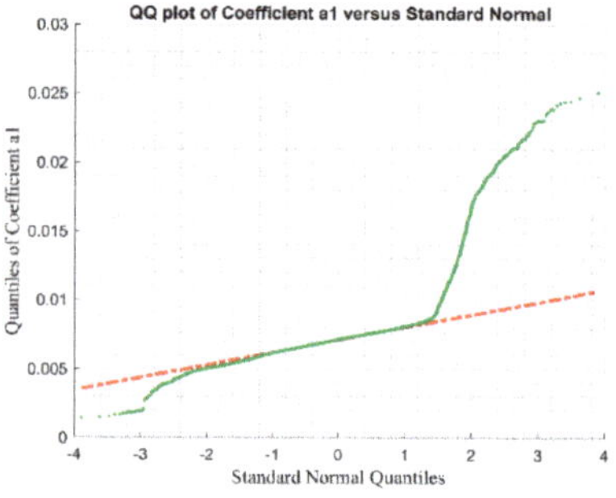

Figure 14. *Cont.*

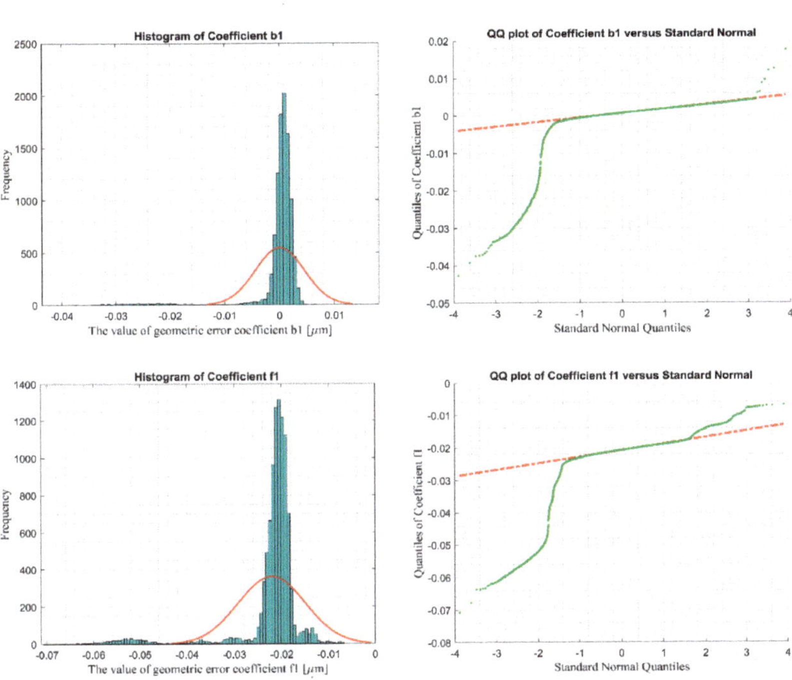

Figure 14. Distributions of the uncertainties of some geometric error correction coefficients.

5. Experimental Study of the Applications of Determined Coefficients

This section aims to verify the applicability of the determined error correction coefficients and the residuals of measured and corrected points with respect to certified points are compared.

This experiment measures the dot grid target standard used in Section 4. The measured area of this standard is different from those in Section 4. All the environmental and operational parameters are the same as those in Section 4. The measurement data are processed using the same algorithms and procedures, until the fitted cylinder centers are approximated parallel with the x-axis. Then the fitted centers are corrected by the error correction coefficients determined. By comparing the mean residuals, the sum of all squared residuals, and the standard deviation of residuals of measured points and corrected points with respect to certified positions, the effectiveness of the calculated coefficients and our model can be observed.

Measurement data are processed using our developed and validated algorithms. The raw measurement data with form removed are shown in Figure 15, implementing the form removal method presented in the pair publication Part I [18] Section 4.1. The cylinders are separated from the base with our developed algorithms, with results demonstrated in Figure 16. Figure 16a shows the initial separated cylinders, Figure 16b shows the cylinders with outliers removed and with clusters classified, and Figure 17 shows the histograms of the three coordinates of the cylinders.

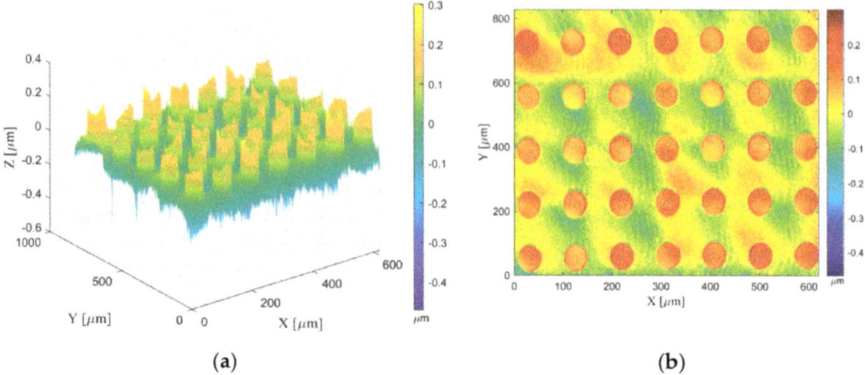

Figure 15. The measured surface with form removed: (**a**) 3D reconstruction and (**b**) 2D reconstruction.

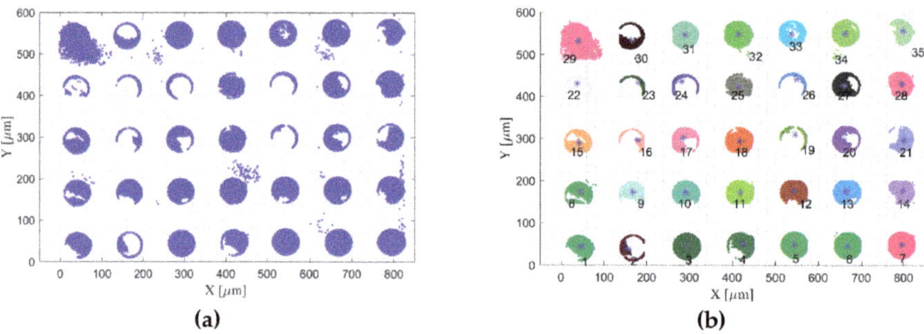

Figure 16. Separation of the cylinders: (**a**) The separated cylinders from the base and (**b**) separate cylinders into individual clusters.

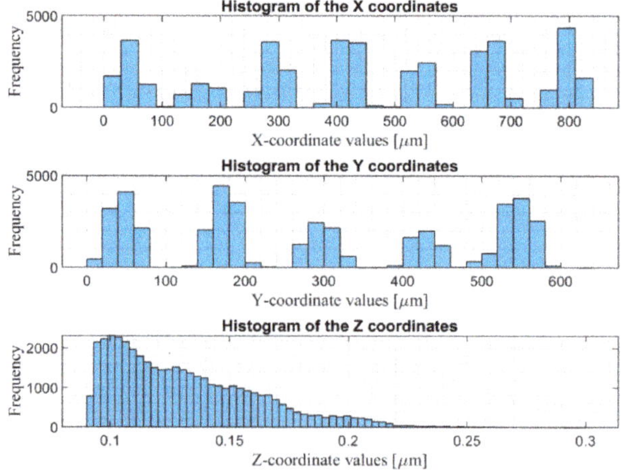

Figure 17. Histogram of the three coordinates of the separated cylinders.

The separated clusters are fitted for their centers. The coordinate values of the fitted cylinder centers are shown in Table 4. Figure 18 shows the fitted cylinder centers, as well as the centroids of each cluster data.

Table 4. The raw coordinate values of the fitted cylinder centers.

Cluster N°	X Coordinate [μm]	Y Coordinate [μm]	Cluster N°	X Coordinate [μm]	Y Coordinate [μm]
1	47.3	42.6	19	541.8	298.1
2	172.3	42.9	20	667.7	301.3
3	294.8	44.0	21	796.8	298.4
4	425.1	49.5	22	40.2	420.6
5	544.9	48.1	23	163.3	417.8
6	669.2	46.6	24	291.1	421.0
7	796.4	48.3	25	415.9	424.6
8	42.4	166.5	26	538.9	423.9
9	167.7	171.9	27	666.1	430.0
10	292.6	172.3	28	794.6	430.3
11	419.3	172.6	29	41.2	530.7
12	544.1	175.2	30	163.3	548.3
13	666.6	174.6	31	289.5	546.6
14	797.0	176.0	32	415.0	548.2
15	40.1	295.9	33	539.9	551.7
16	163.4	294.6	34	662.6	550.7
17	293.0	294.5	35	796.9	550.8
18	417.9	295.3			

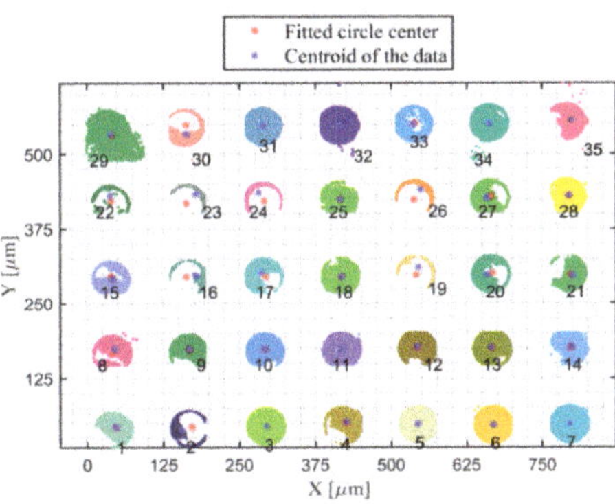

Figure 18. Comparison of the fitted circle center and the centroid of the data.

As shown in Figure 18, the cluster numbered 29 has too many outliers. Here, we only choose the first 28 clusters for the kinematic geometric error correction. The correction employs the mathematical model Equations (4) and (5), as well as our calculated error coefficients. Before correction, the data are aligned to be parallel with the X coordinate as much as possible. Table 5 shows the aligned measurement data and the corrected data.

Table 5. The aligned and corrected measurement data.

Cluster N°	Rotated Measurement Data		Corrected Data by Error Coefficients	
	X Coordinate [μm]	Y Coordinate [μm]	X Coordinate [μm]	Y Coordinate [μm]
1	47.2	42.8	47.5	41.6
2	172.2	43.4	173.4	41.7
3	294.7	44.8	296.7	42.7
4	425.1	50.6	427.7	48.2
5	544.8	49.5	548.2	46.8
6	669.0	48.3	673.1	45.2

Table 5. *Cont.*

Cluster N°	Rotated Measurement Data		Corrected Data by Error Coefficients	
	X Coordinate [μm]	Y Coordinate [μm]	X Coordinate [μm]	Y Coordinate [μm]
7	796.2	50.3	800.9	46.4
8	42.0	166.6	42.1	163.2
9	167.2	172.3	168.2	168.3
10	292.1	173.1	293.9	168.9
11	418.9	173.6	421.3	169.3
12	543.7	176.6	546.8	172.1
13	666.2	176.3	669.9	171.5
14	796.6	178.1	801.0	172.6
15	39.3	296.0	39.1	290.6
16	162.6	295.0	163.2	289.3
17	292.3	296.4	293.7	289.6
18	417.2	299.4	419.3	290.7
19	541.0	303.0	543.8	293.7
20	666.9	300.4	670.4	297.2
21	796.1	420.7	800.2	294.2
22	39.1	418.2	38.6	413.4
23	162.2	418.2	162.5	410.9
24	290.0	421.7	291.1	414.6
25	414.8	425.7	416.7	418.8
26	537.9	425.3	540.4	418.6
27	665.1	431.7	668.3	425.1
28	793.5	432.2	797.4	425.5

The aligned and corrected measurement data are both adjusted to a beginning of $(0, 0)$. The results of the measured and corrected positions are compared with the certified positions in Table 6. Figure 19 illustrates the measured, corrected, and the certified positions. The mean error, the maximum error, the sum of the squared errors, and the standard deviations of the errors are indicated in Table 7. The mean error and residual between the measured positions and the certified positions is 8.1 μm, while the mean error and residual between corrected positions and the certified positions is 5.7 μm, improved 29.6%. The maximum error between the measured positions and the certified positions is 15.6 μm, while the maximum error between corrected positions and the certified positions is 11.5 μm, reduced 26.3%. The sum of squared errors reduced from 2173.3 μm^2 to 1136.2 μm^2, which is 47.7%. It can be observed that all four types of errors are much smaller after correction with the error coefficients.

Table 6. Comparison of the certified, measured, and corrected positions.

Cluster N°	Certified Position		Measured Position (Alignment Rotated)		Corrected Position	
	X Coordinate [μm]	Y Coordinate [μm]	X Coordinate [μm]	Y Coordinate [μm]	X Coordinate [μm]	Y Coordinate [μm]
1	0.0	0.0	0.0	0.0	0.0	0.0
2	125.0	0.0	125.1	0.6	125.9	0.1
3	250.0	0.0	247.6	2.0	249.2	1.1
4	375.0	0.0	377.8	7.8	380.2	6.6
5	500.0	0.0	497.7	6.7	500.7	5.2
6	625.0	0.0	621.9	5.5	625.6	3.6
7	750.0	0.0	749.1	7.5	753.4	4.9
8	0.0	125.0	−5.1	123.8	−5.4	121.6
9	125.0	125.0	120.1	129.5	120.7	126.8
10	250.0	125.0	245.0	130.3	246.4	127.3
11	375.0	125.0	371.7	130.9	373.8	127.8
12	500.0	125.0	496.5	133.8	499.3	130.5
13	625.0	125.0	619.0	133.5	622.4	129.9
14	750.0	125.0	749.4	135.2	753.5	131.0
15	0.0	250.0	−7.8	253.2	−8.4	249.0

Table 6. *Cont.*

Cluster N°	Certified Position		Measured Position (Alignment Rotated)		Corrected Position	
	X Coordinate [μm]	Y Coordinate [μm]	X Coordinate [μm]	Y Coordinate [μm]	X Coordinate [μm]	Y Coordinate [μm]
16	125.0	250.0	115.5	252.2	115.7	247.8
17	250.0	250.0	245.1	252.5	246.2	248.0
18	375.0	250.0	370.1	253.6	371.8	249.1
19	500.0	250.0	493.9	256.6	496.3	252.2
20	625.0	250.0	619.8	260.2	622.9	255.6
21	750.0	250.0	748.9	257.7	752.7	252.6
22	0.0	375.0	−8.0	377.9	−8.9	371.8
23	125.0	375.0	115.1	375.4	115.1	369.3
24	250.0	375.0	242.9	378.9	243.7	373.0
25	375.0	375.0	367.7	382.9	369.2	377.2
26	500.0	375.0	490.7	382.5	492.9	377.0
27	625.0	375.0	617.9	388.9	620.8	383.5
28	750.0	375.0	746.4	389.5	749.9	383.9

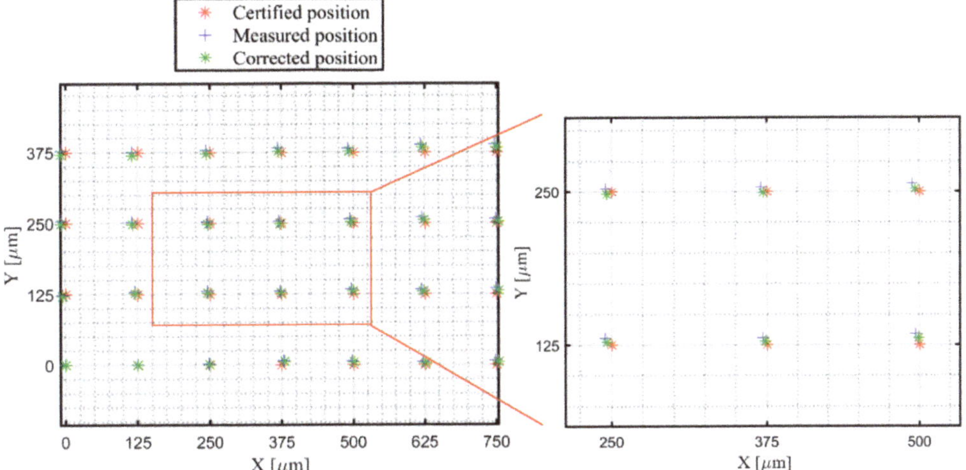

Figure 19. Comparison of the certified, measured, and corrected positions.

Table 7. Errors with respect to the certified positions before correction and after correction.

Data Types	Mean Error [μm]	Maximum Error [μm]	Sum of Squared Errors [μm²]	Standard Deviations of the Errors [μm]
Measured points	8.1	15.6	2173.3	3.5
Corrected points	5.7	11.5	1136.2	2.8

Figures 20 and 21 show the contour of the mean errors of the measured data and the corrected data individually. Figure 22 shows the comparison of the error vectors from the certified positions to the measured positions and the vectors from the certified positions to the corrected positions.

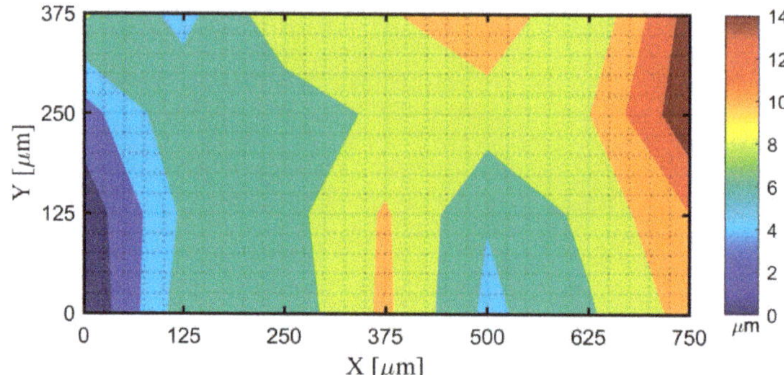

Figure 20. Contour of the Euclidean residuals between each certified and measured points.

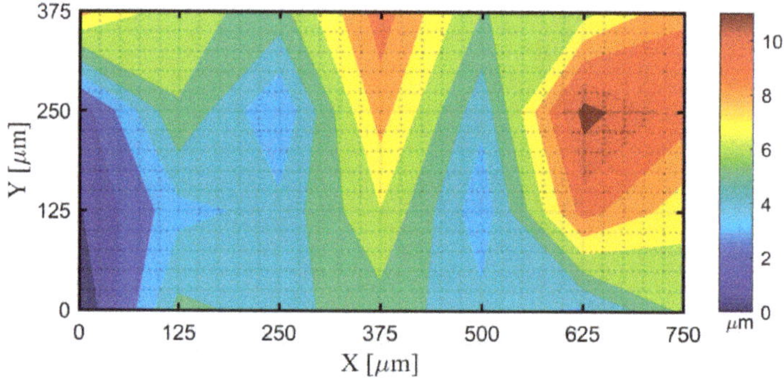

Figure 21. Contour of the Euclidean residuals between each certified and corrected points.

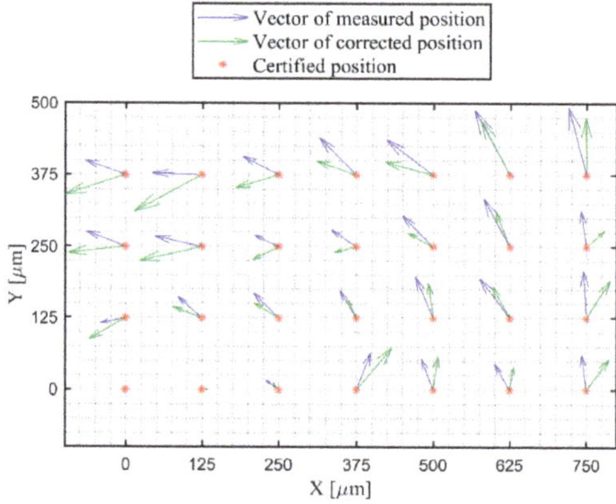

Figure 22. Comparison of the error vectors between the measured and corrected points.

According to the above results, we found that the errors and residuals between the corrected positions and the certified positions are much smaller than the errors and residuals between the measured positions and certified positions. This indicates that our method for the X and Y coordinate calibration and correction is effective and useful.

6. Conclusions and Future Work

This paper implemented two experiments for the illustration and the verification of our proposed method for the correction of the kinematic geometric errors produced by the movement of the lateral stage of confocal microscopes. The experimental results indicate that the mean residual reduced 29.6%, the maximum error reduced 26.3%, and the sum of squared errors reduced 47.7%.

The first experiment measured the dot grid targets with extended topography. After processing the measurement data, the error correction coefficients defined in the mathematical model, i.e., Equations (4) and (5) presented in Part I [18] were determined. The uncertainties of the values of those coefficients were also evaluated using the Monte Carlo method. The simulation number was 1×10^4. Distributions of the uncertainties of each coefficient, as well as their lower and upper boundaries of a 95% interval, were indicated.

The second experiment measured a different area of the same standard artefact. By correcting the measurement data using our mathematical model and the determined coefficients, the corrected results were obtained. The residuals between the raw measured points and their corresponding certified values were compared to those between the corrected points and the certified values.

The data processing algorithms and procedures, such as separations of the flats and cylinders, data partitions, outlier eliminations, K-means clustering, cylinder centers fittings, etc. were the same for the two experimental studies. The difference between the data processing for the two experiments was in the final procedures. The first experiment fitted the values of the coefficients used the nonlinear least squared method. The second method applied the mathematical models and the determined values of the coefficients to the measured data for obtaining the corrected coordinate values of the points.

Results of the experiments demonstrated that our proposed method for lateral stage kinematic geometric error correction is efficient and useful.

Among the next practical steps for improving the proposed method is a focus on the stitching algorithm of the optical element [26], which highly needs a calibration and correction of the stitching result.

Author Contributions: Conceptualization and methodology, C.W. and E.G.; algorithms, experiments, data, manuscript writing, C.W.; resources, funding, supervision, E.G. and Y.Y.; review and editing, C.W., E.G., and Y.Y.

Funding: This work is funded by the Spanish State Programme of Promotion of Scientific Research and Technique of Excellence, State Sub-programme of Generation of Knowledge. Project DPI2016-78476-P "Desarrollo Colaborativo de Patrones de Software y Estudios de Trazabilidad e Intercomparación en la Caracterización Metrológica de Superficies", belonging to the 2016 call for R & D Projects. The authors acknowledge the support from the National Natural Science Foundation of China (NSFC) project no. 51775326 and the Major State Research Development Program of China (2016YFF0101905).

Acknowledgments: Sincere thanks to the computer resources, technical expertise, and assistance provided by the Supercomputing and Visualization Center of Madrid (CeSViMa). The authors thankfully acknowledge the Chinese Scholarship Council (CSC) for funding the first author's doctoral study.

Conflicts of Interest: The authors declare no conflict of interest.

References

1. Jassby, D.; Cath, T.Y.; Buisson, H. The role of nanotechnology in industrial water treatment. *Nat. Nanotechnol.* **2018**, *13*, 670–672. [CrossRef]
2. Pfeifer, T.; Freudenberg, R.; Dussler, G.; Brocher, B. Quality control and process observation for the micro assembly process. *Measurement* **2001**, *30*, 1–18. [CrossRef]
3. Bose, S.; Vahabzadeh, S.; Bandyopadhyay, A. Bone tissue engineering using 3D printing. *Mater. Today* **2013**, *16*, 496–504. [CrossRef]
4. Stout, K.J.; Blunt, L. A contribution to the debate on surface classifications—Random, systematic, unstructured, structured and engineered. *Int. J. Mach. Tools Manuf.* **2001**, *41*, 2039–2044. [CrossRef]
5. Krolczyk, G.M.; Krolczyk, J.B.; Maruda, R.W.; Legutko, S.; Tomaszewski, M. Metrological changes in surface morphology of high-strength steels in manufacturing processes. *Measurement* **2016**, *88*, 176–185. [CrossRef]

6. Wang, C.; D'Amato, R.; Gómez, E. Confidence Distance Matrix for outlier identification: A new method to improve the characterizations of surfaces measured by confocal microscopy. *Measurement* **2019**, *137*, 484–500. [CrossRef]

7. Wang, C.; Caja, J.; Gomez, E. Comparison of methods for outlier identification in surface characterization. *Measurement* **2018**, *117*, 312–325. [CrossRef]

8. Sekiya, F.; Sugimoto, A. Fitting discrete polynomial curve and surface to noisy data. *Ann. Math. Artif. Intell.* **2015**, *75*, 135–162. [CrossRef]

9. Khac, B.C.T.; Chung, K.H. Quantitative assessment of contact and non-contact lateral force calibration methods for atomic force microscopy. *Ultramicroscopy* **2016**, *161*, 41–50. [CrossRef]

10. Ekberg, P.; Mattsson, L. Traceable X,Y self-calibration at single nm level of an optical microscope used for coherence scanning interferometry. *Meas. Sci. Technol.* **2018**, *29*, 035005. [CrossRef]

11. Wilson, T.; Carlini, A.R. Size of the detector in confocal imaging systems. *Opt. Lett.* **1987**, *12*, 227–229. [CrossRef] [PubMed]

12. Kim, T.; Gweon, D.; Lee, J. Enhancement of fluorescence confocal scanning microscopy lateral resolution by use of structured illumination. *Meas. Sci. Technol.* **2009**, *20*, 055501. [CrossRef]

13. Ni, H.; Zou, L.; Guo, Q.; Ding, X. Lateral resolution enhancement of confocal microscopy based on structured detection method with spatial light modulator. *Opt. Express* **2017**, *25*, 2872–2882. [CrossRef] [PubMed]

14. Wang, B.; Zou, L.; Zhang, S.; Tan, J. Super-resolution confocal microscopy with structured detection. *Opt. Commun.* **2016**, *381*, 277–281. [CrossRef]

15. Lee, K.; Lee, J.C.; Yang, S.H. The optimal design of a measurement system to measure the geometric errors of linear axes. *Int. J. Adv. Manuf. Technol.* **2013**, *66*, 141–149. [CrossRef]

16. Ibaraki, S.; Kimura, Y.; Nagai, Y.; Nishikawa, S. Formulation of Influence of Machine Geometric Errors on Five-Axis On-Machine Scanning Measurement by Using a Laser Displacement Sensor. *J. Manuf. Sci. Eng.* **2015**, *137*, 021013. [CrossRef]

17. Daemi, B.; Ekberg, P.; Mattsson, L. Lateral performance evaluation of laser micromachining by high precision optical metrology and image analysis. *Precis. Eng.* **2017**, *50*, 8–19. [CrossRef]

18. Wang, C.; Gómez, E.; Yu, Y. Characterization and correction of the geometric errors in using confocal microscope for extended topography measurement. Part I: Models, Algorithms Development and Validation. *Electronics* **2019**, *8*, 733. [CrossRef]

19. Cox, M.G.; Siebert, B.R.L. The use of a Monte Carlo method for evaluating uncertainty and expanded uncertainty. *Metrologia* **2006**, *43*, S178–S188. [CrossRef]

20. *JCGM 100:2008 Evaluation of Measurement Data—Guide to the Expression of Uncertainty in Measurement (GUM)*; JCGM: Paris, France, 2008.

21. Harris, P.M.; Cox, M.G. On a Monte Carlo method for measurement uncertainty evaluation and its implementation. *Metrologia* **2014**, *51*, S176–S182. [CrossRef]

22. *JCGM 101:2008 Evaluation of Measurement Data—Supplement 1 to the "Guide to the Expression of Uncertainty in Measurement"—Propagation of Distributions Using a Monte Carlo Method*; JCGM: Paris, France, 2008.

23. *JCGM 102: 2011 Evaluation of Measurement Data—Supplement 2 to the "Guide to the Expression of Uncertainty in Measurement"—Extension to Any Number of Output Quantities*; JCGM: Paris, France, 2011.

24. Wubbeler, G.; Krystek, M.; Elster, C. Evaluation of measurement uncertainty and its numerical calculation by a Monte Carlo method. *Meas. Sci. Technol.* **2008**, *19*, 084009. [CrossRef]

25. Bich, W.; Cox, M.G.; Dybkaer, R.; Elster, C.; Estler, W.T.; Hibbert, B.; Imai, H.; Kool, W.; Michotte, C.; Nielsen, L.; et al. Revision of the "Guide to the Expression of Uncertainty in Measurement". *Metrologia* **2012**, *49*, 702–705. [CrossRef]

26. Chen, D.; Peng, J.; Valyukh, S.; Asundi, A.; Yu, Y. Measurement of High Numerical Aperture Cylindrical Surface with Iterative Stitching Algorithm. *Appl. Sci.* **2018**, *8*, 2092. [CrossRef]

 *electronics*

Article

Deep Transfer HSI Classification Method Based on Information Measure and Optimal Neighborhood Noise Reduction

Lianlei Lin, Cailu Chen, Jingli Yang * and Shanshan Zhang

Department of Test and Control Engineering, School of Electronics and Information Engineering, Harbin Institute of Technology, Harbin 150001, China; linlianlei@hit.edu.cn (L.L.); chencailu_hit@163.com (C.C.); zhangshanhit@163.com (S.Z.)
* Correspondence: jinglidg@hit.edu.cn; Tel.: +86-0451-86413532

Received: 5 August 2019; Accepted: 30 September 2019; Published: 2 October 2019

Abstract: Land environment is one of the most commonly and importantly used synthetical natural environments in a virtual test. To recognize the ground truth for the construction of virtual land environment, a deep transfer hyperspectral image (HSI) classification method based on information measure and optimal neighborhood noise reduction was proposed in this article. Firstly, the information measure method was used to select the most valuable spectrum. Specifically, three representative bands were selected using the combination of entropy, color matching function, and mutual information. Based on the selected bands, a patch containing spatial-spectral information was constructed and used as the input of the convolutional neural networks (CNN) network. Then, in order to address the problem that a large number of labeled samples were required in deep learning method, the HSI classification method based on deep transfer learning was proposed. In the proposed method, the transfer of parameters ensured the classification performance with small training samples and reduced the training cost. Moreover, the optimal neighborhood noise reduction was used as the post-processing method to effectively eliminate the salt-and-pepper noise and further improve the classification performance. Experiments on two datasets demonstrated that the proposed method in this article had higher classification accuracy than similar methods.

Keywords: CNN; hyperspectral image classification; information measure; transfer learning; neighborhood noise reduction

---

1. Introduction

In recent years, hyperspectral image (HSI) analysis has been widely used in various fields [1], such as the monitoring of land cover change [2,3] and the environmental science and mineral development [4]. As an advanced machine learning technology, deep learning has been widely used in image classification to learn the hierarchical features of a deep neural network from low-level to high-level [5]. The image classification methods based on the convolutional neural networks (CNN) have shown the ability to detect local features of the hyperspectral input data and obtain the classification results with high accuracy and stability. Rachmadi et al. proposed an adaptation of a convolutional neural network (CNN) scheme proposed for segmenting brain lesions with considerable mass-effect, to segment white matter hyperintensities (WMH) characteristic of brains with none or mild vascular pathology in routine clinical brain magnetic resonance images (MRI) [6]. Krizhevsky et al. proposed a large, deep convolutional neural network to classify the 1.2 million-high resolution images in the ImageNet LSVRC-2010 contest into the 1000 different classes [7]. In addition to the application in common image classification, many CNN-based hyperspectral image classification methods have been developed in recent years. Ma et al. proposed a context deep-learning algorithm for learning the features, which can better characterize the information than the extraction algorithms with predefined features [8]. Zhang et al. proposed a region-based diversified CNN, which can semantically encode context-aware representations to obtain valuable features and improve the classification accuracy [9]. Chen et al. first proposed the Stack Automated Encoder (SAE) framework to incorporate the features with the special-spectral information. Firstly, the validity of SAE was verified by the classical spectral information-based classification method. Secondly, a new classification method based on spatial principal component information was proposed [10]. Slavkovikj et al. proposed a CNN framework for hyperspectral image classification in which spectral features were extracted from a small neighborhood [11]. Makantasis et al. proposed an R-PCA CNN classification method [12], in which PCA was first used to extract the spatial information, and then CNN was used to encode spectral and spatial information. The CNN classification methods with the combined spectral and spatial information have shown better classification performance.

The hyperspectral image contains all of the spectral information of the ground objects, with the typical high-dimensional features. However, the spectral information with high redundancy may interfere with the classification process and reduce the classification accuracy. Therefore, it is very important to reduce the spectral dimensionality of the hyperspectral image before the classification process. Two typical dimensionality-reduction methods have been reported, i.e., feature extraction and band selection methods. The feature extraction method mainly included principal component analysis (PCA) [13], linear discriminant analysis (LDA) [14], and multidimensional scaling [15]. The band selection method mainly included examining correlations [16], calculating mutual information [17,18], etc.. In recent years, information-based band selection has been a very popular research topic, in which the Shannon entropy or its changes, such as mutual information (MI), were usually used as the basis of image information measure. Bajcsy proposed a band selection method under the constraints of classification accuracy and computational requirements, in which the optimal number of bands was determined by the unsupervised method of entropy [19]. Adolfo proposed a clustering method for automatic band selection based on the mutual information in multispectral images [20]. Wang proposed a supervised classification method for band selection based on spatial entropy [21]. Le Moan et al. proposed a new spectral image visualization method to achieve band selection by the first-order, second-order, and third-order information measure [22]. Manel proposed a frequency band selection method based on the hierarchical clustering of spectral layer, and used mutual information measure to reduce the dimensions of the image. Then a new c-means clustering algorithm was proposed to integrate the spatial spectrum information [23]. Hossain et al. proposed a dimensionality reduction method (PCA-nMI) that combined principal component analysis (PCA) with normalized mutual information (nMI) under two constraints [24]. The proposed method maximized the general correlation and minimized the redundancy in selected subspaces.

In the application of the deep neural networks to the classification of hyperspectral images, there were some difficulties in the process of model training, such as the high demand for training samples and the time complexity of computing models. Transfer learning can address the above difficulties to some extent. In transfer learning, the learned knowledge or experience from the source tasks is applied to the object tasks. Based on transfer learning, Li et al. proposed a CNN framework for the anomaly detection [25] and the training of multi-layer CNNs using the differences between adjacent pixels generated from the source image dataset. The experimental performance showed that the proposed algorithm was superior to the classical Reed-Xiaoli [26] and the most advanced representation-based detectors, such as sparse representation-based detectors (SRD) [27] and cooperative representation-based detectors [28]. Wang proposed two architectures to extract the general features of remote scene classification from the pre-trained CNNs [29]. Wang Liwei et al. proposed a hyperspectral image classification method by applying transfer learning in deep residual networks [30], and shared the shallow network weight parameters of deep residual networks. At the same time, in order to address the over-fitting problem in the process of transfer shallow network parameters from the large source dataset to the small object dataset, the strategy of fine-tuning the residual network was proposed, in which the deep layer of deep residual network was randomly initialized and retrained in the object dataset.

The method studied in this article is to construct a virtual land environment for virtual test, which is also one of the research fields of the author's team. In order to build a virtual land environment which can be used as the basis of other natural environment modeling and sensor sensing, the most important thing is to accurately acquire the information of ground truth. So, we propose a scheme to construct the virtual land environment using multi-source satellite's earth observation data. The method proposed in this paper is the key step of ground truth recognition. In this article, the hyperspectral image classification method based on CNN was investigated. In addition, the information measure was used to select the bands and reduce the dimensionality of hyperspectral images, thus, reducing the redundant information. At the same time, the spatial-spectral features were extracted to improve the classification performance of hyperspectral images. Furthermore, the classification method based on deep transfer learning and neighborhood noise reduction was used to achieve high classification accuracy for small-sized samples and reduce the training complexity of the object dataset.

2. The Related Work

The method studied in this article is to build a virtual land environment, which belongs to the field of virtual tests. Virtual test technology is a new test technology. Because of its low cost, high efficiency and ability to support test in a complex environment, virtual test technology has been more and more widely used, such as automobile test [31], building design [32], weapon system test [33] and so on. The virtual test is inseparable from the virtual environment, so in order to meet the needs of the virtual test in a complex environment, it is an important research content to study the modeling theory and method of all kinds of complex environments. The synthetic natural environment, which is composed of land, atmosphere, ocean and space environment, is the space for various human activities [34]. The virtual land environment is one of the important components of the virtual natural environment. The current virtual land environment, which is often used in virtual reality [35] and visual simulation [36], is mainly concerned about visualization and pays attention to the immersion. However, the new virtual test puts higher demands on the virtual land environment, as follows.

(1) Providing support for the construction of other types of natural environment. The synthetic natural environment is an interrelated and interactive organic whole, especially the interaction between the land environment and the atmospheric environment. For example, the MM5 atmospheric model needs six types of land environmental data, such as land surface type, water body, vegetation composition, etc. [37].

(2) Providing a sensing basis for virtual sensors. For example, when using infrared imaging sensors to detect the ground, it is necessary to support the high-precision three-dimensional land environment with the information of ground truth, material, etc.

Based on the above analysis, the virtual land environment used in the virtual test is more important to be used as the modeling basis of other natural environments and the sensing basis of various sensors. What is more, it is a three-dimensional land environment which contains various information, such as ground truth, objects material, color, texture, etc. Therefore, as shown in Figure 1, we propose a scheme to construct a virtual land environment using multi-source satellite earth observation data, which is a joint application of multi-sensor data [38]. Based on hyperspectral, multispectral, panchromatic, and other optical earth observation data and radar earth observation data, such as InSAR, the construction of virtual land environment is completed through four steps: Temporal-spatial-spectral fusion (A), ground truth recognition (B), elevation extraction (C) and synthesis (D). Among them, ground truth recognition (B) is a key step, which is responsible for providing accurate information, such as ground truth, material, etc for the virtual land environment. The HSI classification method proposed in this article belongs to this step, and the ground truth information is obtained through the classification of HSI.

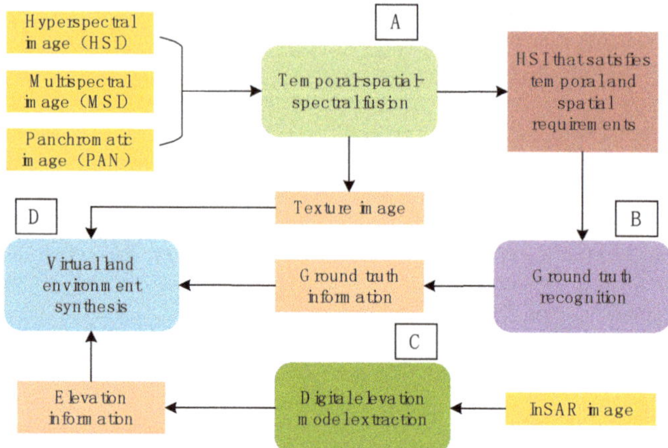

Figure 1. The process of constructing a virtual land environment using multi-source satellite Earth observation data.

In the virtual test, according to the test requirements, a specific space range of virtual land environment is usually constructed, which may use the earth observation data obtained by different sensors. At the same time, the method of ground truth classification based on machine learning needs a large number of marked data as training samples. It is not easy to prepare a large amount of marked data for the images of different scenes obtained by different sensors. Therefore, transfer learning is adopted in this article. After training the classification network in all kinds of typical scenes, we can transfer to a specific scene and do a little training.

3. Dimensionality Reduction of Hyperspectral Image Based on Information Measure

3.1. Spectral Band Preprocessing Based on Entropy and Color Matching Function (CMF)

Based on the information theory, Shannon first proposed the concept of entropy in 1948 [39], in which the amount of information with uncertainty, i.e., the probability of the occurrence of discrete random events was measured. The greater amount of information indicated a smaller redundancy. The information measure based on Shannon's communication theory has been proven very effective in identifying the redundancy of high-dimensional datasets.

The entropy of a random variable is defined as

$$H(X) = -\sum_{i=1}^{n} p_X(x_i) \log_b [p_X(x_i)], \tag{1}$$

where x_i is the event of X, $p_X(x_i)$ is the probability density function of X, and b is the logarithmic order.

Assume X and Y are two random variables, where X has n values and Y has m values. Then their joint entropy is

$$H(X, Y) = -\sum_{i=1}^{n} \sum_{j=1}^{m} p_{X,Y}(x_i, y_j) \log p_{X,Y}(x_i, y_j), \tag{2}$$

where $p_{X,Y}(x_i, y_j)$ is the joint probability density function of X and Y.

When these measurements are applied to hyperspectral images, it is generally assumed that each channel (spectral band) is equivalent to a random variable X, and each pixel in the channel is an event x_i. In the preprocessing step for the spectral band of the hyperspectral image, the channel with less information is eliminated based on Shannon entropy, which are described as follows:

Firstly, the entropy $H(B_i)$ of each spectral band of the hyperspectral image is calculated. The random variable B_i is the i-th spectral band ($i = 1, 2, \ldots n$), x_i is the pixel of the i-th spectral band, and $p_{B_i}(x_i)$ is the probability density function of the band B_i.

$$H(B_i) = -\sum_{i=1}^{n} p_{B_i}(x_i) \log_b\left[p_{B_i}(x_i)\right]. \tag{3}$$

Secondly, the local average entropy of each spectral band is defined as Equation (4), where m is the window size, indicating the size of the neighborhood.

$$\overline{H_m(B_i)} = \frac{1}{m} \sum_{p=-m/2}^{m/2} H\left(B_{i+p}\right). \tag{4}$$

Finally, the bands B_i that meet the following conditions are retained.

$$H(B_i) \in \left[\overline{H_m(B_i)} \times (1 - \sigma), \overline{H_m(B_i)} \times (1 + \sigma)\right], \tag{5}$$

where σ is the threshold factor. The spectral band with higher or lower entropy than the range of σ of the local average entropy $\overline{H_m(B_i)}$ is considered to be irrelevant.

The blue line in Figure 2 is the entropy curve, in which the horizontal axis represents the spectral dimension and the vertical axis represents the entropy of each spectral band. The smoothness of the entropy curve determines the window size m and the threshold factor σ. If the entropy curve is smooth, the change of the adjacent spectral band information, i.e., the uncertainty of the band information is small; thus, the number of bands outside the relevant range, i.e., the probability of having an uncorrelated band, is also small. In this condition, few spectral bands are redundant, thus, smaller σ and m values should be chosen to improve the ability of excluding redundant bands. On the contrary, if the entropy curve fluctuates greatly, the change of the adjacent spectral band information, i.e., the uncertainty of the band information, is large; thus, the number of bands outside the correlation range, i.e., the probability of having an uncorrelated band, is also large. In this case, more spectral bands are redundant, thus, larger σ and m values should be chosen to exclude the redundant bands and retain the bands with valid information.

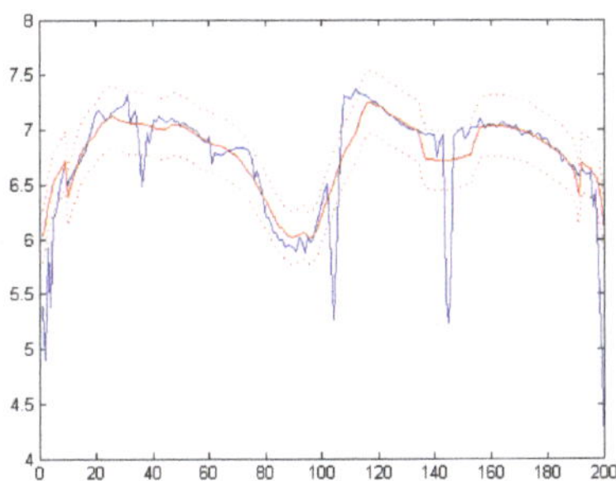

Figure 2. Exclusion of the unrelated spectral bands of the Indian pines dataset.

The color matching function (CMF) in CIE 1931 standard chromaticity observer [40] described the human eye visual color characteristics. This function was used to achieve the complete process of preliminary selection of spectral bands based on the calculation of entropy. At a specific wavelength, the CMF determined the number of the three primary lights (red, blue and green), which must be mixed in a certain order to achieve the same visual effect as the corresponding monochrome light at that wavelength. By applying the CIE color matching to hyperspectral images in the visible range, hyperspectral images can be represented by the corresponding color matching [41].

As shown in Figure 3, the wavelength of the first effective spectral band was $\lambda = 360$ nm, and the wavelength of the last effective spectral band was $\lambda = 830$ nm. Then a linear interpolation was performed between the first and last effective spectral band.

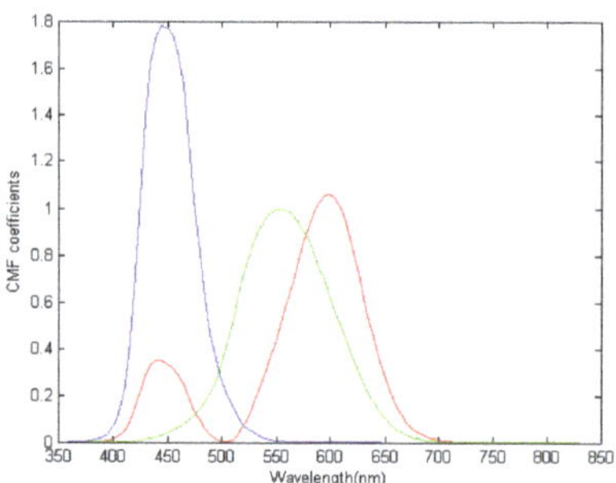

Figure 3. CIE 1931 color matching curve between 360 nm and 830 nm.

In order to obtain the band selection sets, the thresholds t for the CMF coefficients of the three primary colors were set, i.e., Set_R^t, Set_B^t and Set_G^t were set based on the optical channels of the three primary colors.

In Figure 4, two spectral thresholds ($t = 0.1$, $t = 0.5$) were set for the coefficient variation curve of the CMF of the red light. When the CMF coefficient is above the threshold, the corresponding spectral bands were preserved. It was challenging to set the value of the parameter t without a specific application. In this article, an automatic threshold method was used to define the optimal threshold t to maximize the amount of discarded information. Use $S_{discard}^t$ to label the set of channels which are discarded by threshold processing of the CMF and $S_{selected}^t$ to label the complementary set of $S_{discard}^t$. The optimal threshold t_{opt} is defined as

$$t_{opt} = \mathrm{argmax}(t), H\left(S_{discard}^t\right) < H\left(S_{selected}^t\right),\tag{6}$$

where $H\left(S_{discard}^t\right)$ is the total entropy of the discarded spectral bands obtained by the above derivation, and $H\left(S_{selected}^t\right)$ is the total entropy of the selected spectral bands. Using the above describe method, the initially selected spectral bands can be obtained.

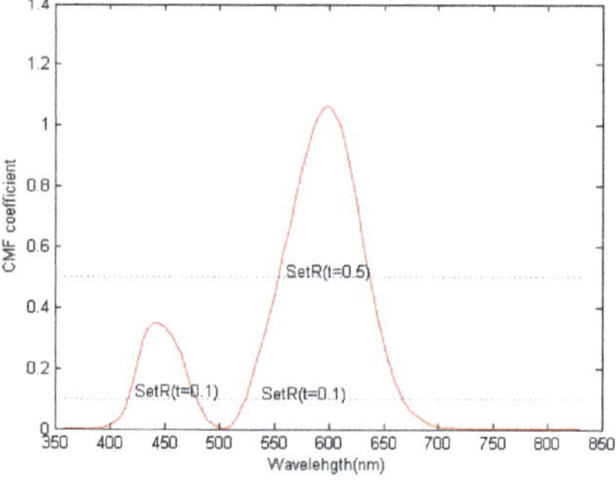

Figure 4. Color matching function (CMF) coefficient changes of red light.

3.2. Band Selection Based on Mutual Information

Mutual Information (MI) is a measure of the useful information, which is defined as the amount of the information contained in a random variable about another random variable. The MI between two random variables X and Y is defined as

$$
\begin{aligned}
I(X,Y) &= \sum_{\substack{i=1\dots n \\ j=1\dots m}} p_{X,Y}(x_i, y_j) \log \frac{p_{X,Y}(x_i,y_j)}{p_X(x_i)\cdot p_Y(y_j)} \\
&= H(X) + H(Y) - H(X,Y)
\end{aligned}
\tag{7}
$$

where $p_X(x_i)$ is the probability density function of X, $p_Y(y_i)$ is the probability density function of Y, and $p_{X,Y}(x_i, y_j)$ is the joint probability density function of X and Y. $H(X)$ is the entropy of the random variable X, and $H(Y)$ is the entropy of the random variable Y, which is calculated by Equation (1). $H(X,Y)$ is the joint entropy of two random variables X and Y, which can be calculated by Equation (2).

Furthermore, Bell proposed the mutual information of three random variables X, Y and Z, as shown in the following [42]:

$$
\begin{aligned}
I(X,Y,Z) &= H(X,Z) + H(Y,Z) - H(X,Y,Z) \\
&= -H(Z) - I(X,Y)
\end{aligned}
\tag{8}
$$

where $H(X,Y,Z)$ is the third-order joint entropy of three random variables X, Y and Z.

The above principle is equally applicable in hyperspectral images. The information of one channel can increase the mutual information between the other two channels. In this case, as the overlapped information between the two channels is less, the interdependence degree between the two random variables is lower, and the amount of contained information is greater. In the dimensionality reduction of hyperspectral images, it is necessary to consider both criteria, i.e., the largest amount of information and the least amount of redundancy.

Pla proposed the application of standardized mutual information [43]. In this article, we used the k-th order normalized information (NI) of the band $S = \{B_1, \cdots, B_k\}$ as the standardized mutual information, in which NI is defined as

$$
NI_k(S) = \frac{K \times I(S)}{\sum\limits_{i=1}^{k} H(B_i)},
\tag{9}
$$

where $I(S)$ is the mutual information of the bands B_1 to B_k, and $H(B_i)$ is the entropy of the band B_i.

Three parts, Set_R^t, Set_B^t and Set_G^t, were obtained by setting threshold t for the CMF coefficients of the three primary colors. As the value of the mutual information $NI_3(S)$ was smaller, the amount of contained information in the selected spectral bands was larger, and the dimensionality reduction effect on the hyperspectral image was better. The spectral bands, x_R, y_B, z_G ($x_R \in Set_R^t, y_B \in Set_B^t, z_G \in Set_G^t$), were obtained to minimize $NI_3(x_{RN^*}, y_{BN^*}, z_{GN^*})$ and selected as the most valuable bands.

3.3. Two Strategies for CNN Inputs

Three spectral bands, x_R, y_B, z_G, were selected for the dimensionality reduction of hyperspectral image based on the information measure, and three grayscale images were obtained. There were two strategies to enter the CNN network, as shown in Figure 5. In the first strategy, the three spectral bands x_R, y_B, z_G were directly used to extract the neighborhood around the pixel and classify the pixel into a patch of $m \times m \times 3$, which was put into the CNN for classification. This method was also called the information measure classification method (IM for short) because it directly used the dimensionality reduction results based on the information measure. In the second strategy, all the spectral information of the pixel was superimposed and classified on the above patch to form a patch of $m \times m \times 6$. All the spectral information of the central pixel of the patch was extracted, combined, intercepted, and reshaped. Then three layers of spectral information were obtained, which had the same shape and size as the three-dimensional spatial-spectral information extracted by the IM method. Then, the patch generated by the IM method and the new $m \times m \times 6$ patch were superimposed and put into CNN for classification. In this article, the proposed method was called the information measure-spectral (IM-SPE for short) classification method.

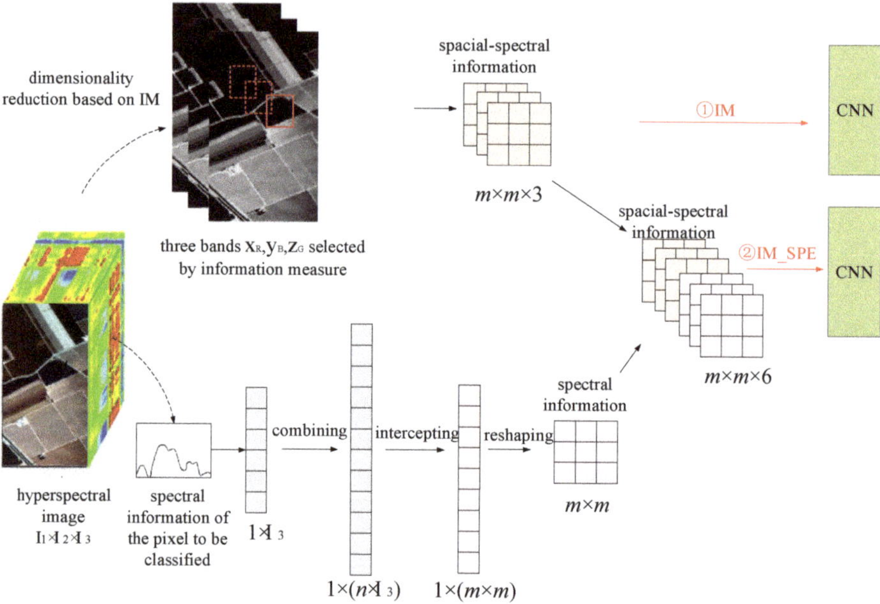

Figure 5. The principle of dimensionality reduction based on the integration of information measure and spectral information.

The spectral information was processed as follows: Assume that the size of hyperspectral data was $I_1 \times I_2 \times I_3$. At first, the one-dimensional spectral information of the central pixel with the size of $1 \times 1 \times I_3$ was superimposed by n times to obtain the one-dimensional spectral information of $1 \times (n \times I_3)$. Then a one-dimensional spectral vector equal to $m \times m$ was intercepted and reshaped to a two-dimensional spectral matrix of $m \times m$. Then a $m \times m \times 6$ spatial-spectral patch was obtained by superimposing three spectral information layers and combining the superimposed information with the three spatial-spectral information extracted by the IM method.

4. Hyperspectral Image Classification Method Based on Deep Transfer Learning

The classification of hyperspectral image based on deep transfer learning can be used to better solve the problem that the sample data is insufficient or relatively small. The specific classification principle is shown in Figure 6.

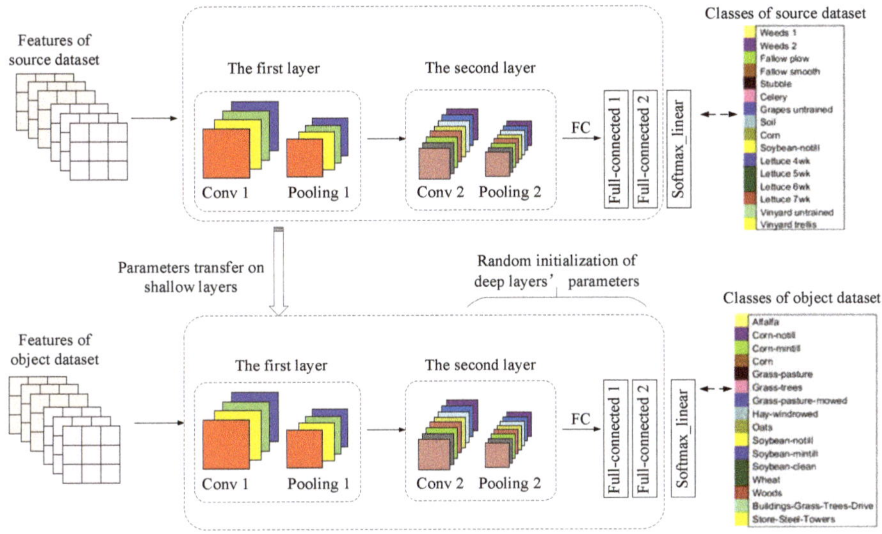

Figure 6. Principle of hyperspectral image classification based on deep transfer learning.

Training was performed on the source dataset to obtain the network model and parameters. The shallow layer structure and parameters were directly transferred to the object dataset, and the deep layer parameters were randomly initialized. Taking the CNN network structure of Figure 6 as an example (including two convolutions and pooling layers, and two fully connected layers), if the source hyperspectral dataset is highly similar to the object hyperspectral dataset, the adjustment of the deep parameters should be in the following order: The last fully connected layer (full-connected2), the first fully connected layer (full-connected1). If the source hyperspectral dataset is not highly similar to the object hyperspectral dataset, the convolutions and pooling layers (conv2 and pooling2) that extract the deep features may also need the random re-initialization of the weighting parameters. Specifically, in the proposed hyperspectral image classification process based on deep transfer learning in this article, the following two situations were mainly considered:

In the first situation, the object dataset has a small number of samples and is similar to the source dataset. In this case, first, the last fully connected layer of the pre-trained layers should be removed, and then a fully connected layer that matches the number of feature classes of the object dataset is added. The weight parameters of other pre-training layers are kept unchanged, and only the weights of the newly added layers are randomly initialized. When the sample size of the object dataset is small, only the new-added fully connected layer is trained using the object dataset to avoid the problem of over-fitting. In details, the learning rate of the previous layers of CNN to 0 and only the last fully connected layer on the object dataset is trained.

In the second situation, the object dataset has a large number of samples, but the number of samples relative to the source dataset is small, and the relative dataset are similar to the source dataset. In this case, first, the last fully connected layer of the pre-training network layers should be removed, and then a fully connected layer that matches the number of feature classes of the object dataset is added. Only the weights of the newly added layers are randomly initialized, while the weight parameters of other pre-training layers are kept unchanged. Since the object dataset has a large amount of data and is not prone to overfitting, the entire network can be retrained. Meanwhile, the features extracted by the original convolutional layer can be used to speed up the training for the object dataset. The entire network can be trained by the specific method through setting the learning rate of the front layeres of the CNN to 0.001.

The classification process of hyperspectral image based on deep transfer learning is shown in Figure 7.

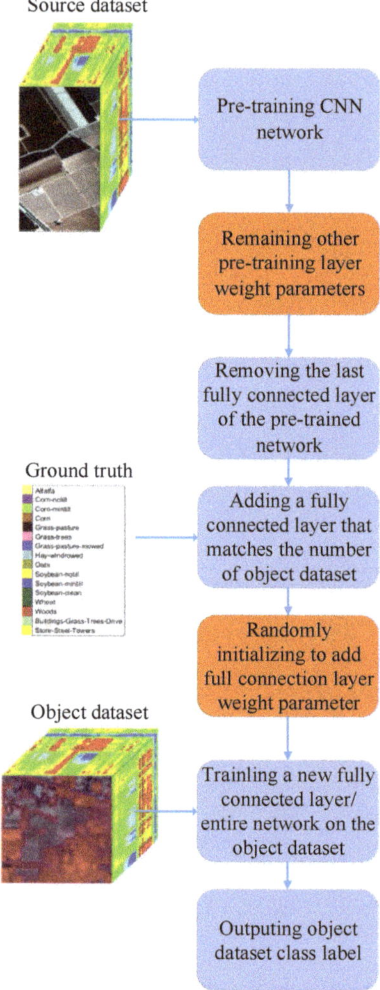

Figure 7. Hyperspectral image classification process based on deep transfer learning.

5. Optimal Neighborhood Noise-Reduction Method

In order to reduce the salt-and-pepper noise in the initial classification results, an optimal neighborhood noise reduction method based on the eight-neighbor mode label was proposed in this article. The optimal neighborhood noise reduction method was used to reprocess the initial classification result of the hyperspectral image. In the method, the hyperspectral image classification label data was used as the input, and the central pixel label was compared to the label of its eight-neighborhood pixel.

If $L_{(i,j)}$ is used to represent the classification result label of a central pixel $p_{(i,j)}$ of the hyperspectral image, the class labels of the central pixel $p_{(i,j)}$ and the eight neighborhood pixels are shown in Figure 8.

$L_{(i-1,j-1)}$	$L_{(i-1,j)}$	$L_{(i-1,j+1)}$
$L_{(i,j-1)}$	$L_{(i,j)}$	$L_{(i,j+1)}$
$L_{(i+1,j-1)}$	$L_{(i+1,j)}$	$L_{(i+1,j+1)}$

Figure 8. Class labels of the central pixel and its eight neighborhood pixels.

As shown in Figure 8, based on the traversal of all the pixel labels of the hyperspectral image, the class label $L_{(i,j)}$ of one central pixel $p_{(i,j)}$ and the class labels of the eight neighborhood pixels were combined into a 3×3 matrix. The 3×3 matrix was transformed into a 1×9 one-dimensional vector. Then the mode M of the nine labels and the number m of the mode labels were calculated. The threshold was set to N ($0 \leq N \leq 9$), and the pixel label of the hyperspectral image that did not need to be classified was set to 0. The central pixel is considered to be noise when the following conditions are met: The class label $L_{(i,j)}$ of the central pixel is not equal to the mode of the nine pixel labels, the mode label is not 0, and the number of the mode labels is $m \geq N$. Since the label of 0 means that the pixel does not need to be classified, the mode label with the value of 0 is excluded to avoid the edge misjudgment. If the central pixel $p_{(i,j)}$ is confirmed to be noise, it is replaced with the mode label of the eight neighborhood pixels. The initial value of the threshold N is generally set to 5. When the central pixel label $L_{(i,j)}$ is not equal to the mode label of the nine pixels, the mode label is not 0, and the number of the mode labels is $m \geq 5$, the central pixel is considered to be noise. The threshold N can be modified according to actual conditions. If the threshold is too large, the noise-reduction effect may not be obvious; if the threshold is too small, the actual information may be misjudged as noise.

The pseudo-code of the optimal neighborhood noise reduction process is shown as follows:

Input: The classification result dataset X of the hyperspectral image, the label of 0 for the pixels that do not need to be classified, and the threshold $N=5$
Output: The noise-reduced classification result data set X of the hyperspectral image and classified image with noise reduction.
Load classification result dataset X
for $i=1$; $i<$ **X.shape[0]**; $i++$
for $j=1$; $j<X.shape[1]$; $j++$
loop traversing each pixel label $L_{(i,j)}$
Transforming the 3×3 matrix composed of the central pixel label $L_{(i,j)}$ and its eight neighborhood pixel labels into a 1×9 one-dimensional vector
Calculating the mode M and its number m of central pixel label and the eight neighborhood pixels labels.
if the class label $L_{(i,j)}$ of the central pixel is not equal to the mode number M of the nine labels
if the mode label is not 0 and the number of the mode label is $m \geq N$
The central pixel $p_{(i,j)}$ is noise
Replace the central pixel label $L_{(i,j)}$ with the mode label M to remove the noise
end if
end if
Update the hyperspectral image classification result label $L_{(i,j)}$ of each pixel $p_{(i,j)}$
end loop $L_{(i,j)}$
end for
end for

---

6. Experiments and Analysis

6.1. Dataset

In order to verify the proposed method in this article, the experiments were conducted on the datasets with similar characteristics. The selected dataset included the Indian Pines dataset, the Salinas dataset, the Pavia University dataset, and the Pavia Center dataset. Both the Indian Pines dataset and the Salinas dataset were acquired by AVIRIS sensors. The corrected spectral dimensions for both datasets were 200 and 204, respectively, which were very close. The real ground objects were divided into 16 classes [44]. The Pavia University dataset and the Pavia Center dataset were collected by ROSIS sensors. The corrected spectral dimensions of these two datasets were 103 and 102, respectively, and the real ground objects were divided into nine classes [45]. The datasets are gotten from the website (http://www.ehu.eus/ccwintco/index.php?title=Hyperspectral_Remote_Sensing_Scenes).

The Indian Pines dataset and the Salinas dataset were similar, and the Pavia University dataset and the Pavia Center dataset were similar. In addition, the former two datasets had a smaller sample size than the latter two datasets. Therefore, the Salinas dataset and the Pavia Center dataset were used as the source datasets in the transfer learning method, while the Indian Pines dataset and the Pavia University dataset were used as the corresponding object datasets. First, the structures and parameters of the shallow network were obtained and validated from the training on the Salinas dataset and Pavia Center dataset. Then the obtained structure and parameters were transferred to the Indian Pines dataset and Pavia University dataset with relatively small sample sizes. Finally, the structure and parameters of the network were fine-tuned.

There was a big difference between the Indian Pines dataset and the Pavia University dataset, which could be used to validate the proposed classification method fully. The Indian Pines dataset had a small sample size and can mainly reflect the vegetation information, including rich species and mostly regular block distribution, rich spectral information and low spatial resolution. The Pavia University dataset had a large image size and can mainly reflect the landscape information of the urban landscape. Although there were few species, the shape of the object was irregular, and the spatial resolution was high.

6.2. Experiments and Result Analysis

6.2.1. Experiments of Dimensionality Reduction Methods

In order to test the effectiveness of the proposed methods in the dimensionality reduction and noise reduction for the hyperspectral images, the classification performance of IM, IM_SPE methods, and these methods superimposed with optimal neighborhood noise-reduction method (IM_DN, IM_SPE_DN methods) were tested and compared with that of the spectral information based CNN classification method (SPE) [46], the spatial information based CNN classification method (PCA1, PCA first principal component) [47], the CNN classification method based on the integration of the spectral information and the first principal component of the spatial information (PCA1_SPE), and the CNN classification method based on PCA's first three principal components of the spectral information (PCA3) [48]. We use OA, AA and Kappa coefficients to evaluate the performance of different methods.

(1) Experiments on Indian pines dataset

The classification performance on the Indian pines dataset is shown in Table 1. From Table 1, for the Indian pines dataset, the PCA1_SPE, PCA3, IM, and IM_SPE classification methods provided the best classification accuracy. Moreover, the OA, AA and Kappa coefficients of IM and IM_SPE methods were superior to those of SPE, PCA1, PCA1_SPE, and PCA3 methods, which indicated that the information measure-based classification method (IM and IM_SPE) had better performance than the spectral information-based classification method (SPE) and the PCA-based classification method (PCA1, PCA1_SPE and PCA3). Among all classification methods, the IM_DN and IM_SPE_DN methods had the best OA, AA, and Kappa coefficients, due to the combination of the dimensionality reduction based on information measure and the optimal neighborhood noise reduction. It has been demonstrated that the optimal neighborhood noise reduction method had a significant effect on the treatment of salt-pepper-noise of the classification results, and can greatly improve the classification accuracy.

Table 1. The classification results on Indian pines dataset (%).

Class	SPE	PCA1	PCA1_SPE	PCA3	IM	IM_SPE	IM_DN	IM_SPE_DN
1	92.30	79.33	99.97	97.80	92.38	98.91	93.48	98.91
2	89.10	94.05	91.85	97.18	97.78	98.20	98.76	98.71
3	87.43	94.02	94.90	**98.34**	96.17	96.91	96.90	97.46
4	88.39	94.98	90.80	**98.27**	97.51	95.44	97.74	96.56
5	96.59	95.96	97.13	97.01	95.54	99.36	96.17	**99.58**
6	97.81	98.47	98.35	98.20	96.84	99.01	97.41	**99.08**
7	91.07	98.18	98.19	98.21	99.95	98.18	**99.96**	98.18
8	98.06	99.76	99.95	99.77	99.89	99.78	**100.00**	99.89
9	79.98	77.49	**100.00**	74.97	79.95	72.48	82.46	72.48
10	90.28	96.37	95.01	96.73	98.17	95.97	**98.96**	96.62
11	91.08	94.67	95.61	97.09	98.50	98.35	99.00	**99.23**
12	87.61	94.18	96.24	95.87	97.78	98.62	98.35	**99.16**
13	99.23	97.52	99.24	98.70	99.45	99.50	99.71	**99.76**
14	95.81	98.24	98.90	99.45	99.47	99.45	99.59	**99.65**
15	78.79	94.08	97.09	98.12	99.69	99.30	99.71	**99.84**
16	96.76	96.74	93.54	98.30	91.92	98.36	93.01	**99.45**
OA	85.31	92.39	93.06	95.79	96.39	96.90	97.36	**97.82**
AA	91.28	94.01	96.67	96.50	96.31	96.74	96.95	**97.16**
Kappa	83.23	91.33	92.08	95.21	95.88	96.46	96.99	**97.52**

Figure 9a–i show the classification results of the SPE, PCA1, PCA1_SPE, PCA3, IM, IM_SPE, IM_DN, and IM_SPE_DN methods on the Indian pines data set. Figure 9i shows the ground truth of Indian pines dataset.

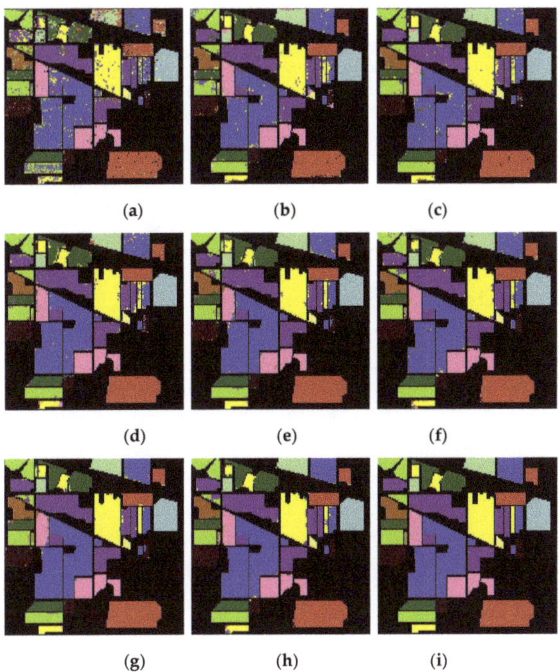

Figure 9. The classification results on Indian Pines dataset: (**a**) SPE; (**b**) PCA1; (**c**) PCA1_SPE; (**d**) PCA3; (**e**) IM; (**f**) IM_SPE; (**g**) IM_DN; (**h**) IM_SPE_DN; (**i**) the ground truth.

From Figure 9, the classification accuracy of the IM and IM_SPE methods was superior to that of the SPE, PCA1, PCA1_SPE and PCA3 methods on the Indian pines dataset. In addition, the dimensionality reduction method based on information measure played an important role in improving the image classification performance. The salt-pepper-noise of the hyperspectral image was significantly reduced by IM_DN and IM_SPE_DN methods, indicating that the optimal neighborhood noise reduction method can provide a more accurate classification effect.

(2) Experiments on Pavia University dataset

The classification performance of the SPE, PCA1, PCA1_SPE, PCA3, IM, IM_SPE, IM_DN, and IM_SPE_DN methods on the Pavia University data set is shown in Table 2. From Table 2, the OA values of the IM and IM_SPE methods were 6.49% and 7.04% higher than the SPE method on the Pavia University dataset, respectively. Compared with the PCA1 method, the OA values of the IM and IM_SPE methods were increased by 3.76% and 4.31%, respectively. Compared with the spatial-spectral fusion methods (PCA1_SPE and PCA3), the OA values were increased by 0.2%~3.43% and 0.75%~3.98%, respectively. Thus, the spectral selection method based on information measure can significantly improve the classification accuracy. Among all classification methods, the IM_DN and IM_SPE_DN methods, which combined the dimensionality reduction based on the information measure and the optimal neighborhood noise reduction, had the best OA, AA, Kappa coefficients. The AA of IM_SPE_DN method reached as high as 99.20%. Therefore, when the sample size is sufficiently large, the optimal neighborhood noise reduction method can improve the classification performance to a large extent.

Table 2. The classification results on Pavia University dataset (%).

Class	SPE	PCA1	PCA1_SPE	PCA3	IM	IM_SPE	IM_DN	IM_SPE_DN
1	94.45	96.13	96.92	97.20	98.50	98.51	99.03	**99.30**
2	93.36	95.27	95.04	97.00	97.17	97.60	98.10	**99.12**
3	90.30	92.73	92.75	96.38	97.93	97.50	**98.77**	98.10
4	95.49	97.95	98.72	97.74	97.83	99.03	98.13	**99.12**
5	99.66	99.83	99.88	98.89	99.71	**100.00**	99.72	**100.00**
6	90.43	94.06	91.66	96.99	96.72	96.67	98.14	**98.83**
7	95.50	92.88	95.13	98.21	98.47	98.90	98.84	**99.39**
8	93.16	96.95	96.59	97.76	98.92	98.82	**99.16**	99.02
9	99.87	98.64	99.88	98.04	98.08	**99.94**	98.31	**99.94**
OA	89.76	92.49	92.82	96.05	96.25	96.80	97.44	**98.56**
AA	94.69	96.05	96.29	97.58	98.15	98.55	98.69	**99.20**
Kappa	86.55	90.18	90.55	95.11	95.05	95.77	96.63	**98.10**

Figure 10a–i show the classification results of the SPE, PCA1, PCA1_SPE, PCA3, IM, IM_SPE, IM_DN, and IM_SPE_DN methods on the Pavia University data set, and Figure 10i is the ground truth of Pavia University dataset.

As can be seen in Figure 10, it is obvious that the information measure-based CNN classification method (IM and IM_SPE) can achieve higher classification accuracy on the Pavia University dataset. Moreover, the optimal neighborhood noise reduction method (IM_DN and IM_SPE_DN) can effectively reduce the salt-pepper-noise.

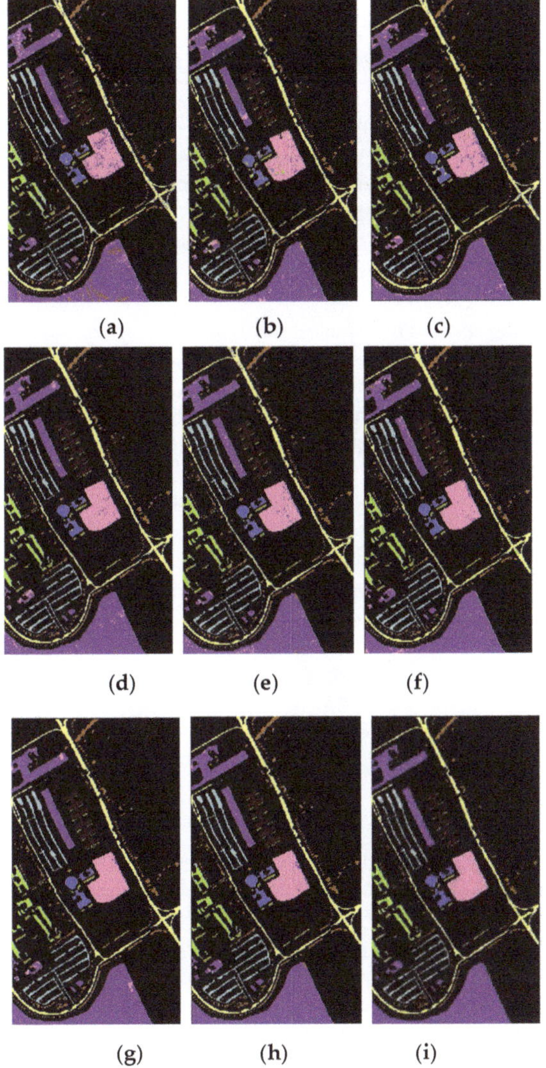

Figure 10. The classification results on Pavia University dataset: (**a**) SPE; (**b**) PCA1; (**c**) PCA1_SPE; (**d**) PCA3; (**e**) IM; (**f**) IM_SPE; (**g**) IM_DN; (**h**) IM_SPE_DN; (**i**) the ground truth.

The experimental results show that the band selection method based on information measure is better than the feature extraction method based on PCA on Indian pines dataset and Pavia University dataset. As we all know, besides PCA, there are some common dimensionality reduction methods, such as Kernel-PCA (KPCA), independent component correlation (ICA) [49], locally linear embedding (LLE) [50], etc. Research shows that machine learning by feature extraction can achieve better generalization performance than that without feature extraction. This demonstrates the fact that dimensionality reduction can improve generalization performance. Generally speaking, KPCA and ICA perform better than PCA—which is explained by the fact that KPCA and ICA can explore higher order information of the original inputs than PCA. Instead of the sample covariance matrix, (the second-order information) as used in PCA, the negentropy in ICA could take into account the higher order information of the original inputs. By using the kernel method to generalize PCA into nonlinear, KPCA also implicitly takes into account the high order information of the original inputs. A higher number of principal components could also be extracted in KPCA, eventually resulting in the best generalization performance. LLE is much better than PCA in dealing with so-called manifold dimensionality reduction. LLE maps its inputs into a single global coordinate system of lower dimensionality, and its optimizations do not involve local minima. By exploiting the local symmetries of linear reconstructions, LLE is able to learn the global structure of nonlinear manifolds, such as those generated by images of faces or documents of text. So, in the future research, we can try to use KPCA, ICA, LLE to replace the methods based on information measurement, or compare our in this paper with the classification methods based on these dimensionality reduction methods for further experimental testing.

Moreover, entropy and mutual information are used to select the most representative band of the hyperspectral image, in order to reduce the dimension. From the introduction, we can see that some hyperspectral image classification methods based on RBMS or DBN have appeared in the past few years. It is known that Restricted Boltzmann Machines (RBMs) based on unsupervised learning can be used to preprocess the data and basically to help the "machine learning" process become more efficient. Mousas et al. used RBMs to preprocess the motion features of a character's hand to enhance the estimation rate [51]. Nam et al. used sparse RBM to encode the preprocessed data into high-dimensional feature vectors in the field of music annotation and retrieval [52]. The classification methods based on RBMS or DBN generally take all spectral information of each pixel as the input of the network, and realize the classification of hyperspectral images only according to spectral information. In recent years, some classification methods using spatial-spectral information have achieved better classification results. For feature extraction of the image after dimensionality reduction, the reason why we use CNN instead of RBMS or DBN is that we want to use spatial-spectral information to classify hyperspectral images in order to improve classification accuracy. The validity of this method is also proved by the experiments.

6.2.2. Experiments of Deep Transfer Learning Methods

(1) Transfer experiment from Salinas to Indian pines

Salinas was used as a source dataset to pre-train CNN. Then the shallow layers' weight parameters were transferred to the object dataset, Indian Pines. In addition, the fine-tuning of the parameters in the network and the optimal neighborhood noise reductions were performed. In this experiment, 5% of the Salinas dataset samples were randomly selected to pre-train CNN, 10% of the Indian Pines dataset were selected as the training set, and the rest of the dataset were used as the test samples.

In order to fully verify the effectiveness of the transfer learning method, the classification method based on all spectrum data of hyperspectral images (that is, No Dimensionality reduction, NDR), the IM method, and the IM_SPE method were combined with the deep transfer learning method (MIG) to obtain the NDR_MIG, IM_MIG, and IM_SPE_MIG methods, respectively. Then the NDR_MIG, IM_MIG, and IM_SPE_MIG methods were compared with the classification methods without transfer (i.e., NDR, IM, IM_SPE). At the same time, in order to further verify the effectiveness of the proposed noise reduction method, NDR_MIG, IM_MIG, IM_SPE_MIG, were combined with the optimal neighborhood noise reduction method (DN) to obtain the NDR_MIG_DN, IM_MIG_DN, and IM_SPE_MIG_DN methods, respectively. The classification result of each method is shown in Table 3.

From Table 3, on the Indian Pines dataset, the three classification evaluation indicators (OA, AA, and Kappa coefficients) of the classification methods combined with transfer learning (NDR_MIG, IM_MIG and IM_SPE_MIG) were better than those of the non-transfer learning classification method. Especially, the classification accuracy of the NDR_MIG method was 1.77% higher than that of the NDR method. Among all classification methods, the classification methods combined with transfer learning (IM_MIG, IM_SPE_MIG and NDR_MIG) showed the best OA, AA, and Kappa coefficients. In particular, the classification accuracy of NDR_MIG_DN method was higher than that of the NDR method by 2.73%. The problem of the low classification accuracy, due to the insufficient training samples and serious noise was addressed to some degree.

Table 3. The classification results on Indian pines (%).

Class	NDR	IM	IM_SPE	NDR_MIG	IM_MIG	IM_SPE_MIG	NDR_MIG_DN	IM_MIG_DN	IM_SPE_MIG_DN
1	93.47	92.38	**98.91**	93.48	93.46	95.65	93.48	94.55	95.65
2	95.08	97.78	**98.20**	96.65	96.99	97.22	97.21	97.62	98.00
3	97.06	96.17	96.91	98.92	98.98	98.47	99.21	**99.32**	99.03
4	95.95	97.51	95.44	99.72	**99.99**	99.53	99.95	**99.99**	99.75
5	98.09	95.54	**99.36**	98.00	98.43	98.70	98.66	98.67	99.23
6	98.35	96.84	99.01	98.78	98.78	98.95	**99.20**	**99.20**	99.17
7	99.95	99.95	98.18	98.18	98.20	99.98	98.19	98.21	**99.99**
8	99.87	**99.89**	99.78	99.47	99.69	99.79	99.68	99.69	99.79
9	**99.98**	79.95	72.48	97.48	99.97	97.48	**99.98**	**99.98**	97.48
10	96.70	**98.17**	95.97	97.34	97.38	97.28	97.90	98.02	97.97
11	97.35	98.50	98.35	98.27	98.71	98.79	99.05	**99.20**	99.19
12	94.22	97.78	**98.62**	95.88	96.07	96.48	96.99	97.36	97.82
13	99.48	99.45	99.50	99.51	99.48	99.74	**99.76**	99.49	99.74
14	98.55	99.47	99.45	99.40	99.64	99.43	99.69	**99.77**	99.63
15	96.74	**99.69**	99.30	98.51	97.51	98.23	98.80	97.78	98.76
16	92.40	91.92	98.36	95.09	98.79	96.72	95.10	**98.81**	97.81
OA	94.89	96.39	96.90	96.66	97.08	97.15	97.62	97.86	**98.02**
AA	97.08	96.31	96.74	97.79	98.25	98.28	98.30	98.60	**98.69**
Kappa	94.18	95.88	96.46	96.20	96.67	96.74	97.29	97.56	97.75

Figure 11a–i are the classification results of the NDR, IM, IM_SPE, NDR_MIG, IM_MIG, IM_SPE_MIG, NDR_MIG_DN, M_MIG_DN, and IM_SPE_MIG_DN methods on the Indian Pines data set

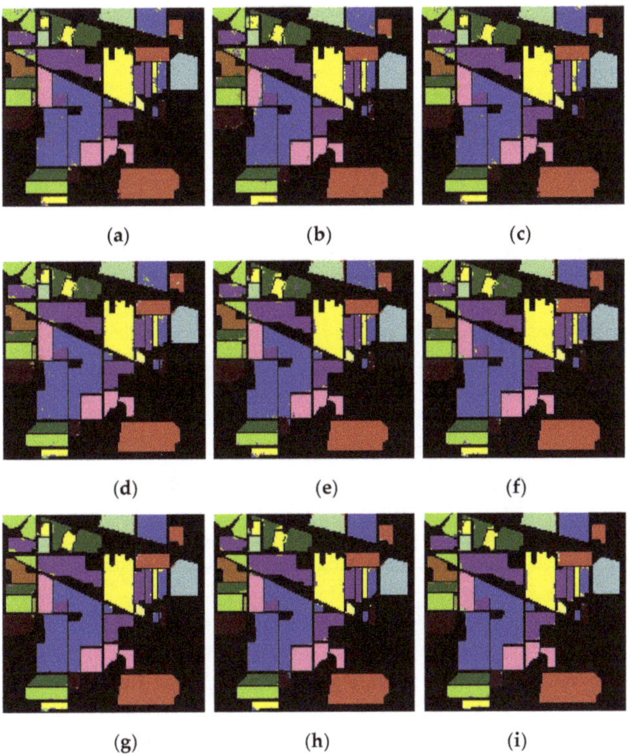

Figure 11. The classification results on Indian Pines dataset: (**a**) NDR; (**b**) IM; (**c**) IM_SPE; (**d**) NDR_MIG; (**e**) IM_MIG; (**f**) IM_SPE_MIG; (**g**) NDR_MIG_DN; (**h**) IM_MIG_DN; (**i**) IM_SPE_MIG_DN methods.

From Figure 11, it is obvious that the overall classification results on the Indian Pines dataset with small sample size using the classification method with transfer learning are significantly better than those using the classification method without transfer learning. The optimal neighborhood noise reduction method can be used to remove the salt-pepper-noise of the hyperspectral image. Among all the methods, the classification method based on both the deep transfer learning and the optimal neighborhood noise reduction showed the best and most stable classification performance.

(2) Transfer Experiments from Pavia Center to Pavia University

Pavia Center was used as the source dataset to pre-train CNN. Then the weight parameters of the shallow layers were transferred to the object dataset, Pavia University. In addition, the fine-tuning of the network and the optimal neighborhood noise reduction were performed. The Pavia Center and Pavia University datasets were used to represent the datasets with general and sufficient sample sizes, respectively. In the experiment, 9% of the source dataset, Pavia Center samples, were randomly selected to pre-train CNN, 9% of the Pavia University samples were selected as the training set of the object dataset, and the rest samples of the datasets were used as the test samples.

The classification results of the NDR, IM, IM_SPE, IM_MIG, IM_SPE_MIG, NDR_MIG, IM_MIG_DN, IM_SPE_MIG_DN, and NDR_MIG_DN methods on the Pavia University dataset are shown in Table 4.

Table 4. The classification results on Pavia University (%).

Class	NDR	IM	IM_SPE	NDR_MIG	IM_MIG	IM_SPE_MIG	NDR_MIG_DN	IM_MIG_DN	IM_SPE_MIG_DN
1	96.95	98.50	98.51	98.42	98.60	98.98	98.74	99.00	**99.17**
2	96.96	97.17	97.60	99.27	99.77	99.84	99.80	99.89	**99.93**
3	97.95	97.93	97.50	98.26	99.13	99.02	98.92	99.34	**99.39**
4	99.53	97.83	99.03	99.57	99.69	99.70	99.60	99.69	**99.73**
5	99.66	99.71	**100.00**	99.85	99.74	99.85	99.85	99.78	99.85
6	98.27	96.72	96.67	99.26	99.75	99.84	99.79	99.92	**99.93**
7	99.17	98.47	98.90	98.68	99.53	98.12	99.33	**99.83**	98.83
8	98.36	98.92	98.82	99.28	99.37	99.35	99.54	99.52	**99.56**
9	99.56	98.08	**99.94**	99.67	99.72	99.77	99.67	99.78	99.77
OA	95.87	96.25	96.80	98.48	99.12	99.22	99.19	99.41	**99.47**
AA	98.49	98.15	98.55	99.14	99.48	99.39	99.47	**99.64**	99.57
Kappa	94.59	95.05	95.77	97.99	98.83	98.97	98.93	99.22	**99.29**

From Table 4, the transfer learning and optimal neighborhood noise reduction methods increased the overall classification accuracy (OA) to above 99% on the Pavia University dataset. In particular, compared with the NDR method, the proposed NDR_MIG and NDR_MIG_DN methods increased the OA by 2.61% and 3.32%, respectively. In addition, the classification and noise-reduction effect of this method was more prominent for the dataset with a larger sample size. The Kappa coefficient of the NDR_MIG_DN method achieved 98.93%, which indicated that almost all the samples were correctly classified based on the consistency check.

Figure 12a–i show the classification results of the NDR, IM, IM_SPE, NDR_MIG, IM_MIG, IM_SPE_MIG, NDR_MIG_DN, M_MIG_DN, and IM_SPE_MIG_DN methods on the Pavia University dataset.

From Figure 12, the classification method with both the deep transfer learning and the optimal neighborhood noise reduction had outstanding performance on the Pavia University dataset. The classification performance of the proposed classification method with both the deep transfer learning and the optimal neighborhood noise reduction was significantly better than that of the non-transfer learning method (IM, IM_SPE) and NDR). In particular, the hyperspectral images processed by the classification method with both deep transfer learning and optimal neighborhood noise reduction (IM_MIG_DN, IM_SPE_MIG_DN, and NDR_MIG_DN) were almost completely noise-free and correctly classified.

From the above two groups of experiments, i.e., transfer learning classification and neighborhood noise reduction experiments, the classification method based on the transfer learning and neighborhood noise reduction exhibited significant advantages in solving the problem of low classification accuracy under the condition of insufficient training samples, thus, can avoid the over-fitting phenomenon in the training of small CNNs. By transfer between two similar datasets with a large sample size, the computational complexity can be reduced, and the accurate and stable classification results can be obtained. At the same time, through the optimal neighborhood noise reduction, the final classification result was almost noiseless. The results indicated that the transfer learning classification method and the optimal neighborhood noise reduction method could significantly improve the classification performance for the hyperspectral image.

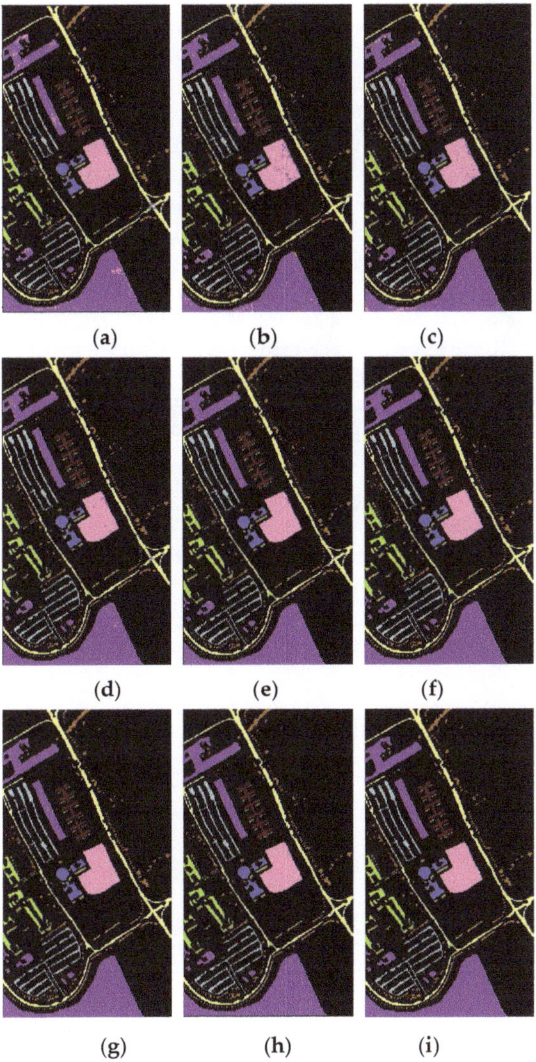

Figure 12. Classification results on the Pavia University dataset: (**a**) NDR; (**b**) IM; (**c**) IM_SPE; (**d**) NDR_MIG; (**e**) IM_MIG; (**f**) IM_SPE_MIG; (**g**) NDR_MIG_DN; (**h**) IM_MIG_DN; (**i**) IM_SPE_MIG_DN methods.

7. Conclusions

In this article, a deep transfer HSI classification method based on information measure and optimal neighborhood noise reduction was proposed. In this method, the information measure was used to reduce the dimension of the hyperspectral image. Then the fusion of the key spectral information and spatial information of the hyperspectral image was achieved, and the redundant spectral information was processed. On this basis, a classification method based on deep transfer learning and neighborhood noise reduction was proposed. The obtained classification accuracy for small samples was higher than 98% on average. Compared with the non-transfer learning method, the total classification accuracy was improved by at least 3%. For the Pavia University dataset with more samples, the classification accuracy of above 99% was obtained. The proposed method can both reduce the computational complexity to some degree and solve the problem of lower classification accuracy caused by insufficient training samples and salt-pepper-noise.

This method is suitable for HSI classification with insufficient training samples. When this situation occurs, we can use labeled samples in similar scenarios to train the network initially, and then adjust the network by transfer learning and a small number of labeled samples to achieve accurate classification of object scenarios. In addition, another advantage of this method is that by dimensionality reduction based on information measure, a pseudo-color image of the hyperspectral image can be obtained, and the hyperspectral image can be visualized. It is worth noting that the core of this method is based on transfer learning, so its limitation is that, at first, we need to get good training on a source scene similar to the target scene, which requires a sufficient number of training samples in the source scene. In the application of constructing a virtual land environment, it is necessary to select some typical scenes for common sensors, mark the samples in these typical scenes, and train the initial network. When constructing the virtual land environment for a specific area in the virtual test, the initial network trained by appropriate scene is selected, and then the high accuracy ground truth of the task area can be obtained by using the proposed method. In this paper, we make the validation by using public datasets. In future work, we will acquire relevant satellite data according to the actual task requirements, and use the proposed method to achieve high-precision ground feature information in the construction of a virtual land environment.

Author Contributions: Conceptualization and methodology, L.L. and C.C.; validation and formal analysis, C.C. and S.Z.; resources, J.Y. and C.C.; writing—original draft preparation, C.C.; writing—review and editing, C.C. and J.Y.; supervision, L.L.

Funding: This work was supported by the National Natural Science Foundation of China (Grant No. 61671170) and National Key R&D Plan (Grant No.2017YFB1302701).

Conflicts of Interest: The authors declare no conflict of interest.

References

1. Fang, B.; Li, Y.; Zhang, H.; Chan, J.; Bei, F.; Ying, L.; Haokui, Z. Semi-Supervised Deep Learning Classification for Hyperspectral Image Based on Dual-Strategy Sample Selection. *Remote Sens.* **2018**, *10*, 574. [CrossRef]
2. Lacar, F.M.; Lewis, M.M.; Grierson, I.T. Use of hyperspectral imagery for mapping grape varieties in the Barossa Valley, South Australia IGARSS 2001. Scanning the Present and Resolving the Future. In Proceedings of the IEEE 2001 International Geoscience and Remote Sensing Symposium (Cat. No.01CH37217), Sydney, Australia, 9–13 July 2001.
3. Mou, L.; Ghamisi, P.; Zhu, X.X. Unsupervised Spectral-Spatial Feature Learning via Deep Residual Conv-Deconv Network for Hyperspectral Image Classification. *IEEE Trans. Geosci. Remote Sens.* **2018**, *99*, 1–16. [CrossRef]
4. Wu, C.; Du, B.; Zhang, L. Slow Feature Analysis for Change Detection in Multispectral Imagery. *IEEE Trans. Geosci. Remote Sens.* **2014**, *52*, 2858–2874. [CrossRef]
5. Hao, W.; Saurabh, P. Convolutional Recurrent Neural Networks for Hyperspectral Data Classification. *Remote Sens.* **2017**, *9*, 298.

6. Nam, J.; Herrera, J.; Slaney, M.; Smith, J.O. Segmentation of white matter hyperintensities using convolutional neural networks with global spatial information in routine clinical brain MRI with none or mild vascular pathology. *Comput. Med Imaging Graph.* **2018**, *66*, 28–43.

7. Krizhevsky, A.; Sutskever, I.; Hinton, G.E. Imagenet classification with deep convolutional neural networks. *Adv. Neural Inf. Process. Syst.* **2012**, 1097–1105. [CrossRef]

8. Ma, X.; Geng, J.; Wang, H. Hyperspectral image classification via contextual deep learning. *Eurasip. J. Image Video Process.* **2015**, *2015*, 20. [CrossRef]

9. Zhong, Y.; Zhang, L. An Adaptive Artificial Immune Network for Supervised Classification of Multi-/Hyperspectral Remote Sensing Imagery. *IEEE Trans. Geosci. Remote Sens.* **2012**, *50*, 894–909. [CrossRef]

10. Chen, Y.; Lin, Z.; Zhao, X.; Wang, G.; Gu, Y.; Chen, Y.; Lin, Z.; Zhao, X. Deep Learning-Based Classification of Hyperspectral Data. *IEEE J. Sel. Top. Appl. Earth Obs. Remote Sens.* **2014**, *7*, 2094–2107. [CrossRef]

11. Slavkovikj, V.; Verstockt, S.; de Neve, W.; van Hoecke, S.; van de Walle, R.; Slavkovikj, V.; Verstockt, S.; de Neve, W. Hyperspectral image classification with convolutional neural networks. In Proceedings of the 23rd ACM international conference on Multimedia, Brisbane, Australia, 26–30 October 2015.

12. Makantasis, K.; Karantzalos, K.; Doulamis, A.; Doulamis, N. Deep supervised learning for hyperspectral data classification through convolutional neural networks. In Proceedings of the 2015 IEEE International Geoscience and Remote Sensing Symposium (IGARSS), Milan, Italy, 26–31 July 2015.

13. Jiang, J.; Ma, J.; Chen, C.; Wang, Z.; Cai, Z.; Wang, L. SuperPCA: A Superpixelwise PCA Approach for Unsupervised Feature Extraction of Hyperspectral Imagery. *IEEE Trans. Geosci. Remote Sens.* **2018**, 1–13. [CrossRef]

14. Chu, H.S.; Kuo, B.C.; Li, C.H.; Lin, C.T. A semisupervised feature extraction method based on fuzzy-type linear discriminant analysis. In Proceedings of the IEEE International Conference on Fuzzy Systems, Taipei, Taiwan, 27–30 June 2011.

15. Binliang, J.; Chenglong, F.; Zhaohui, W. Multidimensional scaling used for image classification based on binary partition trees. *Comput. Eng. Appl.* **2015**.

16. Zhou, X.; Xiang, B.; Zhang, M. Novel Spectral Interval Selection Method Based on Synchronous Two-Dimensional Correlation Spectroscopy. *Anal. Lett.* **2013**, *46*, 340–348. [CrossRef]

17. Jeng-Shyang, P.; Lingping, K.; Sung, P.; Sung, P.W.; Tsai, S.; Vaclav, S. Alpha-Fraction First Strategy for Hierarchical Wireless Sensor Networks. *J. Internet Technol.* **2018**, *19*, 1717–1726.

18. Guo, B.; Gunn, S.R.; Damper, R.I.; Nelson, J.D. Band Selection for Hyperspectral Image Classification Using Mutual Information. *IEEE Geosci. Remote Sens. Lett.* **2006**, *3*, 522–526. [CrossRef]

19. Groves, P.; Bajcsy, P. Methodology for hyperspectral band and classification model selection. In Proceedings of the IEEE Workshop on Advances in Techniques for Analysis of Remotely Sensed Data, Greenbelt, MD, USA, 27–28 October 2003.

20. Martínez-Usó, A.; Pla, F.; García-Sevilla, P.; Sotoca, J.M. Automatic Band Selection in Multispectral Images Using Mutual Information-Based Clustering. In Proceedings of the Iberoamerican Congress on Pattern Recognition, Cancun, Mexico, 14–17 November 2006; Springer: Berlin/Heidelberg, Germany, 2006.

21. Wang, B.; Wang, X.; Chen, Z. A hybrid framework for reservoir characterization using fuzzy ranking and an artificial neural network. *Comput. Geosci.* **2013**, *57*, 1–10. [CrossRef]

22. Le Moan, S.; Mansouri, A.; Voisin, Y.; Hardeberg, J.Y. A Constrained Band Selection Method Based on Information Measures for Spectral Image Color Visualization. *IEEE Trans. Geosci. Remote Sens.* **2011**, *49*, 5104–5115. [CrossRef]

23. Salem, M.B.; Ettabaa, K.S.; Bouhlel, M.S. Hyperspectral image feature selection for the fuzzy c-means spatial and spectral clustering. In Proceedings of the 2016 International Image Processing, Applications and Systems (IPAS), Hammamet, Tunisia, 5–7 November 2016.

24. Hossain, M.A.; Jia, X.; Pickering, M. Subspace Detection Using a Mutual Information Measure for Hyperspectral Image Classification. *IEEE Geosci. Remote Sens. Lett.* **2014**, *2*, 424–428. [CrossRef]

25. Li, W.; Wu, G.; Du, Q. Transferred Deep Learning for Anomaly Detection in Hyperspectral Imagery. *IEEE Geosci. Remote Sens. Lett.* **2017**, *14*, 597–601. [CrossRef]

26. Mehmood, A.; Nasrabadi, N.M. Kernel wavelet-Reed-Xiaoli. An anomaly detection for forward-looking infrared imagery. *Appl. Opt.* **2011**, *50*, 2744–2751. [CrossRef]

27. Yokoya, N.; Iwasaki, A. Object Detection Based on Sparse Representation and Hough Voting for Optical Remote Sensing Imagery. *IEEE J. Sel. Top. Appl. Earth Obs. Remote Sens.* **2015**, *8*, 2053–2062. [CrossRef]

28. Zhang, W.; Teng, S.; Fu, X. Scan attack detection based on distributed cooperative model. In Proceedings of the IEEE International Conference on Computer Supported Cooperative Work in Design, Xi'an, China, 16–18 April 2008.
29. Wang, J.; Luo, C.; Huang, H.; Zhao, H.; Wang, S. Transferring Pre-Trained Deep CNNs for Remote Scene Classification with General Features Learned from Linear PCA Network. *Remote Sens.* **2017**, *9*, 225. [CrossRef]
30. Wang, L.; Li, J.; Zhou, G.; Yang, D. Application of Deep Transfer Learning in Hyperspectral Image Classification. *Comput. Eng. Appl.* **2019**, *5*, 181–186. [CrossRef]
31. You, S.; Joo, S.G. *Virtual Testing and Correlation with Spindle Coupled Full Vehicle Testing System*; SAE Technical Paper; SAE International: Thousand Oaks, WA, USA, 2006.
32. Whyte, J.; Bouchlaghem, N.; Thorpe, A.; McCaffer, R. From CAD to virtual reality: Modelling approaches, data exchange and interactive 3D building design tools. *Autom. Constr.* **2000**, *10*, 43–55. [CrossRef]
33. Bartoldus, K.; Hartung, D.; Eibl, H.; Boehm, J.; Grieb, M.; Pongratz, H. Autonomous Weapons system simulation system for generating and displaying virtual scenarios on board and in flight. U.S. Patent Application 10/413,569, 20 November 2003.
34. Birkel, P.A. Fall. Synthetic Natural Environment (SNE) Conceptual Reference Model. In Proceedings of the Fall Simulation Interoperability Workshops, Orlando, FL, USA, 14–18 September 1998.
35. Lyons, D.M. System and Method for Permitting Three-Dimensional Navigation through a Virtual Reality Environment Using Camera-Based Gesture Inputs. U.S. Patent 6,181,343, 30 January 2001.
36. Arayici, Y. An approach for real world data modelling with the 3D terrestrial laser scanner for built environment. *Autom. Constr.* **2007**, *16*, 816–829. [CrossRef]
37. Chen, F.; Dudhia, J. Coupling an advanced land surface–hydrology model with the Penn State–NCAR MM5 modeling system. Part I: Model implementation and sensitivity. *Mon. Weather Rev.* **2001**, *129*, 569–585. [CrossRef]
38. Throng-The, N.; Jeng-Shyang, P.; Thi-Kien, D. *An Improved Flower Pollination Algorithm for Optimizing Layouts of Nodes in Wireless Sensor Network*; Digital Object Identifier; IEEE Access: Piscataway, NJ, USA, 2019; Volume 7. [CrossRef]
39. Shannon, C.E. A Mathematical Theory of Communication. *Bell Syst. Tech. J.* **1948**, *27*.
40. Shaw, M.Q.; Fairchild, M.D. Evaluating the 1931 CIE Color Matching Functions. *Res. Appl.* **2002**, *27*, 316–329.
41. Jacobson, N.P.; Gupta, M.R. Design goals and solutions for display of hyperspectral images. *IEEE Trans. Geosci. Remote Sens.* **2005**, *43*, 2684–2692. [CrossRef]
42. Bell, A.J. The co-information lattice. In Proceedings of the 4th International Symposium on Independent Component Analysis and Blind Signal Separation (ICA2003), Nara, Japan, 1–4 April 2003.
43. MartÍnez-UsÓMartinez-Uso, A.; Pla, F.; Sotoca, J.M.; García-Sevilla, P. Clustering-Based Hyperspectral Band Selection Using Information Measures. *IEEE Trans. Geosci. Remote Sens.* **2008**, *45*, 4158–4171. [CrossRef]
44. Palmason, J.A.; Benediktsson, J.A.; Sveinsson, J.R.; Chanussot, J. *Classification of Hyperspectral Data from Urban Areas Using Morphological Preprocessing and Independent Component Analysis*; IEEE: Piscataway, NJ, USA, 2005.
45. Shao, Y.; Sang, N.; Gao, C. Representation Space-Based Discriminative Graph Construction for Semi-supervised Hyperspectral image classification. *IEEE Signal Process. Lett.* **2017**, *25*, 1.
46. Xinyi, S. Hyperspectral Image Classification Based On Convolutional Neural Networks. Master Thesis, Harbin Institute of Technology, Harbin, Heilongjiang, China, July 2016.
47. Chen, Y.; Jiang, H.; Li, C.; Jia, X.; Ghamisi, P. Deep Feature Extraction and Classification of Hyperspectral Images Based on Convolutional Neural Networks. *IEEE Trans. Geosci. Remote Sens.* **2016**, *54*, 6232–6251. [CrossRef]
48. Lishuan, H. Study of Dimensionality Reduction and Spatial-spectral Method for Classification of Hyperspectral Remote Sensing Image. Ph.D. Thesis, China University of Geosciences, Wuhan, Hubei, China, October 2018.
49. Cao, L.J.; Chua, K.S.; Chong, W.K.; Lee, H.P.; Gu, Q.M. A comparison of PCA, KPCA and ICA for dimensionality reduction in support vector machine. *Neurocomputing* **2003**, *55*, 321–336. [CrossRef]

50. Roweis, S.T.; Saul, L.K. Nonlinear dimensionality reduction by locally linear embedding. *Science* **2000**, *290*, 2323–2326. [CrossRef] [PubMed]
51. Mousas, C.; Anagnostopoulos, C.N. Learning Motion Features for Example-Based Finger Motion Estimation for Virtual Characters. *3d Res.* **2017**, *8*, 25. [CrossRef]
52. Nam, J.; Herrera, J.; Slaney, M.; Smith, J.O. Learning Sparse Feature Representations for Music Annotation and Retrieval. In Proceedings of the 13th International Society for Music Information Retrieval Conference, ISMIR 2012, Porto, Portugal, 8–12 October 2012; pp. 565–570.

Article

Quality Assessment of Tire Shearography Images via Ensemble Hybrid Faster Region-Based ConvNets

Chuan-Yu Chang [1,*], **Kathiravan Srinivasan** [2], **Wei-Chun Wang** [1], **Ganapathy Pattukandan Ganapathy** [3], **Durai Raj Vincent** [2] and **N Deepa** [2]

[1] Department of Computer Science and Information Engineering, National Yunlin University of Science and Technology, Yunlin 64002, Taiwan; freesky1952@gmail.com
[2] School of Information Technology and Engineering, Vellore Institute of Technology (VIT), Vellore, Tamil Nadu 632014, India; kathiravan.srinivasan@vit.ac.in (K.S.); pmvincent@vit.ac.in (D.R.V.); deepa.rajesh@vit.ac.in (N.D.)
[3] Centre for Disaster Mitigation and Management, Vellore Institute of Technology (VIT), Vellore, Tamil Nadu 632 014, India; seismogans@yahoo.com
* Correspondence: chuanyu@yuntech.edu.tw

Received: 18 November 2019; Accepted: 25 December 2019; Published: 28 December 2019

Abstract: In recent times, the application of enabling technologies such as digital shearography combined with deep learning approaches in the smart quality assessment of tires, which leads to intelligent tire manufacturing practices with automated defects detection. Digital shearography is a prominent approach that can be employed for identifying the defects in tires, usually not visible to human eyes. In this research, the bubble defects in tire shearography images are detected using a unique ensemble hybrid amalgamation of the convolutional neural networks/ConvNets with high-performance Faster Region-based convolutional neural networks. It can be noticed that the routine of region-proposal generation along with object detection is accomplished using the ConvNets. Primarily, the sliding window based ConvNets are utilized in the proposed model for dividing the input shearography images into regions, in order to identify the bubble defects. Subsequently, this is followed by implementing the Faster Region-based ConvNets for identifying the bubble defects in the tire shearography images and further, it also helps to minimize the false-positive ratio (sometimes referred to as the false alarm ratio). Moreover, it is evident from the experimental results that the proposed hybrid model offers a cent percent detection of bubble defects in the tire shearography images. Also, it can be witnessed that the false-positive ratio gets minimized to 18 percent.

Keywords: intelligent tire manufacturing; digital shearography; faster region-based CNN; tire bubble defects; tire quality assessment

1. Introduction

Industry 4.0 is the novel digital technology meant for industries, and this paradigm enables the communication, collection, and analysis of data through machines, thereby allowing quicker, more agile, and efficient processes for making superior quality goods with minimal expenditure. Moreover, this digital industrial technology will assist in enhancing productivity, enabling industrial development, and revamping the profile of the personnel involved, thereby nurturing changes in the competence of business organizations and states. Further, this paradigm will foster superior efficiencies and will modify the conventional production associations between the suppliers, manufacturers, and clients and also the communication amongst humans and machines. Besides, due to the phenomenal growth in technology and agile expansion of Industry 4.0, several manufacturing firms have embraced automation, thereby replacing labor-intensive tasks in conventional production units [1]. Also, it can be observed that enabling technologies for smart tire quality assessment for realizing intelligent tire manufacturing has been gaining prominence to address challenges such as automated tire defects detection.

Generally, it can be witnessed that in the modern-day scenario, there is a humongous and swift growth in the gadgets, equipment, and devices connected to the internet. These devices, gadgets, and equipment have profound computing characteristics, and at the same time, they are exceedingly performance-oriented [2]. Due to all these facts, the concept of deep learning has evolved into a newer dimension, and it plays a significant part in processing and recognizing images, speech, and video, and so on. Furthermore, the implementation of a deep learning paradigm for automation in conventional manufacturing units significantly minimizes the usage of physical labor, and it also enhances the overall competence and efficacy of these units. Digital shearography is a laser-based measuring approach that relies on the processing of digital data, interferometry, and phase-shifting paradigm [3–5].

A shearography system was developed by applying a spatial light modulator for controlling the amount of shearing and the direction of the phase light automatically and accurately. The system eliminates the nonlinear random error and enhances the efficiency of testing [6]. A system was developed using shearography to examine the exterior heatproof covering of a cylinder, and defect detection was done using an artificial intelligence-based recognition algorithm for deep learning, namely Faster R-CNN, which detects the bounding defects [7]. A model was developed using a deep convolutional neural network for the detection of defects captured in the X-ray images. Moreover, the fully convolutional network was selected for the pixel-wise prediction of defect location and segmentation [8]. A binary classification model was established using a convolutional neural network to classify the defects in pipes into two classes, such as minor defects and major defects, and the case study was conducted from 256 shearography images [9]. An algorithm was developed based on deep learning for the classification of defects in tires using the proposed multi-column convolutional neural network (CNN) by integrating several CNNs [10].

Our research primarily focuses on the deep learning-based bubble defects detection in tire shearography images, which plays a significant role in the automation of the tire manufacturing industries. Figure 1 portrays the digital shearography set-up for capturing tire shearography images utilizing a digital computer.

The tire manufacturing process includes five substantial stages compounding, mixing, shaping, and vulcanizing, testing. Further, in the course of tire production, there is a chance for the bubble defects to appear in the tires, as the air might not be entirely removed from the tires.

Consequently, when the car is driven at high speed, such defective tires seem to suffer a greater chance of bursting, hence this scenario might place the human lives at risk. Therefore, in order to overcome this issue, the tire manufacturing units ensure that the tire is tested successfully by using several quality control mechanisms, prior to its dispatch and delivery. During the testing process, detecting the bubble defects that are present internally within the tires, turns out to be a substantial task. Moreover, the bubble defects in the four different shearography images are portrayed in Figure 2. The operator identifies the bubble detects in the shearography images and assesses whether the size of the bubble is passable. Nevertheless, the quality of inspection significantly relies on the experience and expertise of the operator. As a result, the bubble defect detection process might need substantial human expertise. Besides, the lassitude of the operator might lead to poor judgment and discrepancies in detecting the defects.

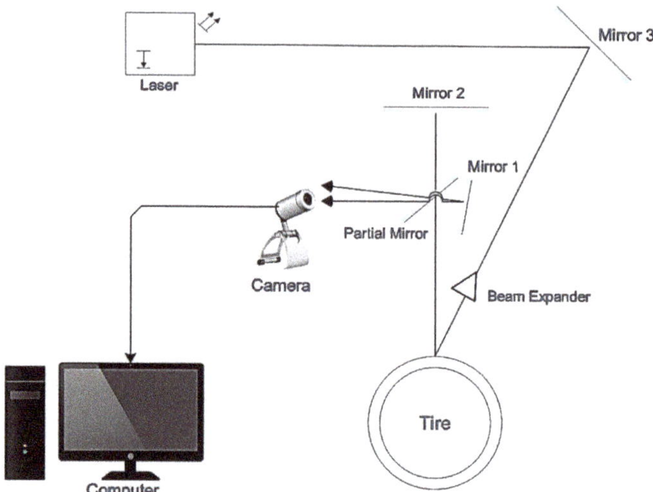

Figure 1. Digital shearography set-up for capturing tire shearography images.

Since 2012, there has been rapid progress in the convolutional neural networks-based research on image, visual, and computer-vision-based tasks [11,12]. The approach presented in the preliminary study [13] establishes two CNN architectures for tire bubble-defects diagnosis. Although, the scheme presented in [13] provides a precise identification of the bubble defects, however, the false alarm ratio, also known as the false positive ratio—more than twenty percent—is significant. The Faster R-CNN comprises two networks, primarily for generating the region proposals it makes use of a region proposal network (RPN) and secondly, a network that utilizes these region proposals for detecting the bubble defects [14].

The key contributions of this work are summarized as follows:

- The substantial contribution of this work lies in improving the architecture established earlier in [13] for effectively realizing intelligent tire manufacturing with automated defects detection.
- A Faster Region-based convolutional neural networks (R-CNN) is combined along with the architecture described in [13] for minimizing the false positive ratio.
- Further, this significant modification to the CNN architecture helps in minimizing the labor cost involved in the tire manufacturing industry.
- The results of the proposed hybrid model indicate that this approach asserts a hundred percent detection of bubble defects in the tire shearography images.
- From the results, it can be perceived that the false alarm ratio can be minimized to 18 percent.

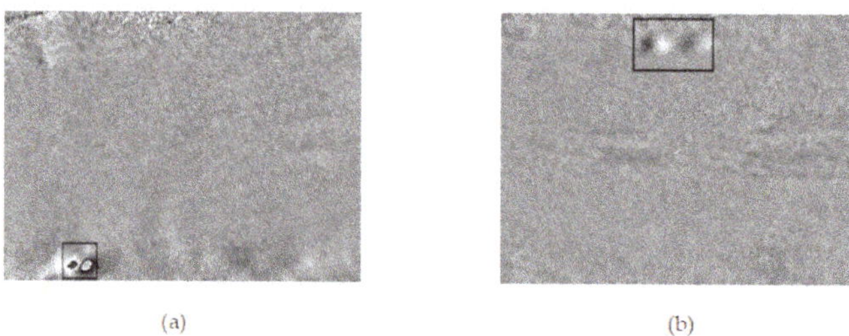

(a)   (b)

Figure 2. *Cont.*

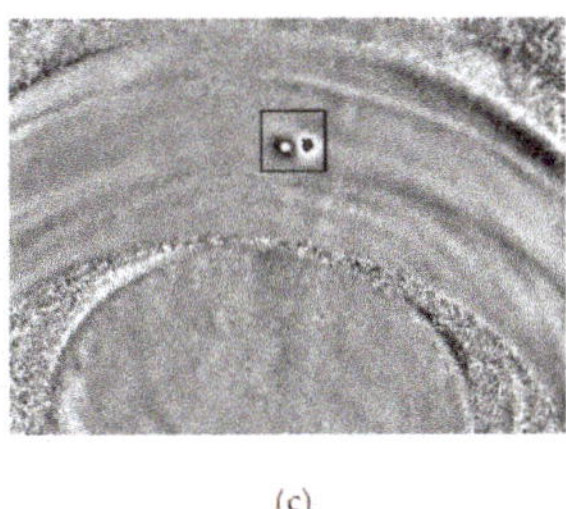 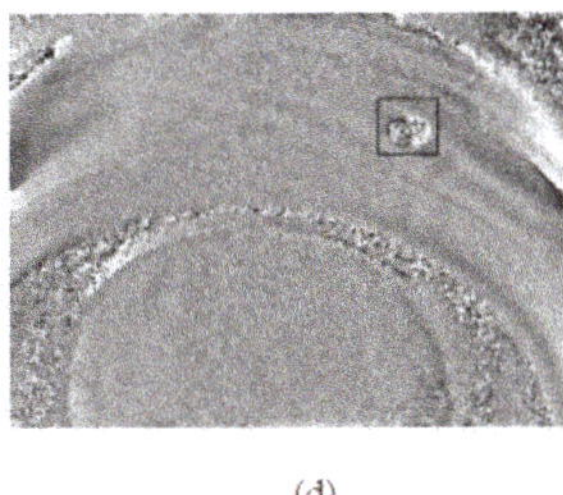

(c) (d)

Figure 2. Shearography images of tires, where (**a,b**) portray bubble defects present in the shearography images of the tire treads, and (**c,d**) illustrate bubble defects present in the shearography images of the tire sidewalls.

2. Materials and Methods

A two-stage hybrid model for detecting bubble defects in tires is proposed in this work. The primary stage includes a CNN architecture for diagnosing tire bubble-defects, and the second stage makes use of a Faster Region-based ConvNets architecture for minimizing the false positive or the false alarm ratio (FAR). A flow diagram of the proposed two-stage ensemble hybrid model is portrayed in Figure 3.

2.1. Faster Region-Based Convolutional Neural Networks

A model referred to as Regions with CNN features (R-CNN), which is a scalable object detection approach that enhances the mean average precision, was established by the research in [15]. In the research desscribed in [16], another improved version of the R-CNN model known as the Fast R-CNN was deployed with various novelties for enhancing the training and testing speed at the same time augmenting the accuracy of detection. Further, the work presented in [17] established a Faster R-CNN that introduced an RPN which shares the convolutional features of the full image with the network responsible for detection; hence, this ensures that the region proposals are achieved at a low cost. It can be observed that the RPN approach is deployed instead of the Selective Search (SS) technique [18] in the case of Faster R-CNN/ Faster Region-based ConvNets. Further, this method considerably reduces the time-period necessary for extracting the candidate regions and also for increasing the overall efficiency. Figure 4 illustrates the architectural model of the Faster R-CNN network.

The Faster R-CNN network with a ZF-net exhibits the detection results with an accuracy of 59.9% for the PASCAL VOC 2007 test set [19–21]. Besides, for the same test set, the Faster R-CNN network with VGG16 architecture accomplishes the detection results with 73.2% accuracy [19–21]. Henceforth, it can be observed that the Faster R-CNN with VGG16 architecture achieves superior accuracy, which makes it the most sought after approach. Moreover, this technique is utilized in this research to enhance the detection accuracy of the tire bubble defects. In Figure 5, the architectural model of the fully convolutional region proposal network [22] is depicted.

Figure 5 portrays the fact that the fully convolutional region proposal network applies a 3×3 window over the feature maps received from the ConvNets. Subsequently, for assessing the candidate regions, we make use of the anchors with various areas and ratios. Additionally, the chosen candidate expanses are placed into the 256-dimensional trajectory, and further, they are passed on as the inputs to the box regression layer (reg) and a box-class layer (cls). For each proposal, the outcome of the box-class layer approximates the target object or the non-target object probabilities. Consequently, a positive label will be allocated for an anchor with an Intersection-over Union (IoU) overlay proportion more significant than the value 0.7 in comparison to some ground truth box. Besides, the negative label will be allocated for the non-positive anchor with an Intersection-over Union proportion lesser than 0.3 for the remaining ground truth boxes. It can be clearly noted that the anchors which are neither positive nor negative have no role in the training for accomplishing the target. In the box regression layer, the positive sample co-ordinates achieved by the box-class layer are modified to suit the ground truth's bounding box aptly.

2.2. Image Enhancement

The classification capability and competence of the convolutional neural networks rely heavily on the two vital parameters, namely, the quality and quantity of the training samples. Nevertheless, the arduous task for this research is the identification of speckle patterns encompassing the bubble defects. In order to overcome this issue; hence, the blocks from the speckle patters encompassing the bubble defects were randomly chosen. Also, the chosen data were rotated horizontally and vertically, and then the resultant dataset helps in achieving the essential dataset required for the training process. The imperfect bubble blocks were detached physically. In this way, this research could achieve about ten times the training data. Hence, this approach makes sure that the patterns of the tire bubble defects were adequate for the training process.

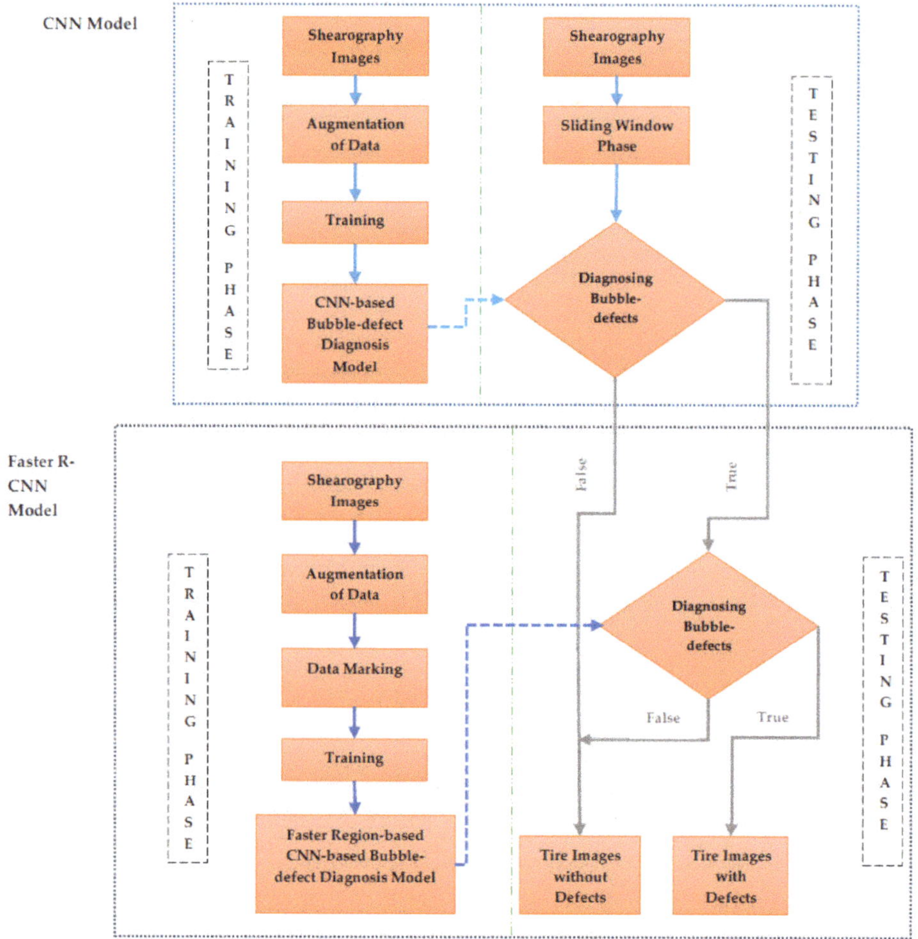

Figure 3. Flow diagram of the proposed two-stage ensemble hybrid model.

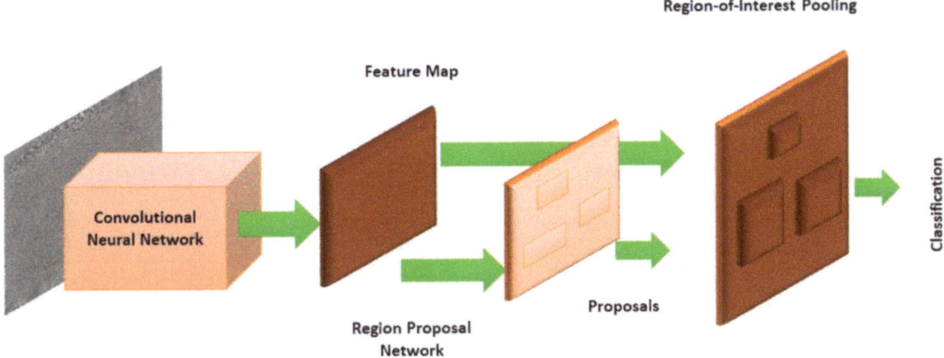

Figure 4. The architectural model of the Faster R-CNN network.

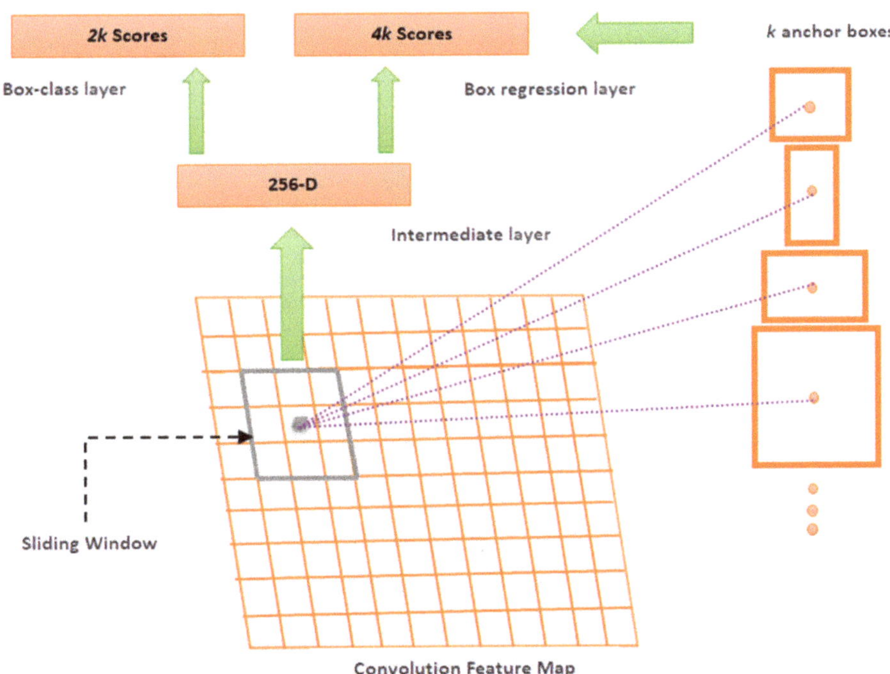

Figure 5. The architectural model of the fully convolutional region proposal network.

2.3. Classification of the Bubble Defects in Tires

It can be noticed from [13] that two convolutional neural network architectures were established for diagnosing the bubble-defects available in the treads and sidewalls of the tires. Though this approach accurately classifies the tire bubble-defects, nevertheless, the FAR seems marginally more significant than 20 percentage. Thus, our work enhances the approach in [13] by incorporating a Faster-RCNN network for reducing the false alarm ratio. The modified hybrid Faster Region-Based Convolutional Neural Networks architecture is illustrated in Figure 6. The various components of the proposed hybrid model are presented in Table 1. In the CNN, the hyper-parameters settings for tire tread are learning rate = 0.01, epoch = 30,000, batch size = 40, gamma = 0.001, power = 0.75, and momentum = 0.9. The hyper-parameters settings for tire sidewall are learning rate = 0.00001, epoch = 18,000, batch size = 18, gamma = 0.001, power = 0.75, and momentum = 0.9. In the Faster-RCNN, the learning rate, step size, and momentum are set as 0.00001, 50,000, and 0.9, respectively, for both tire tread and tire sidewall.

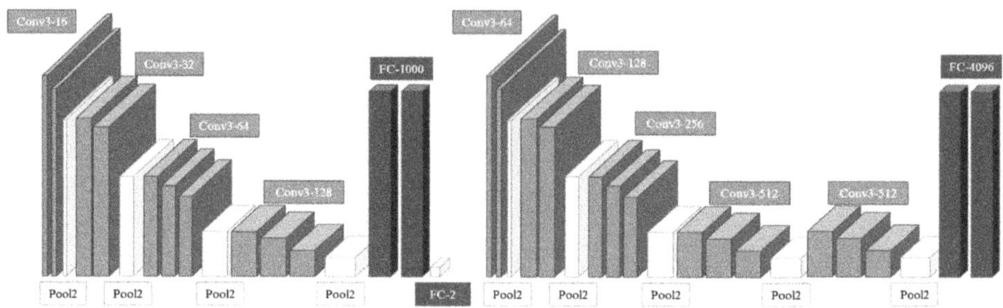

(a)

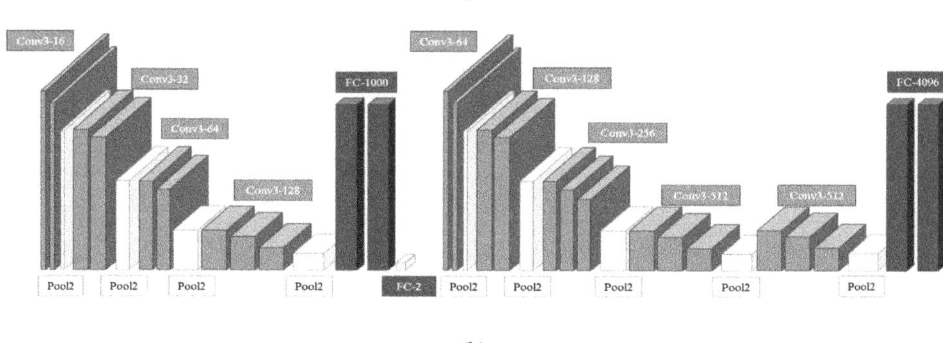

(b)

Figure 6. (a) The proposed hybrid Faster Region-Based ConvNets architecture for diagnosing the bubble-defects in treads and (b) The proposed hybrid Faster Region-Based ConvNets architecture for diagnosing the bubble-defects in sidewalls of tires.

Table 1. The various components of the proposed hybrid model.

Convolutional Neural Networks	
(a) Tread of Tires	**(b) Sidewall of Tires**
ConvNet3-16	ConvNet3-16
ConvNet3-16	ConvNet3-16
Max-pooling process	
ConvNet3-32	ConvNet3-32
ConvNet3-32	ConvNet3-32
Max-pooling process	
ConvNet3-64	ConvNet3-64
ConvNet3-64	ConvNet3-64
	ConvNet3-64
Max-pooling process	
ConvNet3-128	ConvNet3-128
ConvNet3-128	ConvNet3-128
ConvNet3-128	ConvNet3-128

Table 1. *Cont.*

Max-pooling process
Fully Connected-1000
Fully Connected-1000
Fully Connected-2
Softargmax function
Faster Region-based Convolutional Neural Networks
ConvNet3-64
ConvNet3-64
Max-pooling Process
ConvNet3-128
ConvNet3-128
Max-pooling Process
ConvNet3-256
ConvNet3-256
ConvNet3-256
Max-pooling Process
ConvNet3-256
ConvNet3-256
ConvNet3-256
ConvNet3-512
ConvNet3-512
ConvNet3-512
ConvNet3-512
Reshape process
Soft-max function
Reshape process
Proposal
ROI pooling layer
Full-connection
Softmax function
Bbox_pred   Cls_prob

2.4. The Sliding Window Phase

In this work, the original shearography image had a size of 1360×1024 pixels. In order to facilitate bubble defect detection, the shearography tire images are fragmented into a variety of blocks via the sliding window phase. Subsequently, it is evident that the location of the tire bubble defects is not known; consecutive sliding windows with 50% overlapping regions for the extraction of speckle patterns are used to avoid fragmenting the bubble defects and causing erroneous results.

The overlapping threshold has been selected to poise the time-period required for processing and also for the efficient detection of bubble defects. The abstract depiction of the sliding window indicating the overlap is presented in Figure 7.

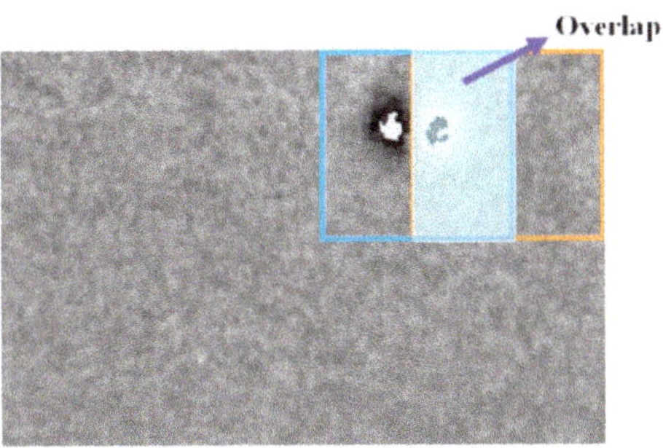

Figure 7. The abstract depiction of the sliding window indicating the overlap.

Further, this research establishes a classifier that performs the process of detection of bubble defects in treads and sidewalls of tires, which is presented in Section 2.3. Moreover, the sliding window described in Section 2.4 is used to check the segmented block images sequentially for bubble defects. If the classifier classifies a block as containing bubbles, the Faster R-CNN is used to determine if the result is a false positive. If the resultant image is not false positive, then in the original image, the respective location of the block is encircled. As a result, this image is passed on to the professional operators for assessing the quality of the tires and also for removing the defective piece. Furthermore, this devised semi-automated assessment process offers cost-leadership when compared with the traditional manual inspection and also improves the reliability of the inspection process.

3. Results

In this work, the diagnosis of bubble defects in tires established in [13] and the Faster Region-based Convolutional Neural Networks approach is amalgamated for obtaining 100% detection of defects and also aiding in reducing the false alarm ratio. The evaluation metrics, such as the accuracy, sensitivity, and specificity, are used for assessing the performance of the proposed hybrid model. These metrics are computed using the following expressions:

$$Accuracy = (TP + TN)/(P + N) \tag{1}$$

$$Sensitivity = TP/P \tag{2}$$

$$Specificity = TN/N \tag{3}$$

where TP stands for true positive, it represents the amount of diagnosed bubble patterns, which really possesses the bubble defects, and FP stands for false-positive and, it indicates the amount of not bubble patterns, which are wrongly diagnosed as bubble defects. True negative (TN) illustrates the amount of not bubble patterns, which are diagnosed as not bubble defects. Positives (P) represents the real bubble defects and negatives (N) denotes the not bubble defects. Among the evaluation metrics, sensitivity is the most necessary measure for achieving the complete detection of bubble defects.

Moreover, a tire company provided the shearography images deployed in this research. Usually, the tire bubble defects were physically delineated with the assistance of experienced professionals. The amount of training images and blocks are clearly organized in Table 2. Further, it is evident that for the training process, the tire manufacturer supplied the 325 tire shearography images with bubble defects. Subsequently, the image enhancement approach is deployed for imitating 8596 and 5052 blocks from 223 tire tread images and 102 tire sidewall images containing bubble defects. Additionally, Table 3 indicates the test dataset, it comprises of 541 tire shearography images deprived of bubble defects and 256 tire shearography images having bubble defects.

Table 2. Training Dataset Details.

	Tire Treads		Tire Sidewalls	
	No. of Images	No. of Blocks	No. of Images	No. of Blocks
Shearography without bubble	1409	8811	1545	10514
Shearography with bubbles	223	8596	102	5052

Table 3. Testing Dataset Details.

	Tire Treads	Tire Sidewalls
	No. of Images	No. of Images
Shearography without bubble	262	279
Shearography with bubbles	136	120

An area with bubble defects is expected to be smaller than the area of the default anchor of the Faster R-CNN. Therefore, in this work, the anchor's ratio and scale are adjusted according to the area of the bubble defects. Table 4 shows the ratio and scale adjustment of the anchors. Twelve anchor configurations are used for candidate regions in the marking of bubbles.

Table 4. Anchors Configuration.

	Original	Proposed Ensemble Hybrid Model
Ratios	[0.5,1,2]	[0.3,0.4,0.5,0.75]
Scale	[8,16,32]	[8,16,32]

The proposed hybrid model has been compared with various classifiers including the Support Vector Machine (SVM) [23], Random Forest Model [24], Haar-like AdaBoost Method [25], Chang's method [13], and the integrated model comprising of SVM, Random Forest Model, AdaBoost method. Besides, the proposed model was compared with these methods for verifying its performance. Table 5 illustrates the diagnosis of bubble-defects in treads of tire shearography images for several existing methods in comparison with the proposed ensemble hybrid model in terms of the evaluation metrics such as accuracy, sensitivity, and specificity. Additionally, Table 6 depicts the diagnosis of bubble-defects in sidewalls of tire shearography images for numerous prevailing approaches in comparison with the proposed ensemble hybrid model in terms of the assessment metrics such as accuracy, sensitivity, and specificity. Further, it can be witnessed from these tables that the work in [13] and the presented ensemble hybrid approach achieve 100 percent sensitivity, by successfully identifying each and every bubble-defect. Also, it can be observed that the presented ensemble hybrid approach surpasses all other existing approaches in terms of specificity. However, the presented ensemble hybrid approach requires a processing time of approximately 7 seconds/image, whereas the approach established in [13] takes only a processing-time of roughly 6 seconds/image. Nevertheless, the presented ensemble hybrid model is superior in other means and also in terms of specificity, when compared with the other existing approaches.

Figure 8a–d illustrate the shearography images or the speckle patterns acquired using digital shearography, and Figure 8e–h depict the detection of bubble defects in tires using the proposed hybrid Faster Region-based convolutional neural networks model. Figure 8e–h indicate the fact that all bubble defects in tires have been detected successfully. Figure 9a–d depict the false positive or the false alarm inspection results in [13], where the shearography images do not have bubble defects; however, they get misrepresented as possessing the bubble defects. Figure 9e–h illustrate the assessment results of the hybrid Faster Region-based convolutional neural networks model using the same set of input images. It can be witnessed in Figure 9e–h that the shearography images have no bubble defects. Besides, it reveals the fact that the proposed hybrid Faster Region-based convolutional neural networks model effectively reduces the false-positive ratio or the false alarm rate.

Table 5. Diagnosis of Bubble-defects in Treads of Tire Shearography Images.

Measurement Methods	Accuracy (%)	Sensitivity (%)	Specificity (%)
Support Vector Machine [23]	55.53	92.65	36.26
Random-Forest Model [24]	59.3	96.32	40.08
Haar-like Ada-Boost Method [25]	62.81	97.06	45.04
Integrated Model comprising of Support Vector Machine, Random-Forest Model, Ada-Boost Method	79.15	96.32	70.23
Chang's method [13]	87.94	100	81.68
Proposed Hybrid Faster Region-based Convolutional Neural Networks Model	89.16	100	83.09

Table 6. Diagnosis of Bubble-defects in Sidewalls of Tire Shearography Images.

Measurement Methods	Accuracy (%)	Sensitivity (%)	Specificity (%)
Support Vector Machine [23]	50.13	81.67	36.56
Random-Forest Model [24]	44.61	85.83	26.88
Haar-like Ada-Boost Method [25]	46.37	82.5	30.82
Integrated Model comprising of Support Vector Machine, Random-Forest Model, Ada-Boost Method	61.9	85	51.97
Chang's method [13]	85.46	100	79.21
Proposed Hybrid Faster Region-based Convolutional Neural Networks Model	86.87	100	80.15

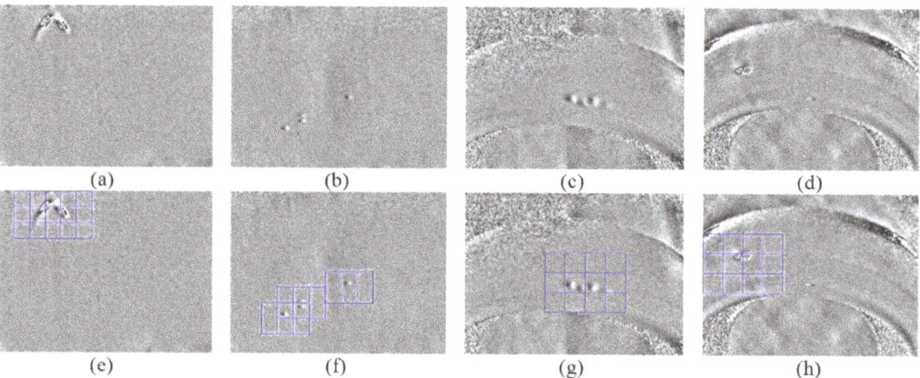

Figure 8. (**a–d**) the shearography images or the speckle patterns acquired using digital shearography, (**e–h**) the detection of bubble defects in tires using the proposed hybrid Faster Region-based convolutional neural networks model.

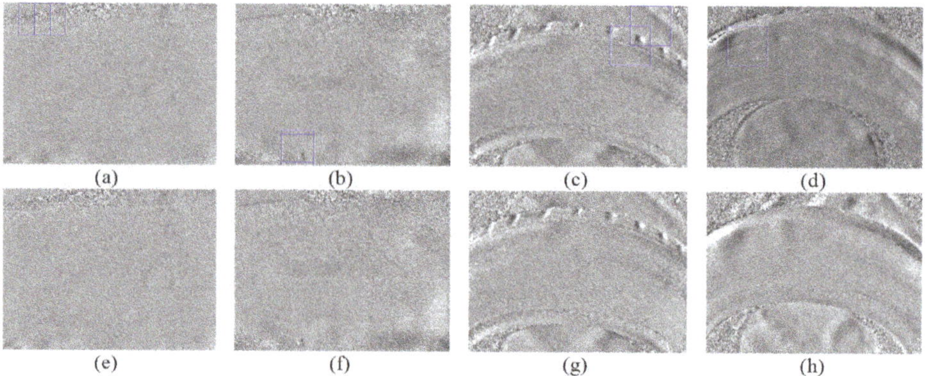

Figure 9. Chang's model misclassification results [13] and the Proposed Hybrid Faster Region-based Convolutional Neural Networks Model. (**a–d**) diagnosis outcomes of tire bubble defects using Chang's model, (**e–h**) the Proposed Hybrid Faster Region-based Convolutional Neural Networks Model's bubble defects detection results.

4. Conclusions

In the tire manufacturing process, the diagnosis of bubble-defects in the treads and sidewalls of shearography tire images represents a significant task. Therefore, enabling smart tire quality assessment seems to be an essential way of realizing intelligent tire manufacturing practices that can ensure automated detection of defects. Further, an ensemble hybrid combination of the CNN with a high-performance Faster Region-based ConvNets for classifying and diagnosing the bubble-defects present in the tire shearography images. The proposed hybrid Faster Region-based convolutional neural networks model reduces misjudgments caused by human errors and achieves high consistency in the quality of bubble-defect detection. It is clearly evident from the results that in addition to thoroughly diagnosing the bubble-defects in tires, the hybrid Faster Region-based convolutional neural networks model decreases the false alarm ratio of not-bubble defects in tires from 20% to a rate of 18%. Also, it has to be noted that this hybrid system model was deployed in a tire manufacturing unit, and it produced efficient results in automatically diagnosing the bubble-defects in treads and sidewalls of tires. In the future work, more advanced CNN enabled approaches can be implemented for automated detection of defects [26–30], thus ensuring and realizing a sustainable tire manufacturing process.

Author Contributions: Conceptualization, C.-Y.C., K.S., and W.-C.W.; methodology, C.-Y.C., K.S., and W.-C.W.; software, K.S., and W.-C.W.; validation, G.P.G., D.R.V., and N.D.; formal analysis, K.S., and W.-C.W.; investigation, K.S., and W.-C.W.; resources, C.-Y.C. and K.S.; data curation, G.P.G., D.R.V. and N.D.; writing—original draft preparation, K.S.; writing—review and editing, C.-Y.C., K.S., W.-C.W., G.P.G., D.R.V. and N.D.; visualization, W.-C.W.; supervision, C.-Y.C.; project administration, C.-Y.C.; funding acquisition, C.-Y.C. All authors have read and agreed to the published version of the manuscript.

Funding: This research was partially funded by "Intelligent Recognition Industry Service Research Center" from The Featured Areas Research Center Program within the framework of the Higher Education Sprout Project by the Ministry of Education (MOE) in Taiwan. Grant number: N/A and the APC was funded by the aforementioned Project.

Conflicts of Interest: The authors declare no conflict of interest.

References

1. Lelli, F. Interoperability of the Time of Industry 4.0 and the Internet of Things. *Future Internet* **2019**, *11*, 36. [CrossRef]
2. Kathiravan, S.; Kanakaraj, J. A Review on Potential Issues and Challenges in MR Imaging. *Sci. World J.* **2013**, *2013*, 783715. [CrossRef] [PubMed]
3. Zhao, Q.; Dan, X.; Sun, F.; Wang, Y.; Wu, S.; Yang, L. Digital Shearography for NDT: Phase Measurement Technique and Recent Developments. *Appl. Sci.* **2018**, *8*, 2662. [CrossRef]
4. Lopato, P. Double-Sided Terahertz Imaging of Multilayered Glass Fiber-Reinforced Polymer. *Appl. Sci.* **2017**, *7*, 661. [CrossRef]
5. Steinchen, W.; Yang, L. *Digital Shearography: Theory and Application of Digital Speckle Pattern Shearing Interferometry*; SPIE Press: Bellingham, WA, USA, 2003.
6. Sun, F.; Wang, Y.; Yan, P.; Zhao, Q.; Yang, L. The application of SLM in shearography detecting system. *Opt. Lasers Eng.* **2019**, *114*, 90–94. [CrossRef]
7. Ye, Y.; Ma, K.; Zhou, H.; Arola, D.; Zhang, D. An automated shearography system for cylindrical surface inspection. *Measurement* **2019**, *135*, 400–405. [CrossRef]
8. Wang, R.; Guo, Q.; Lu, S.; Zhang, C. Tire Defect Detection Using Fully Convolutional Network. *IEEE Access* **2019**, *7*, 43502–43510. [CrossRef]
9. Fröhlich, H.B.; Fantin, A.V.; de Oliveira, B.C.F.; Willemann, D.P.; Iervolino, L.A.; Benedet, M.E.; Jnior, A.A.G. Defect classification in shearography images using convolutional neural networks. In Proceedings of the 2018 International Joint Conference on Neural Networks (IJCNN), Rio de Janeiro, Brazil, 8–13 July 2018; pp. 1–7.
10. Cui, X.; Liu, Y.; Zhang, Y.; Wang, C. Tire defects classification with multi-contrast convolutional neural networks. *Int. J. Pattern Recognit. Artif. Intell.* **2018**, *32*, 1850011. [CrossRef]
11. Srinivasan, K.; Sharma, V.; Jayakody, D.N.K.; Vincent, D.R. D-ConvNet: Deep learning model for enhancement of brain MR images. *Basic Clin. Pharmacol. Toxicol.* **2018**, *124*, 3–4.
12. Srinivasan, K.; Ankur, A.; Sharma, A. Super-resolution of Magnetic Resonance Images using deep Convolutional Neural Networks. In Proceedings of the 2017 IEEE International Conference on Consumer Electronics—Taiwan (ICCE-TW), Taipei, Taiwan, 12–14 June 2017; pp. 41–42. [CrossRef]

13. Chang, C.-Y.; Huang, J.-K. Tires Defects Detection Using Convolutional Neural Networks. In Proceedings of the 2017 International Conference on Visualization, Graphics and Image Processing (CVGIP), Nantou, Taiwan, 21–23 August 2017.

14. Chang, C.-Y.; Wang, W.-C. Integration of CNN and Faster R-CNN for Tire Bubble Defects Detection. In Proceedings of the 13th International Conference on Broadband and Wireless Computing, Communication and Applications (BWCCA-2018), Taichung, Taiwan, 27–29 October 2018. [CrossRef]

15. Girshick, R.; Donahue, J.; Darrell, T.; Malik, J. Rich feature hierarchies for accurate object detection and semantic segmentation. In Proceedings of the IEEE Conference on Computer Vision and Pattern Recognition, Washington, DC, USA, 23–28 June 2014; pp. 580–587.

16. Girshick, R. Fast R-CNN. In Proceedings of the 2015 IEEE International Conference on Computer Vision (ICCV), Washington, DC, USA, 7–13 December 2015; pp. 1440–1448. [CrossRef]

17. Ren, S.; He, K.; Girshick, R.; Sun, J. Faster R-CNN: Towards Real-Time Object Detection with Region Proposal Networks. *IEEE Trans. Pattern Anal. Mach. Intell.* **2017**, *39*, 1137–1149. [CrossRef] [PubMed]

18. Uijlings, J.R.; van de Sande, K.E.; Gevers, T.; Smeulders, A.W. Selective search for object recognition. *Int. J. Comput. Vis.* **2013**, *104*, 154–171. [CrossRef]

19. Matthew, D.Z.; Fergus, R. Visualizing and understanding convolutional neural networks. In Proceedings of the 13th European Conference Computer Vision and Pattern Recognition, Zurich, Switzerland, 6–12 September 2014; pp. 818–833.

20. Everingham, M.; Van Gool, L.; Christopher, K.I.; Williams, J.W.; Zisserman, A. The PASCAL Visual Object Classes(VOC) Challenge. *Int. J. Comput. Vis.* **2010**, *88*, 303–338. [CrossRef]

21. Simonyan, K.; Zisserman, A. Very deep convolutional networks for large-scale image recognition. In Proceedings of the International Conference on Learning Representations, San Diego, CA, USA, 7–9 May 2015.

22. Long, J.; Shelhamer, E.; Darrell, T. Fully convolutional networks for semantic segmentation. In Proceedings of the IEEE Conference Computer Vision and Pattern Recognition, Boston, MA, USA, 7–12 June 2015.

23. Cortes, C.; Vapnik, V. Support-vector Networks. *Mach. Learn.* **1995**, *20*, 273–297. [CrossRef]

24. Breiman, L. Random forests. *Mach. Learn.* **2001**, *45*, 5–32. [CrossRef]

25. Freund, Y.; Robert, E.; Shapire, A. Decision-Theoretic Generalization of On-Line Learning and an Application to Boosting. *J. Comput. Syst. Sci.* **1997**, *55*, 119–139. [CrossRef]

26. Liu, B.; Zou, D.; Feng, L.; Feng, S.; Fu, P.; Li, J. An FPGA-Based CNN Accelerator Integrating Depthwise Separable Convolution. *Electronics* **2019**, *8*, 281. [CrossRef]

27. Sinha, R.S.; Hwang, S.-H. Comparison of CNN Applications for RSSI-Based Fingerprint Indoor Localization. *Electronics* **2019**, *8*, 989. [CrossRef]

28. Rivera-Acosta, M.; Ortega-Cisneros, S.; Rivera, J. Automatic Tool for Fast Generation of Custom Convolutional Neural Networks Accelerators for FPGA. *Electronics* **2019**, *8*, 641. [CrossRef]

29. Li, T.; Zhao, E.; Zhang, J.; Hu, C. Detection of Wildfire Smoke Images Based on a Densely Dilated Convolutional Network. *Electronics* **2019**, *8*, 1131. [CrossRef]

30. Wang, D.; Shen, J.; Wen, M.; Zhang, C. Efficient Implementation of 2D and 3D Sparse Deconvolutional Neural Networks with a Uniform Architecture on FPGAs. *Electronics* **2019**, *8*, 803. [CrossRef]

Article

High-Resolution Image Inpainting Based on Multi-Scale Neural Network

Tingzhu Sun [1,2], Weidong Fang [3], Wei Chen [1,2,4,*], Yanxin Yao [5], Fangming Bi [1,2] and Baolei Wu [1,2]

[1] School of Computer Science and Technology, China University of Mining Technology, Xuzhou 221000, Jiangsu, China; stz0526@163.com (T.S.); bfm@cumt.edu.cn (F.B.); blwu@cumt.edu.cn (B.W.)

[2] Mine Digitization Engineering Research Center of the Ministry of Education, China University of Mining and Technology, Xuzhou 221116, Jiangsu, China

[3] Key Laboratory of Wireless Sensor Network & Communication, Shanghai Institute of Micro-system and Information Technology, Chinese Academy of Sciences, Shanghai 201800, China; weidong.fang@mail.sim.ac.cn

[4] School of Earth and Space Sciences, Peking University, Beijing 100871, China

[5] School of Communication and Information Engineering, Beijing Information Science & Technology University, Beijing 100101, China; yanxin_buaa@126.com

* Correspondence: chenwdavior@163.com; Tel.: +86-1392-176-1978

Received: 10 October 2019; Accepted: 15 November 2019; Published: 19 November 2019

Abstract: Although image inpainting based on the generated adversarial network (GAN) has made great breakthroughs in accuracy and speed in recent years, they can only process low-resolution images because of memory limitations and difficulty in training. For high-resolution images, the inpainted regions become blurred and the unpleasant boundaries become visible. Based on the current advanced image generation network, we proposed a novel high-resolution image inpainting method based on multi-scale neural network. This method is a two-stage network including content reconstruction and texture detail restoration. After holding the visually believable fuzzy texture, we further restore the finer details to produce a smoother, clearer, and more coherent inpainting result. Then we propose a special application scene of image inpainting, that is, to delete the redundant pedestrians in the image and ensure the reality of background restoration. It involves pedestrian detection, identifying redundant pedestrians and filling in them with the seemingly correct content. To improve the accuracy of image inpainting in the application scene, we proposed a new mask dataset, which collected the characters in COCO dataset as a mask. Finally, we evaluated our method on COCO and VOC dataset. the experimental results show that our method can produce clearer and more coherent inpainting results, especially for high-resolution images, and the proposed mask dataset can produce better inpainting results in the special application scene.

Keywords: image inpainting; content reconstruction; instance segmentation

---

1. Introduction

Every day, about 300 million pictures are captured and shared on social networks, and a large part of them are human-centered pictures (including selfies, street photos, travel photos, etc.,). Many human-related research directions have been produced in computer vision and machine learning in recent years. Among them, target tracking [1] (including pedestrian detection [2], pedestrian reidentification [3], human pose estimation [4], etc.,), face recognition [5], and face image inpainting (including pet eye fix [6], eye-closing to eye-opening [7], etc.,) are the research hotspots. Many researchers devote themselves in improving the performance of the existing network. However, integrating existing researches and enabling them to solve common problems in life is also of high practical significance.

In our social networks, we can often see street photos as shown on the left of Figure 1, but actually, the original images seem as shown on the right side of Figure 1. We can see redundant pedestrians in their background destroying beauty and artistic conception of the image. So the purpose of our study is to delete the redundant pedestrian in the image and ensure the reality of background inpainting. It involves pedestrian detection, identifying redundant pedestrians and filling in them with the seemingly correct content. This is a challenging problem because (1) the result largely depends on the accuracy of redundant pedestrian detection; (2) the diversity of background information under the redundant pedestrian area is difficult to recover; (3) the training data lacks real output samples to define the reconstruction loss. We want to deploy our work in the real world as a working application, so we took an interactive approach, removing unnecessary sections by manually selecting unnecessary pedestrian areas after pedestrian detection. After the user removed the unnecessary parts, our algorithm successfully filled the remaining holes with the surrounding background information.

(a) (b)

Figure 1. Street photo of the paper's special application scene. (**a**) The repaired figure, (**b**) the figure that is not repaired.

To counter the problems above, we combine the research of instance segmentation, image inpainting. Firstly, we need to complete the instance segmentation of human which detects the regions existing characters. Then it needs us to identify the target character and "protect" the region existing the target character. Finally, we use an image inpainting algorithm to repair other regions. To further improve the inpainting result of the task, we build a new mask dataset, which collects the characters in COCO dataset as a mask, representing various pose. The new mask dataset can produce a better inpainting result on character filtering tasks.

In recent years, deep network has achieved high-quality results in instance segmentation, image inpainting, and so on. Instance segmentation is the combination of object detection and semantic segmentation. First, it uses an object detection algorithm to locate each object in the image with positioning boxes. And then it adapts a semantic segmentation algorithm to mark the target objects in different positioning boxes to achieve the purpose of instance segmentation. The latest instance segmentation is Mask-R-CNN [8], which adds a mask branch of predictive segmentation for each region of interest based on Faster-R-CNN [9]. The mask branch only adds a small computational overhead but supports rapid systems and quick experiments.

Image inpainting can be defined as entering an incomplete image and filling in the incomplete area with semantically and visually believable content. Since Deepak [10] et al. adapts encoder-decoder to complete the inpainting of face images, image inpainting has two transformations from dealing with a fixed shape region to dealing with any non-central and irregular region [11,12], and from distortion to a smooth and clear inpainting result [13,14]. Image inpainting based on GAN network has made great progress in recent years, but for the high-resolution images, the inpainting results will still appear blurred texture and the unpleasant boundaries that are inconsistent with the surrounding area. We found SRGAN [15] has proved superiority on restoring finer texture details, therefore we put forward a new method based on deep generative model. The methods is a two-stage network consisting of content reconstruction and texture detail restoration. We further restore the finer texture details inspired by the architecture of SRGAN after holding the visually believable fuzzy texture. It can effectively solve the problem of structural distortion and texture blur to improve the quality of image inpainting.

In this paper, we propose a high-resolution image inpainting method based on the multi-scale neural network and build a new mask dataset for the special application scene. The main contributions of this paper are:

(1) Based on the current most advanced image inpainting network, we build a texture detail restoration network to restore the details of high-resolution images inspired by SRGAN. The experimental results show that our method can generate a smoother, clearer and more coherent inpainting result than other methods.

(2) To remove unnecessary pedestrians from the image, we proposed a new mask dataset, which collected various pose and could produce better inpainting results in the task of character filtering.

To train our network, we applied the new mask dataset to simulate the real pedestrian. Although we just built the mask dataset to represent the removed pedestrian area, we achieved good results in the real-world data.

The rest is organized as follows. The second part introduces the research status of instance segmentation and image inpainting at home and abroad. The third part describes the improved image inpainting network, and describes the construction of the mask dataset and the special application scene. The fourth part gives the experimental results. The fifth part gives the conclusion.

2. Related Works

In the past ten years, computer vision has made great progress in image processing tasks such as classification, target detection, segmentation, and so on. The performance of deep network has been greatly improved in these tasks, which lays a foundation for the new research difficult problems of image processing and provides support for image inpainting in this paper. We briefly review the relevant work in various sub-areas related to this article.

Instance segmentation integrates image classification, image segmentation, and target detection in computer vision. The earliest region-based CNN(R-CNN) [16] detecting object with a bounding box is to process a certain number of candidate object regions on each ROI independently. Faster-R-CNN [9] based on R-CNN improves by learning the attention mechanism of Region Proposal Network (RPN). Faster-R-CNN is flexible and robust for many later improvements [17–19], and leads the several current benchmarks. Li [20] et al. combines the two types of score map [21] and the target detection [22] to realize "full convolution instance segmentation" (FCIS). Different from the usual method, which predicts a set of position-sensitive channels with full convolution, this method abandons full connected layers for the shared subtasks of image segmentation and image classification, making the network more lightweight. In addition, no trainable parameters exist in either the integrated score map or the result, only the classifier exists. The Mask-R-CNN [8] used in the paper adds a mask branch of predictive segmentation to each region of interest based on Faster-R-CNN [9]. The mask branch only adds a small computational overhead and supports rapid systems and quick experiments.

Traditional inpainting approaches based on diffusion or patch typically use variational algorithms or patch similarity to spread information from background to holes, such as [23,24]. One of the most advanced methods for image inpainting at present is PatchMatch [25], without the use of deep learning, which fills in holes with statistical data of available images through iterated search for the most suitable patch. Although it produces a smoother result, it assumes the texture of the inpainting area can be found elsewhere in the image. This assumption does not always hold. Therefore, it is good at restoring patterned regions, such as background reconstruction, but has difficulty in reconstructing locally unique patterns.

Generative adversarial network makes the research of image inpainting to a peak. Vanilla GANs [26] shows good performance in generating clear images, but has difficulty extending to higher-resolution images due to the instability of training. Several techniques for stable training processes have been proposed, including DCGAN [27], energy-based GAN [28], Wasserstein GAN (WGAN) [29,30], WGAN-GP [31], BEGAN [32], and LSGAN [33]. A more relevant task of image inpainting is conditional image generation. For example, Pix2Pix [34], Pix2Pix HD [35], and CycleGAN [36] transform images in different domains using paired or unpaired data.

The commonly used loss function of image inpainting based on generative adversarial network is a combination of adversarial loss and L2 loss. L2 loss can excite the output of the generated network with variance computing, but cannot capture the high-frequency details and repair the clear texture structure. So the introduction of adversarial loss can effectively solve the problem.

The basic model of image inpainting based on generative adversarial networks is the encoder-decoder used by Deepak [10] et al. To improve the inpainting result of face images, it combines L2 loss with adversarial loss. The latest and effective image inpainting models based on deep learning are mostly developed on this basis. However, the shape of the repaired region is fixed so it has a strong limit in practical application. In response to this question, Liu [12] et al. introduces partial convolution, which can process any non-central, irregular region. However, the method still needs to establish a mask dataset based on deep neural network and conduct pre-training on the irregular masks of random lines. Iizuka [11] uses dilated convolution to increase the receptive field, which obtains the image information in a larger range as much as possible without missing extra information. This method is suitable for solving the inpainting problem of non-center and irregular region, but it has poor inpainting result on structural objects. In recent years, GAN has made a great breakthrough in the application of image inpainting. In the future, there will be more research progress on image inpainting based on deep learning.

Yu [14] et al. improves the generated network of image inpainting based on Iizuka's research [11], and proposes a unified feedforward generation network with a novel context attention layer. The proposed network consists of two phases. The first phase is to roughly extract the missing content after reconstruction loss training with dilated convolution. The second phase is to integrate the context attention. The core idea of context attention is to use the characteristics of known patch as convolution filters to generate patch. The two generating networks are similar to UNET.

Inspired by [14], the issue is divided into two subtasks: (1) The first subtask uses the context encoder (CE) [10] to fill in the large areas need repaired according to the environmental information. (2) As CE cannot recover high-frequency details, the second subtask uses a network similar to SRGAN to capture high-frequency details.

3. The Approach

3.1. Improved Image Inpainting Network

We first constructed our generated network of image inpainting by copying and improving the most advanced inpainting [14] model recently. The network shows a good inpainting result in natural images.

Our improved image inpainting network is shown in Figure 2. We follow the same input and output configurations as in [14] for training and inference. The improved generated network takes an image with white pixels filled in the holes A and a binary mask indicating the hole regions as the input pair, and it outputs the final repaired image. The size of the input image is 256 × 256 and the size of the output image is also 256 × 256. We trained our improved image inpainting network on two datasets including COCO and VOC datasets, as described in the Section 4.

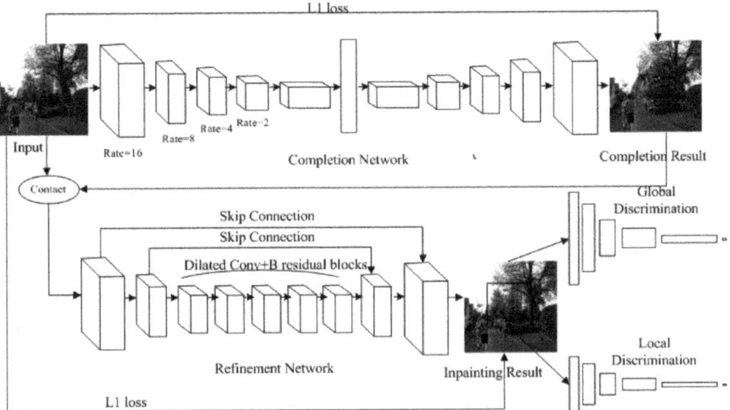

Figure 2. Improved image inpainting network.

To further improve the visual effect of high-resolution image inpainting and reduce the blurred texture and the unpleasant boundaries that inconsistent with the surrounding area, the network we introduced consists of two stages: 1) content reconstruction network, 2) texture details restoration network. The first network is a completion network used to complete the image and obtain the rough prediction results, and it adopts reconstruction loss when training. The second network is a refinement network. It takes coarse prediction results as input to further restore finer texture details of high-resolution images without changing semantic information of coarse prediction results, and it adopts reconstruction loss and adversarial loss when training. The goal of the texture details restoration network is to ensure that the image texture of the holes is "similar" to the surrounding area.

Content Reconstruction Network (Completion Network): Different from [14], we use the VGG network as the encoder, which can better obtain the detailed features of the images. We use continuous 3 × 3 convolution kernels (using small convolution kernels is superior to the use of bigger convolution kernels) for a given receptive field. Also, we alternately use four layers of dilated convolution (rate 16, 8, 4, 2, corresponding feature map size 128, 64, 32, 8) in the intermediate convolution layer. The purpose of dilated convolution is to capture a larger field of view with fewer parameters so that the part under the remaining holes is consistent with its surrounding environment. Then we take the output information of the encoder through the decoder. In our implementation, the content reconstruction network adapts to the context encoder network.

As shown in Figure 2, the five-layer encoder gradually samples down, and each layer of the encoder is composed of Convolution, Relu, BN, and Dilated Convolution. The rate of dilated convolution decreases with the decrease of the size of the feature map. The decoder gradually samples features up to the input image scale. We use transposed convolution instead of convolution in the decoder.

Texture Details Restoration Network (Refinement Network): Inspired by SRGAN, we add multiple residual blocks and skip connections between input and output in the middle layers of the texture detail restoration network. Each residual block uses two 3 × 3 convolution layers, 64 characteristic figures, and the batch normalized layer (BN) after every convolution layer, and uses ReLU as the activation function. The texture detail restoration network uses two sub-pixel convolution layers instead of deconvolution to enlarge the feature size. Reducing invalid information through the sub-pixel convolution layer can make the high-resolution image smoother, reduce the blurred texture and the unpleasant boundaries that inconsistent with the surrounding area, and obtain a better visual result.

3.2. Loss Function

Inspired by Iizuka [11], this paper uses L1 loss while attaching the loss of WGAN-GP [31] to the global and local output of the second-stage network, so as to enhance the consistency between the global and local. The original WGAN used the Wasserstein distance $W(P_r, P_g)$ to compare the distribution differences between the generated data and the actual data. Wasserstein is defined as follows:

$$W(P_r, P_g) = \inf_{Y \in \Pi(P_r, P_g)} E_{(x,y) \sim Y}[\|x - y\|] \tag{1}$$

Of which, $\Pi(P_r, P_g)$ is the set of all possible joint distributions combined by P_r and P_g. For each possible joint distribution Y, we can sample $(x, y) \sim Y$ from it to get a real sample x and a generated sample y. Then we calculate the distance $\|x - y\|$ between the samples. Finally, we calculate the expected value E of the distance between sample pairs in the joint distribution Y. On this basis, the objective function based on WGAN is established:

$$\min_{G} \max_{D \in \mathcal{D}} E_{x \sim P_r}[D(x)] - E_{\tilde{x} \sim P_g}[D(\tilde{x})] \tag{2}$$

where, $\mathcal{D}$ is a set of 1-Lipschitz functions, P_g is the model distribution implicitly defined by $\tilde{x} = G(z)$, and z is the input of the generator. In order to realize the Lipschitz continuity condition, the original WGAN clip the updated parameter of the discriminator to a smaller interval $[-c, c]$, so the parameter gathers at two points of $-c$ and c, which limits the fitting ability to some extent.

WGAN-GP has improved on the basis of WGAN, replacing weight clipping with gradient penalty: $\lambda E_{\tilde{x} - P_{\tilde{x}}}(\|\nabla_{\tilde{x}} D(\tilde{x})\|_2 - 1)^2$. WGAN-GP uses the penalty limits the value of gradient.

For image inpainting, we only try to predict region A need repaired, therefore gradient penalty should only be applied to pixels in region A. We can achieve it by gradient multiplication and mask m. Format is as follows:

$$\lambda E_{\tilde{x} \sim P_{\tilde{x}}}(\|\nabla_{\tilde{x}} D(\tilde{x}) \odot (1 - m)\|_2 - 1)^2 \tag{3}$$

where, the mask value is 0 for missing pixels and 1 for pixels at other locations. λ is set to 10 at all LABS.

The improvement also addresses the problem of the disappearance of training gradient and gradient explosion. Moreover, it has a faster convergence speed in deep learning than the original WGAN. It can also generate higher quality images and reduce the time of parameter adjustment in the training process.

3.3. Generation of Mask Data Sets

In WeChat or other social networks, some photos of scenic spots with a comfortable and clean background are shared by tourists. However, there will be more redundant pedestrians in background destroying the beauty and artistic conception of the images, especially in popular tourist resorts. So we propose an image inpainting task of retaining the target character in the image while filtering out the redundant pedestrians in background. In order to complete the image inpainting task described above, we construct the relevant mask dataset of image inpainting, which must contain various pose to produce a better inpainting result on the character filter task. This paper uses the COCO dataset to construct the mask dataset, which is a large and rich dataset of object detection, segmentation, and caption. The dataset includes 91 types of targets, which contains more than 30,000 images of human, mainly from the complex daily scenes to meet the needs of various pose.

We select images with multiple pedestrians in COCO dataset to construct the irregular mask dataset. As shown in Figure 3, first we take the picture C through the Mask-R-CNN network to find all the people in the image. Mask-R-CNN is a general instance segmentation framework, which can not only find all the target objects in the image but also accurately segment them. We could segment them after finding all the people, and the instance segmentation result is expressed as M. So mask C_{mask} can be expressed as:

$$C_{mask}(x,y) = \begin{cases} 255 & C(x,y) \in M \\ 0 & otherwise \end{cases} \tag{4}$$

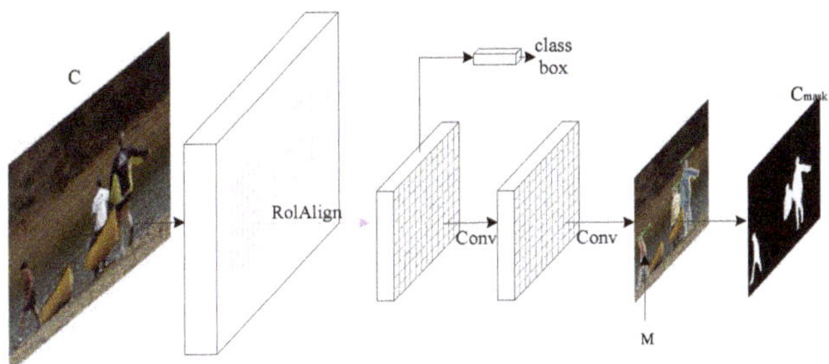

Figure 3. Construction of irregular mask dataset.

We can generate 34,980 irregular masks C_{mask} through the method with COCO dataset. Among the 34,980 irregular masks, we randomly selected 23,500 masks used for training and 11,480 masks for testing. To confirm that our mask dataset is true and reliable, we designed an experimental framework for the image inpainting task. As the entire network is shown in the following figure, the input used for training is a real image named IMAG with one character at most, and the size of the input image is 256×256. Image IMAG first detects the target pedestrian through the Mask-R-CNN network. Then we randomly select a mask C_{mask} applied to image IMAG from the 23,500 masks for training. Finally, image IMAG and Mask C_{mask} can be used as input repair to train the image inpainting network.

However, in the actual application scene, the target character of the street photography is often prominent and unobtrusive. So we should try our best not to destroy the structure of the target character, and simply perform in background. In order to make the training more consistent with the actual application scene, we need to "protect" the target pedestrian when applying the binary mask to the target image IMAG to simulate real pedestrians. The protection mechanism can be defined as:

$$A(x,y) = \begin{cases} 255 & A(x,y) \notin P \quad and \quad C_{mask}(x,y) = 255 \\ A(x,y) & otherwise \end{cases} \tag{5}$$

where, $A(x,y)$ represents an image with the area need repaired, and P is the area of detected target pedestrian.

In this way, the experimental method to test the performance of our irregular mask dataset is complete. The entire network to train is shown in Figure 4. The network integrates the existing research of instance segmentation and image inpainting. It can solve the common problem of more unnecessary pedestrians in image background destroying the beauty and artistic conception of the images in daily life.

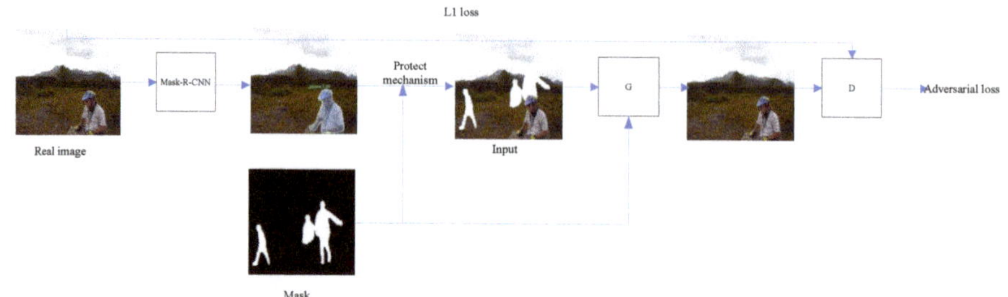

Figure 4. Network framework for pedestrian removal.

4. Result

4.1. Improved Inpainting Network

We evaluate our proposed image inpainting model on VOC2017 and COCO dataset without using tags or other information related to these images. The COCO dataset contains 118,288 images for training and 100 test images. VOC dataset contains 17,126 images for training and 100 test images. These test images are randomly selected from the validation dataset.

We compared the experimental results with PatchMatch [25] and contextual attention (Yu J [14]). PatchMatch [25] is one of the most advanced methods in patch synthesis, and contextual attention (Yu J [14]) is currently a relatively advanced image inpainting network based on deep learning. To be fair, we use all the methods to train on our dataset. Yu J [14] trained the model to handle the fixed hole. Therefore, we used fixed holes on the testing dataset to make it easy to compare the results with PatchMatch [25] and contextual attention (Yu J [14]). The fixed hole is located in the center of the input image, with the size 128 × 128. All results are generated from directly exported training models, and no post-processing is performed.

First of all, the display comparison between our results and PatchMatch [25], contextual attention (Yu J [14]) in high-resolution images is shown in Figure 5. It can be seen, the inpainting results of our model are more realistic, smoother and more similar to the texture of the surrounding area than the other two methods. Next, the quantitative comparison in Table 1 also shows the results of the comparison between our method and PatchMatch [25], contextual attention (Yu J [14]). We use three evaluation indexes: peak signal-to-noise ratio (PSNR), structural similarity index (SSIM), and average error (L1 loss). The unit of PSNR is dB, and the larger the value is, the smaller the image distortion is. The value range of SSIM ranges from 0 to 1, and the larger the value is, the smaller the image distortion is. L1 loss is the sum of the absolute difference between input and output, and the smaller the value is, the smaller the image distortion is. As you can see from the table, the methods based on deep learning have a better performance than the traditional methods based on patch in three indexes including PSNR, SSIM and L1 loss. Our model has improved in terms of data compared with contextual attention (Yu J [14]). And it is obvious in Figure 5 that our model can effectively reduce the blurred texture and the unpleasant boundaries that inconsistent with the surrounding area. Our results are superior to contextual attention, which prove the effectiveness of our model in recovering texture details in image inpainting.

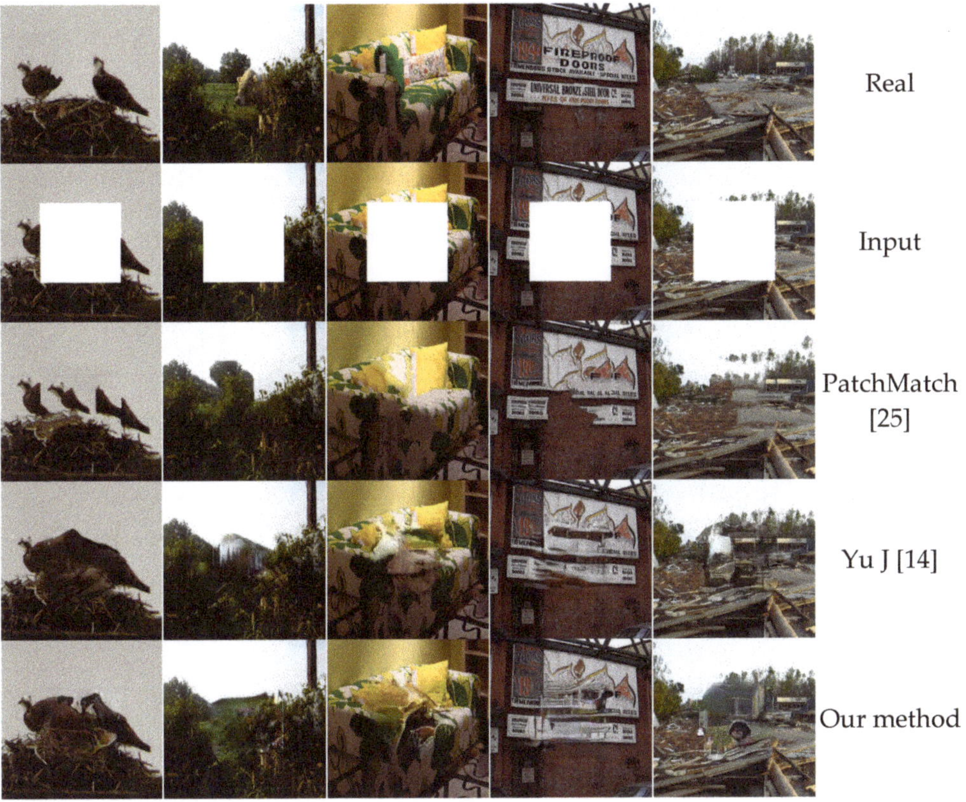

Figure 5. Comparison diagram of the results of our algorithm, PatchMatch [25] algorithm and Yu J [14] algorithm.

Table 1. The results of image inpainting using three methods.

Method	PSNR	SSIM	L1 Loss (%)
PatchMatch [25]	17.36	0.5908	8.78
Yu J [14]	19.14	0.7090	5.06
Our method	19.78	0.7205	5.52

Our full model is implemented on TensorFlow v1.3, CUDNN v7.0, CUDA v9.0 and run on hardware of CPU Intel(R) Xeon(R) gold 5117 (2.00 GHz) and GPU GTX 1080 Ti. We introduced 16 residuals into the texture detail repair network. However, in the training, these 16 residual blocks consume a lot of memory and slow down the training speed. After trying to lessen 16 residual blocks to 5 residual blocks, we found that our full model run 0.2 s per frame on the GPU, with significant improvement in speed and no significant change in performance.

In addition, the proposed inpainting framework can also be applied to conditional image generation, image editing, and computational photography tasks, including image-based rendering, image super-resolution, boot editing, and so on

4.2. Mask Experiment

We also evaluated our proposed mask dataset on two dataset including VOC2017 and COCO dataset. In the previous section, our model is trained to handle fixed holes to make it easy to compare.

While the model in [13] was trained to handle random holes and used the irregular mask dataset proposed by Liu [12], which can meet our experimental requirements. So we used the inpainting model proposed in [13] to prove the reliability of proposed mask dataset in removing redundant pedestrians in the image background.

We compared our mask dataset with the mask dataset proposed by Liu [12], the mixed mask dataset by training with the inpainting model [13]. Our mask dataset contains 23,500 masks for training and 11,480 masks for testing which are randomly generated from COCO dataset. The mask dataset proposed by Liu [12] contains 55,115 masks for training. The mixed mask dataset has a total of 78,615 masks (including 23,500 training masks that we randomly generated and 55,115 masks for training in Liu [12]).

Our comparison results are shown in the Table 2, from which we can see that we have improved the data in processing the image inpainting in a special application scene. We preserved the target character while filtering out redundant pedestrians by using our randomly generated masks. Although using our mask dataset and mixed mask dataset have similar results, it performes poorly without using our mask dataset. The comparison proves the reliability of our mask dataset in removing redundant pedestrians from image background.

Table 2. The results of image inpainting using different mask datasets.

COCO Dataset	PSNR	SSIM	L1 Loss (%)
Liu mask [12]	26.59	0.9146	2.31
Liu mask [12] + our mask	27.55	0.9233	2.01
our mask	27.54	0.9230	2.02

Figure 6 shows the intermediate results of our test, from top to bottom, which are the original picture, the instance segmentation result about people, the result of removing redundant pedestrians, and the result of image inpainting. The visualization results show that our experiment can easily screen out one or more redundant pedestrians in background and remove them.

Figure 6. The result with our proposed mask dataset.

But our experiment still has some limitations. (1) When the target character is "glued" to the other characters, as shown in the left-most figure, it produces poor results even if the redundant pedestrians can be detected. (2) Although the instance segmentation can segment the redundant pedestrians in background, it is not accurate enough to leave the hands or shoes of the redundant pedestrians, affecting the visual result. This requires further study of the experiment.

Finally, we randomly downloaded some travel photos from the internet for testing. The photos contain mountain scenery, buildings, streets, coast, and other areas. As we can see from Figure 7, the method proposed in the paper also has a high visual result in real life. In the future, we can apply it to mobile phone application to detect pedestrians in background of personal travel photos, wedding photos, and other photos. At the same time users filter unnecessary pedestrians with one key and share the beautiful travel photos in real time.

Figure 7. *Cont.*

Figure 7. The actual application effect in network image.

5. Conclusions

Image inpainting based on GAN has made great progress in recent years, but they can only process low-resolution images because of memory limitations and difficulty in training. For high-resolution images, the inpainted regions become blurred and the unpleasant boundaries become visible. Many researchers are committed to improve the existing image inpainting network framework. We propose a novel high-resolution image inpainting method based on deep generative model. It is a two-stage network including content reconstruction and texture detail restoration. After obtaining the visually believable fuzzy texture, we further restore the finer texture details improve the image inpainting quality. Meanwhile, we integrate the existing research of instance segmentation and image inpainting to delete the unnecessary pedestrians in background and ensure the reality of background restoration. To improve the accuracy of image inpainting in the special application scene, we proposed a new mask dataset, which collected the characters in COCO dataset as a mask, and could produce better inpainting results for the special application scene.

In our future work, we will experiment with convolutional deep belief network (CDBN) [37] and PCANET [38] based on the paper. Like the CNN, CDBN can extract the high-frequency features of images. According to the latest research, CDBN performs better than CNN in the classification task of large-size images, so it may be better to use CDBN instead of CNN for high-resolution images. In addition, PCANET can conduct feature fusion of feature maps of different sizes in encoder, which strengthens the correlation between input and output. However, the better result of PCANET may come at the cost of speed.

Author Contributions: Conceptualization, T.S. and W.C.; methodology, T.S., W.C. and W.F.; validation, F.B. and B.W.; formal analysis, T.S. and W.C.; investigation, T.S.; resources, W.C.; writing—original draft preparation, T.S.; writing—review and editing, T.S., W.C. and Y.Y.; visualization, T.S.; supervision, W.C.; project administration, W.C.; funding acquisition, W.C.

Funding: This word was funded by National Natural Science Foundation of China, grant number 51874300, National Natural Science Foundation of China and Shanxi Provincial People's Government Jointly Funded Project of China for Coal Base and Low Carbon, grant number U1510115, and the Open Research Fund of Key Laboratory of Wireless Sensor Network & Communication, Shanghai Institute of Microsystem and Information Technology, Chinese Academy of Sciences, grant numbers 20190902 and 20190913. The APC was funded by 51874300, U1510115, 20190902, and 20190913.

Conflicts of Interest: The authors declare no conflict of interest.

References

1. Danelljan, M.; Bhat, G.; Khan, F.S.; Felsberg, M. ECO: Efficient Convolution Operators for Tracking. In Proceedings of the 2017 IEEE Conference on Computer Vision and Pattern Recognition (CVPR), Honolulu, HI, USA, 21–26 July 2017; pp. 6931–6939.
2. Mao, J.; Xiao, T.; Jiang, Y.; Cao, Z. What Can Help Pedestrian Detection. In Proceedings of the IEEE Conference on Computer Vision and Pattern Recognition, Honolulu, HI, USA, 21–26 July 2017; pp. 6034–6043.
3. Xiao, T.; Li, S.; Wang, B.; Lin, L.; Wang, X. Joint Detection and Identification Feature Learning for Person Search. In Proceedings of the 2017 IEEE Conference on Computer Vision and Pattern Recognition (CVPR), Honolulu, HI, USA, 21–26 July 2017; pp. 3376–3385.
4. Cao, Z.; Simon, T.; Wei, S.-E.; Sheikh, Y. Realtime Multi-person 2D Pose Estimation Using Part Affinity Fields. In Proceedings of the 2017 IEEE Conference on Computer Vision and Pattern Recognition (CVPR), Honolulu, HI, USA, 21–26 July 2017; pp. 1302–1310.
5. Schroff, F.; Kalenichenko, D.; Philbin, J. FaceNet: A unified embedding for face recognition and clustering. In Proceedings of the 2015 IEEE Conference on Computer Vision and Pattern Recognition (CVPR), Boston, MA, USA, 7–12 June 2015; pp. 815–823.
6. Yoo, S.; Park, R.-H. Red-eye detection and correction using inpainting in digital photographs. *IEEE Trans. Consum. Electron.* **2009**, *55*, 1006–1014. [CrossRef]
7. Dolhansky, B.; Ferrer, C.C. Eye In-painting with Exemplar Generative Adversarial Networks. In Proceedings of the 2018 IEEE/CVF Conference on Computer Vision and Pattern Recognition, Salt Lake City, UT, USA, 18–23 June 2018; pp. 7902–7911.
8. He, K.; Gkioxari, G.; Dollar, P.; Girshick, R.B. Mask R-CNN. In Proceedings of the IEEE International Conference on Computer Vision, Venice, Italy, 22–29 October 2017; pp. 2980–2988.
9. Ren, S.; He, K.; Girshick, R.; Sun, J. Faster R-CNN: Towards Real-Time Object Detection with Region Proposal Networks. In Proceedings of the 28th International Conference on Neural Information Processing Systems (NIPS), Montreal, Canada, 7–12 December 2015; MIT: Cambridge, MA, USA, 2015; pp. 91–99.
10. Pathak, D.; Krahenbuhl, P.; Donahue, J.; Darrell, T.; Efros, A.A. Context Encoders: Feature Learning by Inpainting. In Proceedings of the 2016 IEEE Conference on Computer Vision and Pattern Recognition (CVPR), Las Vegas, NV, USA, 26 June–1 July 2016; pp. 2536–2544.
11. Iizuka, S.; Simo-Serra, E.; Ishikawa, H. Globally and locally consistent image completion. *ACM Trans. Graph.* **2017**, *36*, 107. [CrossRef]
12. Liu, G.; Reda, F.A.; Shih, K.J.; Wang, T.-C.; Tao, A.; Catanzaro, B. Image Inpainting for Irregular Holes Using Partial Convolutions. In Proceedings of the Computer Vision—ECCV 2018, Munich, Germany, 8–14 September 2018; pp. 89–105.
13. Nazeri, K.; Ng, E.; Joseph, T.; Qureshi, F.; Ebrahimi, M. EdgeConnect: Generative Image Inpainting with Adversarial Edge Learning. *arXiv* **2019**, arXiv:1901.00212.
14. Yu, J.; Lin, Z.; Yang, J.; Shen, X.; Lu, X.; Huang, T.S. Generative Image Inpainting with Contextual Attention. In Proceedings of the 2018 IEEE/CVF Conference on Computer Vision and Pattern Recognition, Salt Lake City, UT, USA, 18–23 June 2018; pp. 5505–5514.
15. Ledig, C.; Theis, L.; Huszár, F.; Caballero, J.; Aitken, A.; Tejani, A.; Totz, J.; Wang, Z.; Shi, W. Photo-Realistic Single Image Super-Resolution Using a Generative Adversarial Network. In Proceedings of the IEEE Conference on Computer Vision and Pattern Recognition, Las Vegas, NV, USA, 26 June–1 July 2016.
16. Girshick, R.; Donahue, J.; Darrell, T.; Malik, J.; Malik, J. Rich Feature Hierarchies for Accurate Object Detection and Semantic Segmentation. In Proceedings of the 2014 IEEE Conference on Computer Vision and Pattern Recognition, Columbus, OH, USA, 24–27 June 2014; pp. 580–587.
17. Lin, T.-Y.; Dollar, P.; Girshick, R.; He, K.; Hariharan, B.; Belongie, S. Feature Pyramid Networks for Object Detection. In Proceedings of the 2017 IEEE Conference on Computer Vision and Pattern Recognition (CVPR), Honolulu, HI, USA, 21–26 July 2017; pp. 936–944.
18. Huang, J.; Rathod, V.; Sun, C.; Zhu, M.; Korattikara, A.; Fathi, A.; Fischer, I.; Wojna, Z.; Song, Y.; Guadarrama, S.; et al. Speed/Accuracy Trade-Offs for Modern Convolutional Object Detectors. In Proceedings of the 2017 IEEE Conference on Computer Vision and Pattern Recognition (CVPR), Honolulu, HI, USA, 21–26 July 2017; pp. 3296–3297.

19. Shrivastava, A.; Gupta, A.; Girshick, R. Training Region-Based Object Detectors with Online Hard Example Mining. In Proceedings of the 2016 IEEE Conference on Computer Vision and Pattern Recognition (CVPR), Las Vegas, NV, USA, 26 June–1 July 2016; pp. 761–769.

20. Li, Y.; Qi, H.; Dai, J.; Ji, X.; Wei, Y. Fully Convolutional Instance-Aware Semantic Segmentation. In Proceedings of the 2017 IEEE Conference on Computer Vision and Pattern Recognition (CVPR), Honolulu, HI, USA, 21–26 July 2017; pp. 4438–4446.

21. Dai, J.; He, K.; Li, Y.; Ren, S.; Sun, J. Instance-Sensitive Fully Convolutional Networks. In *European Conference on Computer Vision*; Springer: Cham, Switzerland, 2016; pp. 534–549.

22. Dai, J.; Li, Y.; He, K.; Sun, J. R-FCN: Object Detection via Region-based Fully Convolutional Networks. In Proceedings of the Advances in Neural Information Processing Systems; Barcelona, Spain, 5–10 December 2016; pp. 379–387.

23. Fedorov, V.V.; Facciolo, G.; Arias, P. Variational Framework for Non-Local Inpainting. *Image Process. Line* **2015**, *5*, 362–386. [CrossRef]

24. Newson, A.; Almansa, A.; Gousseau, Y.; Pérez, P. Non-Local Patch-Based Image Inpainting. *Image Process. Line* **2017**, *7*, 373–385. [CrossRef]

25. Barnes, C.; Shechtman, E.; Finkelstein, A.; Dan, B.G. PatchMatch: A Randomized Correspondence Algorithm for Structural Image Editing. *ACM Trans. Graph.* **2009**, *28*, 24. [CrossRef]

26. Goodfellow, I.J.; Pougetabadie, J.; Mirza, M.; Xu, B.; Wardefarley, D.; Ozair, S.; Courville, A.C.; Bengio, Y. Generative Adversarial Nets. In Proceedings of the Neural Information Processing Systems, Montreal, QC, Canada, 8–13 December 2014; pp. 2672–2680.

27. Radford, A.; Metz, L.; Chintala, S. Unsupervised Representation Learning with Deep Convolutional Generative Adversarial Networks. *arXiv* **2015**, arXiv:1511.06434.

28. Zhao, J.; Mathieu, M.; LeCun, Y. Energy-based Generative Adversarial Network. *ArXiv* **2016**, arXiv:1609.03126.

29. Salimans, T.; Goodfellow, I.; Zaremba, W.; Cheung, V.; Radford, A.; Chen, X. Improved Techniques for Training GANs. *ArXiv* **2016**, arXiv:1606.03498.

30. Arjovsky, M.; Chintala, S.; Bottou, L. Wasserstein GAN. *arXiv* **2017**, arXiv:1704.00028.

31. Gulrajani, I.; Ahmed, F.; Arjovsky, M.; Dumoulin, V.; Courville, A. Improved Training of Wasserstein GANs. *ArXiv* **2017**, arXiv:1704.00028.

32. Berthelot, D.; Schumm, T.; Metz, L. BEGAN: Boundary Equilibrium Generative Adversarial Networks. *ArXiv* **2017**, arXiv:1703.10717.

33. Mao, X.; Li, Q.; Xie, H.; Lau, R.Y.; Wang, Z.; Smolley, S.P. Least Squares Generative Adversarial Networks. In Proceedings of the 2017 IEEE International Conference on Computer Vision (ICCV), Venice, Italy, 22–29 October 2017; pp. 2813–2821.

34. Isola, P.; Zhu, J.; Zhou, T.; Efros, A.A. Image-to-Image Translation with Conditional Adversarial Networks. *arXiv* **2016**, arXiv:1611.07004.

35. Wang, T.-C.; Liu, M.-Y.; Zhu, J.-Y.; Tao, A.; Kautz, J.; Catanzaro, B. High-Resolution Image Synthesis and Semantic Manipulation with Conditional GANs. *ArXiv* **2017**, arXiv:1711.11585.

36. Zhu, J.-Y.; Park, T.; Isola, P.; Efros, A.A. Unpaired Image-to-Image Translation Using Cycle-Consistent Adversarial Networks. In Proceedings of the 2017 IEEE International Conference on Computer Vision (ICCV), Venice, Italy, 22–29 October 2017; pp. 2242–2251.

37. Lee, H.; Grosse, R.; Ranganath, R.; Ng, A.Y. Convolutional deep belief networks for scalable unsupervised learning of hierarchical representations. In Proceedings of the 26th Annual International Conference, Montreal, QC, Canada, 14–18 June 2009; pp. 609–616.

38. Chan, T.-H.; Jia, K.; Gao, S.; Lu, J.; Zeng, Z.; Ma, Y. PCANet: A Simple Deep Learning Baseline for Image Classification? *IEEE Trans. Image Process.* **2015**, *24*, 5017–5032. [CrossRef] [PubMed]

Article

Deep Learning-Enhanced Framework for Performance Evaluation of a Recommending Interface with Varied Recommendation Position and Intensity Based on Eye-Tracking Equipment Data Processing

Piotr Sulikowski [1,*] and Tomasz Zdziebko [2]

[1] Faculty of Information Technology and Computer Science, West Pomeranian University of Technology, ul. Zolnierska 49, 71-210 Szczecin, Poland

[2] Faculty of Economics, Finance and Management, University of Szczecin, ul. Mickiewicza 64, 71-101 Szczecin, Poland; tomasz.zdziebko@usz.edu.pl

* Correspondence: psulikowski@wi.zut.edu.pl

Received: 7 January 2020; Accepted: 30 January 2020; Published: 5 February 2020

Abstract: The increasing amount of marketing content in e-commerce websites results in the limited attention of users. For recommender systems, the way recommended items are presented becomes as important as the underlying algorithms for product selection. In order to improve the effectiveness of content presentation, marketing experts experiment with the layout and other visual aspects of website elements to find the most suitable solution. This study investigates those aspects for a recommending interface. We propose a framework for performance evaluation of a recommending interface, which takes into consideration individual user characteristics and goals. At the heart of the proposed solution is a deep neutral network trained to predict the efficiency a particular recommendation presented in a selected position and with a chosen degree of intensity. The proposed Performance Evaluation of a Recommending Interface (PERI) framework can be used to automate an optimal recommending interface adjustment according to the characteristics of the user and their goals. The experimental results from the study are based on research-grade measurement electronics equipment *Gazepoint* GP3 eye-tracker data, together with synthetic data that were used to perform pre-assessment training of the neural network.

Keywords: recommender system; human computer interaction; eye-tracking device; deep learning

---

1. Introduction

Fast e-commerce development inspires increasing attention to sales-boosting solutions, especially recommending systems, which aim to replace salespeople from traditional shops. Shopping online offers the benefit of convenience, but on the other hand it is lacking the personal touch of salespeople, especially when a customer has to select from a very large number of alternatives. Thus, the optimization of user experience, including personalization and implementing recommending interfaces, has a crucial role in e-commerce website design. While, in a physical store, a salesperson may directly recommend products, in an online shopping environment it is the recommending interface that helps promote products which may be interesting to the customer. Recommender systems play a vital role in motivating purchase decisions and usually prove successful in enhancing sales [1].

In a recommender system, a user model is usually created, constituting a description of a user, in order to facilitate interactions between the user and the system [2]. A digital representation of a user model is a user profile, which reflects their preferences, transactions, online behavior, etc. [3]. Online systems process a wide stream of user data [4–7] essential to build user profiles and recommend items which are optimal in terms of fit and, as a consequence, resulting sales. A lot of effort has been made to analyze that data spectrum and discover user preferences and needs [8,9]. Early solutions were founded on content-based and collaborative filtering algorithms [10], which were then extended towards explanation interfaces [11] with the use of context [12] and other approaches such as social media data inclusion [13,14].

The final performance of a recommending system, however, depends on factors that go beyond the recommendation algorithms themselves [15]. While there is substantial research in the area of those algorithms, there are substantially fewer studies in the area of the stages which follow in the online recommending process, such as item recommendation presentation. Human-computer interaction with recommending interfaces can be analyzed using DOM-events-based solutions [4] or gaze tracking [16–18]. Results from eye-tracking studies show that gaze data are a valuable source for inferring user interest, and the examination of the visual aspects of organizing a recommending interface may allow to better integrate those interfaces in e-commerce platforms [1,19,20]. In order to optimize the interface, a number of factors can be analyzed, such as the number of recommendations, recommendation item images, descriptions and layouts [21,22]. Since customers are inundated with information, especially marketing content, the habituation effect usually appears, which ends in the banner blindness phenomenon. As a result, even recommendations that are optimal from the algorithm perspective may provide insignificant results unless they are shown to the user in a wise way [23–25]: in the right part of a website, at the right moment of the selection and purchase process, with the right level of content intrusiveness [26–28], and considering personal preferences [29].

This paper is a substantial extension of a conference paper [30] and proposes a validated framework for the performance evaluation of a recommending interface, to optimize its efficiency considering individual user characteristics. The evaluation is based on a deep learning neural network trained on experimental data from an eye-tracking study on the varying visual intensity and position of a recommendation and enhanced with data from implicit user tracking and synthetic data for missing measures. The framework can be implemented as part of e-commerce personalization engine responsible for recommending interface adjustment.

The remainder of the article is structured as follows: the conceptual framework is presented in Section 2. The structure of the experiment and empirical results are provided in Section 3 and conclusions are presented in Section 4.

2. Conceptual Framework

The main objective of this paper is to present a framework for performance evaluation of the positioning of a recommendation within a recommending interface of a website and the varying visual intensity of a recommendation with regard to attracting customer interest. In order to evaluate the viability and usefulness of the framework in terms of user experience and marketing goals, a pre-assessment study is performed. This evaluation is based on a deep neural network model built on data from a study performed with research-grade measurement electronics equipment *Gazepoint* GP3 eye-tracker and synthetic data to perform pre-assessment training of the neural network.

The main assumption behind our proposed framework for Performance Evaluation of a Recommending Interface (PERI) is that different variants of a recommendation interface can have different impact on different users depending on their cognitive abilities [31,32], their way of interacting with a website and their goals of the visit to an e-commerce website. These assumptions have been confirmed by several studies [22,33–35].

In order to determine user interest, one can ask the user explicitly or observe them implicitly. While explicit questioning often disrupts natural behavior and constitutes an extra burden on the user [3,36,37], implicit measures are unobtrusive and therefore better suited to the purpose of the study. The subjects may focus on normally performed tasks, no extraneous cognitive load is generated and no additional motivation is required to provide explicit ratings [38–41].

The methodology of the research assumes the use of gaze tracking for user behavior observation. Eye tracking is a powerful method used to generate implicit feedback and one of the most popular techniques of observing human–computer interaction. Within the scope of the study, gaze-based data are analyzed and interpreted in a basic e-commerce scenario. Eye movements are used to discover which areas of an e-commerce website are most looked at, and which of them are the most relevant to the user, attracting user attention the most. Raw data collected by the eye-tracker device are processed with eye-tracking software and analytics algorithms.

Eye movements may be unordered in nature and unconscious, yet they are generally tightly connected with cognitive processes [42]. Therefore, inference about user attention and interest is possible based on gaze data. A literature review by Buscher et al. confirms that data from gaze-tracking equipment is an excellent source of information on how much attention is paid to particular content on the screen [43].

For the pre-assessment study, total fixation duration is the main gaze-based measure, used together with the buying action. Total fixation is used as an indicator of attractiveness by a number of research studies [35,44–47]. It is calculated as the sum of fixation durations aggregated on a section of a website, in particular the recommendation content (*RC*) section and the main section, with editorial content (*EC*). In the study, in addition to experimenting with the position of a recommending interface on a website and the location of a particular recommendation item (*RI*) within that interface, changes in visual intensity are also taken into account. Three basic levels of intensity are used. Changing the visual intensity of an item is a popular marketing technique used to counteract habituation and attract more attention [48]. Data from the eye-tracking study have been supplemented with features generated on the basis of those data.

Figure 1 depicts the architecture of the framework for the performance evaluation of a recommending interface utilizing certain recommendation positions and intensities. Its key components include the following:

- *User demographic data.* Demographic data about users (i.e., age, education, interests) which can be used to identify user cognitive abilities. These data can be gathered through registration questionnaires;
- *User activity implicit and explicit data gathering.* This module is responsible for collecting data about user behavior and preferences in an untrobusive way by implicitly tracking their activity, and explicitly by gathering opinions expressed mainly in the form of rating stars;
- *User goal identification.* This module is responsible for the identification of the user's goal. In the case of e-commerce websites, visitors can represent different stages of the purchase funnel. A user may be exploring the offer without having buying in mind. User goals can be identified based on a phrase typed in a search engine, the redirections source, and the relation between the items visited by user, usage of product filter utility and history of previous visits;
- *User cognitive abilities identification.* The role of this module is to assess user's cognitive abilities and classify them at one of a number of selected levels. As current cognitive abilities can influence the way a user interacts with a website and processes the provided information, presentation methods should be tailored to user abilities;
- *User preference reasoning.* The role of this module is to infer user personal preferences about particular products, product features and product categories in general. Those preferences are used to construct a user model which is the input for the recommender system;
- *Personalized recommendation engine.* This module is responsible for generating the most accurate personalized product recommendations for individuals, which fit their preferences and also can reach website goals;
- *Performance Evaluation of a Recommending Interface (PERI).* This module is the core of the proposed framework. It is responsible for the evaluation of the performance of a possible set of different ways in which recommendations can be presented. The process of evaluation is carried from the perspective of individual user's goals, cognitive abilities and website goals. The heart of

this module is a prediction model based on a multi-layer deep neural network, which is trained preliminarily on the basis of eye-tracking data.

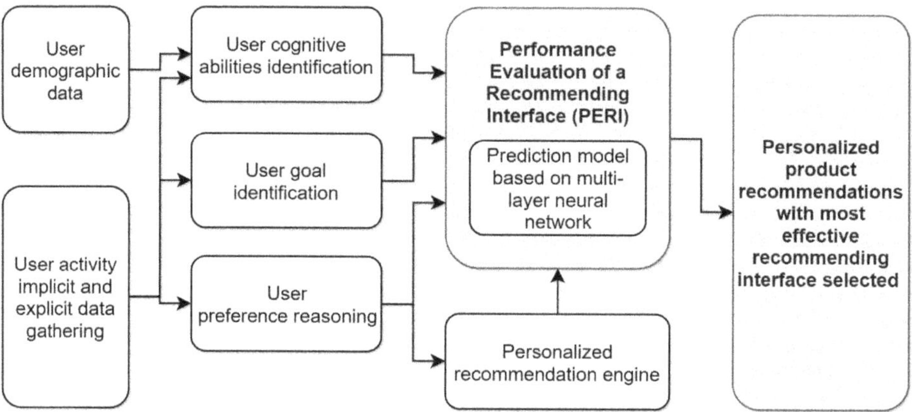

Figure 1. The framework for performance evaluation of a recommending interface.

The proposed framework can be used for any e-commerce site to automatically adjust the recommending interface to the needs, preferences, goals, etc., of individuals and optimize the interface performance, optimally setting up the positions and visual intensities. The prediction model is based on a deep neural network, due to the multi-dimensionality of the preference evaluation task, as this modeling technique handles such sophisticated regression problems in the most accurate way. In real-world solutions, PERI may produce complex evaluation measures by incorporating different user goals. For example, in a scenario where a user is only browsing, without having buying in mind, the success of *RC* can be defined as clicking on an *RC* and then exploring a product page, or just by looking at the product description. Moreover, simply attracting user interest to *RC*, represented by fixation time, can also be of huge importance, as users rely on recommender systems to enhance their confidence in purchase decisions [1].

3. Experimental Results

3.1. Eye-Tracking Experiment Structure and Procedure

This section describes the experiment performed to collect the eye tracking and behavior data used to train the neural network responsible for the evaluation of recommending interfaces.

Task. Each participant was given the task to shop online in order to furnish a studio apartment with six types of furniture. Each subject was asked to move between product categories and select one item from each category, according to their individual preference.

Website. The experiment was composed of a recommending interface within a dedicated e-commerce website, developed using Drupal CMS. The website was available in Polish and consisted of a title, menu, product images and short descriptive text. It covered functions such as product list, buying cart and recommendations.

The editorial content (*EC*) was placed in the central area of the screen, under the main menu. It contained product lists about three screens long with 10 products in each product category. Each product had three unique features: name, product image and price. There were six product categories (*PCj*): wardrobes, chests of drawers, beds, bedside cabinets, tables and chairs. Products in a category were quite similar visually and similarly priced. In addition, under the furniture description there was an 'Add to Cart' button that stored customer choices in a database. Upon selection of a product, its short description was available in the cart preview and on the main cart page. Of course, it was possible to remove the product from the cart in order to allow the user to make changes to the final selection of purchases.

Recommending interface. There were two alternative recommendation interface layouts, i.e., horizontal and vertical recommending mode. This means that the recommendation content (*RC*) section was anchored in one of two dedicated parts of the screen below the main menu: either on the left side of the page, next to the general product list (in vertical mode), or at the top of the page, above the general product list (in horizontal mode). Only one recommendation layout was available at a time, so, when horizontal mode was on, the vertical one was deactivated and vice versa. Figure 2 shows variants of the recommendation content (*RC*) location.

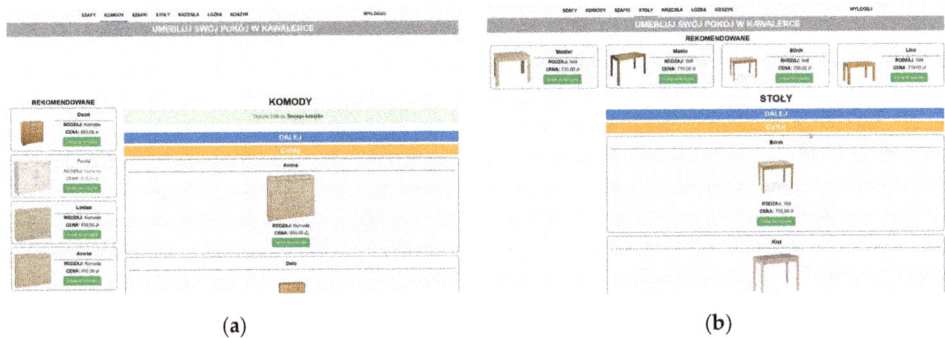

(a)   (b)

Figure 2. Recommendation layouts of the recommending interface: (**a**) vertical; (**b**) horizontal.

The *RC* section consisted of four recommendation items—RC_1 to RC_4,—randomly selected from all products in a category. The section in each variant did not change its location on the screen when browsing products in the product category, regardless of the user scrolling the *EC* section. In fact, only general product lists were made scrollable to ensure reliable subject exposure to the recommendation interface.

It was ensured that product features, i.e., name, image and price, would not stand out from other products in the category. It was assumed that the possible distinction of a particular RC_i location would be achieved only by means of visual intensity VI. Three levels of intensity were used: standard (without any highlight)—VI1, flickering (slowly disappears and reappears every 1–2 s)—VI2 and background in red—VI3. There was a maximum of one RC_i at VI2 or VI3 for each product category. An example of visual intensity of the last kind (VI3) is shown in Figure 3.

Measurement equipment. Research-grade *Gazepoint* GP3 eye tracker, a 60 Hz update rate system, was utilized. The device's nominal accuracy is 0.5–1 degree of visual angle. It allows for ±15 cm range of depth movement and offers 5- and 9-point calibration. It is powered by USB.

Procedure. The experiment proceeded as follows. First, the test person was sitting at the test stand in such a way that their eyes were in the optimal range of the eye-tracking device's camera. It was explained what the device for tracking eyeball movements is, and then the eye tracker was calibrated with *Gazepoint Control* software and a 9-point calibration method. For greater accuracy, calibration was always performed twice, the first time just to familiarize the subject with the process. There was a dual monitor setup with the operator screen invisible to the participant. Thanks to the correct calibration, the device was able to determine the coordinates of the place where the user was looking.

The participant was then informed of their task but was not told about the purpose of the study. After this introduction, the subject had to furnish the apartment. After choosing one item from a category, the subject clicked 'Next' and was automatically moved to the next category. Category by category, the visual intensity of recommendation items changed every time. In addition, for the first three categories, the layout of *RC* was vertical and, after moving to the fourth category, it changed to horizontal and remained thus for the following categories. In general, each participant was presented with at least six subsequent webpages with different recommendation options.

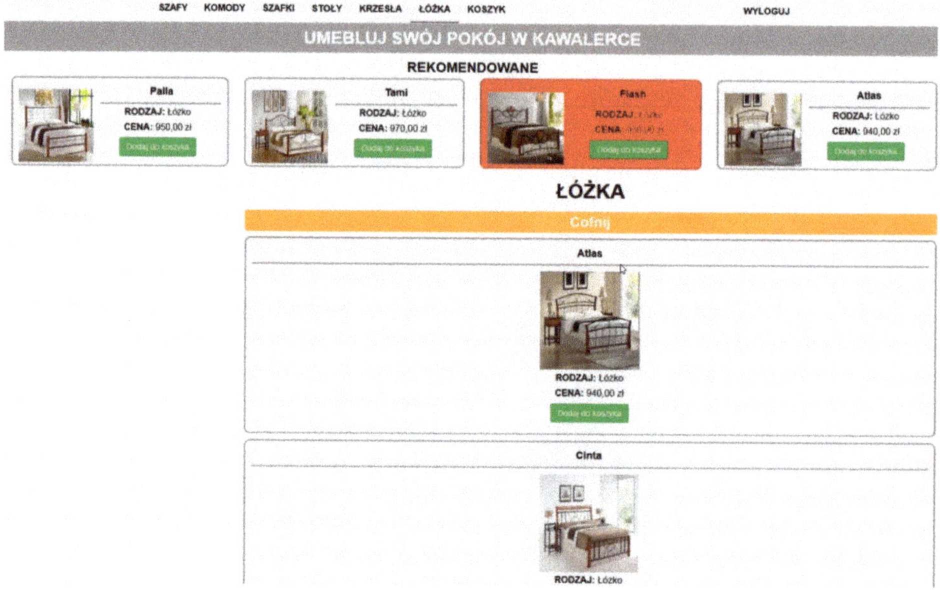

Figure 3. Example of visual intensity VI3 (red background) in the vertical recommending interface.

Each session was monitored live and recorded using *Gazepoint Analysis* software. We constantly double-checked the operator's monitor to ensure the eyes of the subject were in the optimal position relative to the camera, etc. After the participant had completed the task, basic data such as age were collected, and a question was asked about whether the subject felt they were influenced by the recommendations. Finally, all data were saved and stored by the eye-tracking system for further analysis. One experimental run typically lasted about 12 min.

Participants. The initial experimental group of users consisted of 52 people who produced valid eye-tracking data. Most of them were undergraduate or graduate students invited in person or attracted to advertisements for the study, and they were native Polish speakers. They ranged in age from 14 to 54 years (mean = 25.2, σ = 8.0).

3.2. Performance Evaluation of a Recommending Interface Experiment Structure and Procedure

This section relates to the next stage of the experiment necessary to preliminarily implement the proposed framework for Performance Evaluation of a Recommending Interface (PERI). In line with the character of the study, the presented implementation does not cover the full spectrum of data described in the proposal, related to goal identification and preference reasoning modules which were not used since participants were given only one particular task. For the ultimate measure of interface performance, the add-to-cart action was chosen in this implementation. As mentioned in the framework proposal, other performance measures could alternatively be employed, e.g., fixation time on the recommending interface, time spent on a product page accessed via the recommending interface, etc.

Data. Data collected using the eye-tracking device were used to build a deep learning solution and perform our pre-assessment study. Fixation data collected with *Gazepoint Analysis* software constitute lines containing information about all fixations performed by participants. In total, 15,922 fixation records were generated.

Preprocessing. Data were preprocessed in order to extract fixations concerning individual RC_i locations for every product category PC_j and every user who was efficiently involved in the study. As a result, 593 rows were generated, each containing the following features: RC layout (horizontal/vertical)—*rc_layout*, RC_i location (1-4)—*rc_location*, recommendation position intensity level (1-3)—*rc_location_intensity*, total fixation time for RC layout—*fixation_time_layout*, total fixation time for RC_i location—*fixation_time_location*, total time spent on product category page—*fixation_time_category*, percentage of time while fixation was registered inside the RC layout in relation to total time spent on category page—*share_time_layout_category*, percentage of time while fixation was registered inside RC_i location in relation to total time spent on category page—*share_time_location_category*, percentage of time while fixation was registered inside RC_i location in relation to total time spent on RC layout—*share_time_location_layout*, user age—*user_age*, level of user's cognitive abilities—*user_cognitive_ability_level*, adding the product to cart action (and its purchase) from RC—*add_to_cart*. The features concerning the time spent looking at RC were introduced to measure interest in the recommending interface.

All the features beside the last one were used to predict the add-to-cart action, which, in the case of our study, was selected as the ultimate efficiency measure. This measure was selected due to the purchase task given to participants. In another scenario, a different efficiency measure could be applied, for example, interest level generated by recommending interface, measured as time spent on recommended product pages.

Neural network. The preprocessed data were used to train a neural network responsible for the evaluation of recommending interfaces. Multi-layer perceptron deep neural network architecture was chosen as most suitable for the classification problem with a low number of features and training records. It allowed for the deep learning of the relationship between interactions with different recommending interfaces and their efficiency, where success was measured as the add-to-cart action. *IBM SPSS Statistics* was utilized for building the deep learning network.

4. Results

4.1. Eye-Tracking Results of Recommending Interface Efficiency

After completing the task, 33% of participants responded that they felt their selection was influenced by the RC areas of the site (6% felt strongly about it), while others claimed the opposite, including 52% who strongly felt they did not care about recommendations on the website. The last group did indeed seem to show strong resistance to the recommendations—some of those participants, when shown the RC sections after the test, were surprised that they might have neglected most of them at all, treating them comparably to adverts, which confirms the prevalence of the habituation effect.

The analysis of eye-tracking data shows that the task took, on average, 2.3 min to complete. In the study, 312 products were selected for purchase in total. Fixation time on the recommending interface was, on average, 16.3 s per person, which is 12% of the average task completion time. The mean amount of time devoted by subjects to observing RC was 8.2 s and 8.1 s for the vertical and horizontal layouts, respectively. Thus, in terms of fixation time, the two presented variants of the recommending interface layout offered equal performance.

Table 1 shows in more detail the distribution of these times for all locations of recommendation items. It was found that the first three locations, RC_i, were the most favorable, irrespective of the layout. The least eye-catching locations took fourth place on the list, next to the bottom bar of the website (vertical layout) or next to the right edge of the screen (horizontal layout). The most popular of all was the RC_3 location in the horizontal arrangement (3.9 s). This was probably influenced by the fact that this recommendation item was placed directly above the general product list. The second most popular location was RC_2 and the third was RC_1, both in the vertical layout. The apparent popularity of RC_2 in this arrangement was impacted by the fact that, in one product category, this item was shown as flickering (VI2), and the popularity of RC_1, although always shown with standard visual intensity

(VI1), may be influenced by the fact that a lot of people perceive the first location on a list as the best one. It should be noted that, in the case of the vertical layout, this first position still worked better than RC_3, which, for one product category, was presented with dazzling intensity VI3. Item RC_3 in vertical mode performed on a par with item RC_2 in the horizontal layout, the latter being supported by flickering effect (VI2) for one product category.

Table 1. Average fixation time(s) for each recommendation location.

Recommendation Location	Time (s)	
	Vertical *RC*	**Horizontal *RC***
RC_1	2.4	1.3
RC_2	3.1	2.1
RC_3	2.1	3.9
RC_4	0.6	0.8
Total	8.2	8.1

An aggregated heatmap for all participants is presented in Figure 4. It illustrates the views of users in website areas. The areas that received the most attention have a warmer color, while those that were less attractive have a colder one. This map shows that the recommending interface received some attention in relation to the total time spent on completing the task, but less than the main product list. We can also notice some differences in the attractiveness of recommendation items in different locations to the disadvantage of RC_4 for both layout options.

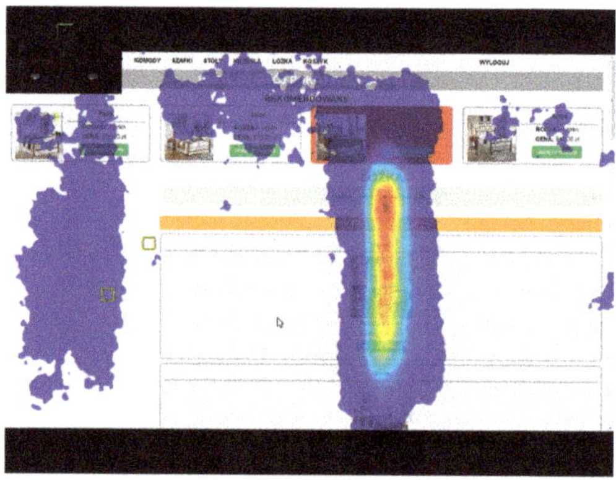

Figure 4. Aggregated heatmap for all participants in the study.

From a sales perspective, 12% of products in all carts were selected directly from the recommendation items. Oddly, this is exactly the same proportion as the one of the recommending interface fixation time to task completion time, which shows the importance of focusing attention on recommended items. Vertical *RC* layout was responsible for 62% of product selections, while the others were due to the horizontal *RC* layout—the vertical layout turned out to be almost twice as effective as the other. This may be related to banner blindness, where banners have historically often been placed in the very same area of a website as horizontal recommendations in the experiment. In the case of the vertical layout, for *RC* with all RC_i at the standard intensity level (VI1), the recommendation-driven purchases (RDPs) were evenly distributed among the recommended products. In the case of *RC* with RC_2 at the flickering intensity level (VI2), the item attracted four out of nine RDPs in the product category; in the case of *RC* with RC_3 on a red background (VI3), the item surprisingly attracted only one out of eight RDPs in the product category. On the whole, RC_2 was the most effective, which means that the second recommendation on the vertical list brought the most sales (48% of RDP's for vertical

RC, and 30% of all RDP's). The recommendation-driven purchase volume is presented in more detail in Table 2.

Table 2. Recommendation driven purchase and visual intensity for each recommendation location (RC_i) and product category (PC_j).

Recommendation Location	Vertical *RC*			Horizontal *RC*		
	PC_1	PC_2	PC_3	PC_4	PC_5	PC_6
RC_1	1 (VI1)	1 (VI1)	0 (VI1)	2 (VI1)	0 (VI1)	4 (VI1)
RC_2	2 (VI1)	4 (VI2)	5 (VI1)	0 (VI1)	3 (VI2)	0 (VI1)
RC_3	1 (VI1)	4 (VI1)	1 (VI3)	0 (VI1)	1 (VI1)	2 (VI3)
RC_4	2 (VI1)	0 (VI1)	2 (VI1)	0 (VI1)	2 (VI1)	0 (VI1)
Total	6	9	8	2	6	6

It has to be noted that only direct recommendation driven purchases were considered, that is, purchases initiated directly from *RC*. It was not feasible to reliably assess non-direct RDPs, that is, the amount of purchases committed from the general product list, yet inspired by recommendation items. Therefore, non-direct RDPs were not analyzed in this study. However, it was noticed in the visual analysis that a few subjects glanced at a recommendation item and, sometime later, decided to select the same product from the general product list, with causation not confirmed.

Another side remark after visual analysis is connected with the fact that the flickering effect (VI2) of a recommendation item seemed to have a prolonged effect on fixation after moving to the next product category. This means that, despite the visual intensity changing to standard, this recommendation location continued to attract attention.

4.2. Results of the Pre-assesment Study of the Proposed Framework for Performance Evaluation of a Recommending Interface (PERI)

Using data described in Section 3.2, the deep neural network was trained for the goal of predicting the performance of recommending interfaces. As a performance measure, the action of adding a product to cart from the RC_i location was used. In total, 40 products were selected directly from RC_i locations. A custom multilayer perceptron with two hidden layers for the binary classification of adding a product to cart was built, the number of neurons being computed automatically. The resulting neural network consisted of four layers (one input, two hidden and one output). The parameters of the neural network are presented in Table 3. Variables *rc_location* and *user_cognitive_ability_level* were treated as categorical variables and, thus, one-hot encoding was performed, resulting in one input neuron for each variable value. In both hidden layers and the output layer, sigmoid function was used as activation function. For training the neural network, the gradient descent algorithm was used, with an initial learning rate of 0.4 and momentum of 0.9. The number of neurons in each hidden layer was determined automatically by using iterative estimation algorithms (*IBM SPSS Statistics*). All input variables were normalized before training of the network.

A test sample of 168 records (around 28.3%) was put aside for the accuracy validation of the neural network. Due to unbalanced data there, were ten positive samples randomly selected. The confusion matrix on the training and testing sample is shown in Table 4. Overall classification accuracy is high for both training and testing datasets, at 98.4% and 98.2%, respectively. The best results are achieved for the not-buying action, with 98.7% and 99.4% of accuracy for both training and testing sets. Regarding predicting the buying action, the accuracy is also quite high—92.9% and 80.0% for the same sets, respectively. Precision and recall accuracy equal 80% and 89%, respectively, and they are the most appropriate metrics for the accuracy evaluation of the model.

Table 3. Parameters of multilayer perceptron neural network.

		Network Information	
		1	*rc_location*
	Factors	2	*rc_layout*
		3	*rc_location_intensity*
		1	*fixation_time_category*
		2	*fixation_time_layout*
Input Layer		3	*fixation_time_location*
	Covariates	4	*share_time_layout_category*
		5	*share_time_location_category*
		6	*share_time_location_layout*
		7	*user_age*
		8	*user_cognitive_ability_level*
	Number of Units		17
	Rescaling Method for Covariates		Normalized
	Number of Hidden Layers		2
Hidden Layer(s)	Number of Units in Hidden Layer 1		8
	Number of Units in Hidden Layer 2		6
	Activation Function		Sigmoid
	Dependent Variables	1	*add_to_cart*
Output Layer	Number of Units		2
	Activation Function		Sigmoid
	Error Function		Sum of Squares

Table 4. Confusion matrix for multilayer perceptron for predicting efficiency of recommending interface.

		Classification		
			Predicted	
Sample	**Observed**	0	1	**Percent Correct**
	0	392	5	98.7%
Training	1	2	26	92.9%
	Overall Percent	92.7%	7.3%	98.4%
	0	157	1	99.4%
Testing	1	2	8	80.0%
	Overall Percent	94.6%	5.4%	98.2%

Other metrics show overall good accuracy of the resulting network, with AUC 0.991 for both actions (buying and not-buying) with high sensitivity and specificity (Figure 5).

The most important variables for the deep neural network are *fixation_time_location*, *fixation_time_layout*, *share_time_location_layout*, *share_time_location_category* and *rc_location* (Table 5). The importance of each predictor was calculated with the SLRM algorithm by removing each predictor variable in turn from the model and verifying how that affects the model's accuracy.

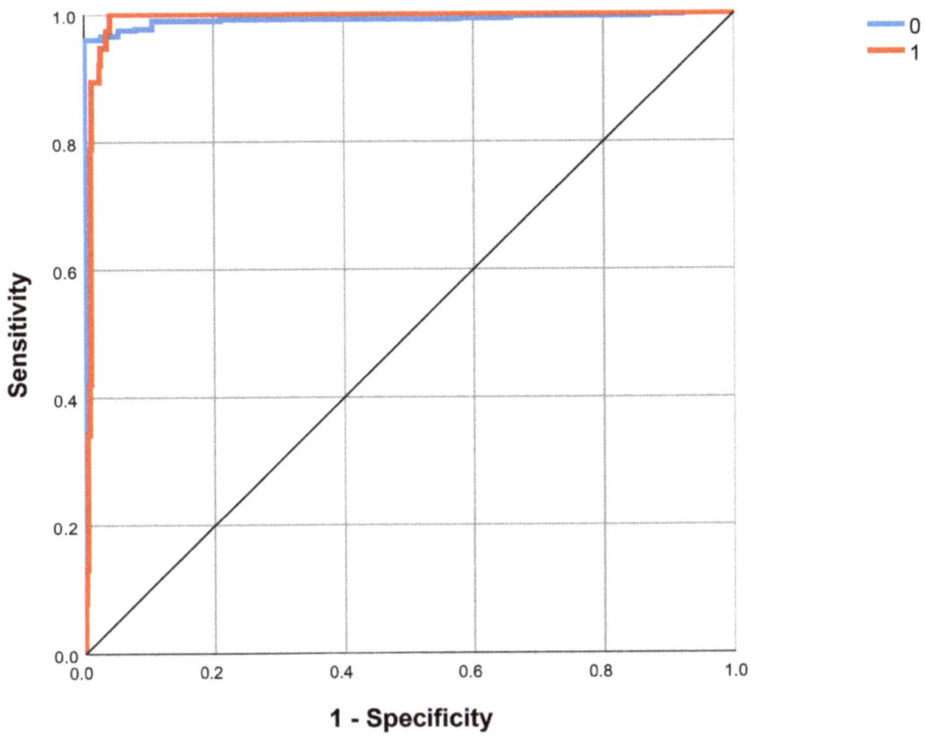

Figure 5. Sensitivity and specificity for the multilayer perceptron.

Table 5. Confusion matrix for multilayer perceptron for predicting the efficiency of the recommending interface.

Independent Variable	Normalized Importance
fixation_time_location	100%
fixation_time_layout	42%
share_time_location_layout	12%
share_time_location_category	8%
rc_location	7%
rc_layout	4%
rc_location_intensity	4%
user_cognitive_ability_level	2%
fixation_time_category	1%
user_age	1%
share_time_layout_category	1%

5. Conclusions

E-commerce platform designers, together with marketers, seek ways of attracting the attention of web users and encouraging them to commit to purchases, in particular with the use of recommending interfaces. The presented study showed the influence of the layout of a recommending interface, the position of a recommendation item and various levels of visual intensity applied to it, on user behavior in a simply structured shopping website. Thanks to the research-grade measurement electronics equipment *Gazepoint* GP3 eye tracker, as well as tracking participants' purchase decisions, the attractiveness of selected website areas was analyzed. A framework for the Performance Evaluation of a Recommending Interface (PERI) was proposed.

There are several major conclusions. In the experiment, an average of 12% of task completion time was used to look at the recommending interfaces and, coincidentally, exactly the same percentage of goods were purchased directly from recommendations. While comparing the vertical and horizonal recommending interface modes, in terms of fixation time, they performed equally, but from the point of view of purchase commitments, the vertical layout proved to be almost twice as effective as the horizontal one. It is speculated that the worse sales performance of the horizontal layout is related to banner blindness, because banners usually occupy a similar rectangular space at the top of the screen. In the better performing vertical arrangement, the most attractive in terms of fixation time was the position on the list, where the effect of slow flickering was used to increase visual intensity. On the other hand, the high visual attractiveness of the first item on the list, despite the lack of any visual distinction, may be due to the preconception that the first is always the best (similar to search engines). The level of attractiveness of the dazzling red back background was relatively low, probably due to the excessively high content intrusiveness that turned out to be counterproductive. It was also found that the first three locations in a recommending interface were the most eye-catching, regardless of the layout, with the least popular locations being the last ones, bordering the bottom or right edge of the website, respectively, for vertical and horizontal layouts. The study justifies considering a vertical rather than horizontal layout when designing a recommending interface and suggests that it is necessary to search for balanced rather than radical visual intensity solutions to counteract the habituation effect without adversely affecting buyers.

The results, based on deep learning solutions used to implement the framework for Performance Evaluation of a Recommending Interface (PERI), showed that the obtained multilayer perceptron has a very good overall prediction accuracy (precision: 80%, recall: 89%) and can be used to assess the performance of different recommending interfaces for users with different characteristics. The prediction accuracy of the adding a product to basket action is a little lower but still high, which is understandable, considering the preliminary character of PERI implementation and the fact that the results were obtained based on a relatively small dataset with a selected number of features. Nevertheless, we showed that the PERI framework can be used to automate an optimal recommending interface adjustment, including adjusting the recommendation position and visual intensity, according to the characteristics of the user. We are planning to perform an extended research with more complex e-commerce stores' websites and subsamples of users of those stores in order to get a wider representation of user characteristics; users will also be given different tasks, from searching to buying, in order to include the goal identification and preference reasoning modules, and further validate the framework. We are also planning to test more types of deep learning networks with more hidden layers and neurons, as well as other machine learning techniques, in order to seek the best-performing architectures for this sophisticated and multidimensional problem.

Author Contributions: Conceptualization, P.S.; methodology, P.S., T.Z.; software, P.S., T.Z.; supervision, P.S.; validation, P.S., T.Z.; formal analysis, P.S., T.Z.; investigation, P.S., T.Z.; resources, P.S.; data curation, P.S., T.Z.; writing—original draft preparation, P.S.; writing—review and editing, P.S., T.Z.; visualization, P.S., T.Z.; project administration, P.S.; funding acquisition, P.S., T.Z. All authors have read and agreed to the published version of the manuscript.

Funding: This research project was supported by the National Science Centre of Poland, Grant No. 2017/27/B/HS4/01216 coordinated by Prof. Jaroslaw Jankowski.

Acknowledgments: We would like to thank Chair of Information Systems Engineering Jaroslaw Jankowski (West Pomeranian University of Technology) for project funding acquisition, administration and co-ordination. We would also like to thank PhD student Kamil Bortko (West Pomeranian University of Technology) for help with experimental lab setup and graduate student Lukasz Dobrowolski (West Pomeranian University of Technology) for support, initial gaze data being presented in his master's thesis with the permission granted by the first author and under his guidance.

Conflicts of Interest: The authors declare no conflict of interest.

References

1. Smith, B.; Linden, G. Two decades of recommender systems at Amazon. com. *IEEE Internet Comput.* **2017**, *21*, 12–18. [CrossRef]
2. Bagher, R.C.; Hassanpour, H.; Mashayekhi, H. User trends modeling for a content-based recommender system. *Expert Syst. Appl.* **2017**, *87*, 209–219. [CrossRef]
3. Jannach, D.; Lerche, L.; Zanker, M. Recommending based on implicit feedback. In *Social Information Access*; Springer: Cham, Switzerland, 2018; pp. 510–569.
4. Sulikowski, P.; Zdziebko, T.; Turzyński, D.; Kańtoch, E. Human-website interaction monitoring in recommender systems. *Procedia Comput. Sci.* **2018**, *126*, 1587–1596. [CrossRef]
5. Sulikowski, P.; Zdziebko, T.; Turzyński, D. Modeling online user product interest for recommender systems and ergonomics studies. *Concurr. Comput. Pract. Exp.* **2019**, *31*, e4301. [CrossRef]
6. Wątróbski, J.; Jankowski, J.; Karczmarczyk, A.; Ziemba, P. Integration of Eye-Tracking Based Studies into e-Commerce Websites Evaluation Process with eQual and TOPSIS Methods. In *Information Systems: Research, Development, Applications, Education, Proceedings of the 10th SIGSAND/PLAIS EuroSymposium 2017, Gdańsk, Poland, 22 September 2017*; Lecture Notes in Business Information Processing; Wrycza, S., Maślankowski, J., Eds.; Springer: Cham, Switzerland, 2017; Volume 300, pp. 56–80.
7. Jankowski, J.; Ziemba, P.; Wątróbski, J.; Kazienko, P. Towards the tradeoff between online marketing resources exploitation and the user experience with the use of eye tracking. In *Intelligent Information and Database Systems, Proceedings of the 8th Asian Conference, ACIIDS 2016, Da Nang, Vietnam, 14–16 March 2016*; Part I. Lecture Notes in Artificial Intelligence; Nguyen, N.T., Trawiński, B., Fujita, H., Hong, T.P., Eds.; Springer: Berlin, Germany, 2016; Volume 9621, pp. 330–343.
8. Melville, P.; Sindhwani, V. Recommender Systems. In *Encyclopedia of Machine Learning and Data Mining*; Springer Publishing Company: New York, NY, USA, 2017; pp. 1056–1066.
9. Yi, B.; Shen, X.; Liu, H.; Zhang, Z.; Zhang, W.; Liu, S.; Xiong, N. Deep matrix factorization with implicit feedback embedding for recommendation system. *IEEE Trans. Ind. Inf.* **2019**, *15*, 4591–4601. [CrossRef]
10. Yang, X.; Guo, Y.; Liu, Y.; Steck, H. A survey of collaborative filtering based social recommender systems. *Comput. Commun.* **2014**, *41*, 1–10. [CrossRef]
11. Zhang, Y.; Chen, X. Explainable recommendation: A survey and new perspectives. *arXiv* **2018**, arXiv:1804.11192.
12. Verbert, K.; Manouselis, N.; Ochoa, X.; Wolpers, M.; Drachsler, H.; Bosnic, I.; Duval, E. Context-aware recommender systems for learning: A survey and future challenges. *IEEE Trans. Learn. Technol.* **2012**, *5*, 318–335. [CrossRef]
13. Lu, J.; Wu, D.; Mao, M.; Wang, W.; Zhang, G. Recommender system application developments: A survey. *Decis. Support Syst.* **2015**, *74*, 12–32. [CrossRef]
14. Seo, Y.D.; Kim, Y.G.; Lee, E.; Baik, D.K. Personalized recommender system based on friendship strength in social network services. *Expert Syst. Appl.* **2017**, *69*, 135–148. [CrossRef]
15. Cremonesi, P.; Elahi, M.; Garzotto, F. User interface patterns in recommendation-empowered content intensive multimedia applications. *Multimed. Tools Appl.* **2017**, *76*, 5275–5309. [CrossRef]
16. Li, Y.; Xu, P.; Lagun, D.; Navalpakkam, V. Towards measuring and inferring user interest from gaze. In Proceedings of the 26th International Conference on World Wide Web Companion, Perth, Australia, 3–7 April 2017; International World Wide Web Conferences Steering Committee: Geneva, Switzerland, 2017; pp. 525–533.
17. Zhao, Q.; Chang, S.; Harper, F.M.; Konstan, J.A. Gaze prediction for recommender systems. In Proceedings of the 10th ACM Conference on Recommender Systems, Boston, MA, USA, 15–19 September 2016; pp. 131–138.
18. Chen, L.; Wang, F. An eye-tracking study: Implication to implicit critiquing feedback elicitation in recommender systems. In Proceedings of the 2016 Conference on User Modeling Adaptation and Personalization, Halifax, NS, Canada, 13–16 July 2016; pp. 163–167.
19. Chen, L.; Wang, F.; Pu, P. Investigating users' eye movement behavior in critiquing-based recommender systems. *AI Commun.* **2017**, *30*, 207–222. [CrossRef]
20. Jin, Y.; Tintarev, N.; Verbert, K. Effects of personal characteristics on music recommender systems with different levels of controllability. In Proceedings of the 12th ACM Conference on Recommender Systems (RecSys'18), Vancouver, BC, Canada, 2–7 October 2018; pp. 13–21.

21. Bortko, K.; Bartków, P.; Jankowski, J.; Kuras, D.; Sulikowski, P. Multi-criteria Evaluation of Recommending Interfaces towards Habituation Reduction and Limited Negative Impact on User Experience. *Procedia Comput. Sci.* **2019**, *159*, 2240–2248. [CrossRef]

22. Hu, R.; Pu, P. Enhancing recommendation diversity with organization interfaces. In Proceedings of the 16th International Conference on Intelligent User Interfaces, Palo Alto, CA, USA, 13–16 February 2011; pp. 347–350.

23. Portnoy, F.; Marchionini, G. Modeling the effect of habituation on banner blindness as a function of repetition and search type: Gap analysis for future work. In Proceedings of the CHI'10 Extended Abstracts on Human Factors in Computing Systems, Atlanta, GA, USA, 10–15 April 2010; pp. 4297–4302.

24. Ha, L. Digital advertising clutter in an age of mobile media. In *Digital Advertising*; Routledge: Abington, UK, 2017; pp. 69–85.

25. Hussein, D.; Han, S.N.; Lee, G.M.; Crespi, N. Social cloud-based cognitive reasoning for task-oriented recommendation in the social internet of things. *IEEE Cloud Comput.* **2015**, *2*, 10–19. [CrossRef]

26. Jankowski, J.; Hamari, J.; Watróbski, J. A gradual approach for maximising user conversion without compromising experience with high visual intensity website elements. *Internet Res.* **2019**, *29*, 194–217. [CrossRef]

27. Resnick, M.; Albert, W. The Impact of Advertising Location and User Task on the Emergence of Banner Ad Blindness: An Eye-Tracking Study. *Int. J. Hum. Comput. Interact.* **2014**, *30*, 206–219. [CrossRef]

28. Jankowski, J. Modeling the Structure of Recommending Interfaces with Adjustable Influence on Users. In *Intelligent Information and Database Systems, Proceedings of the 5th Asian Conference, ACIIDS 2013, Kuala Lumpur, Malaysia, 18–20 March 2013*; Lecture Notes in Computer Science 7803; Springer: Berlin/Heidelberg, Germany, 2013; pp. 429–438.

29. Cheng, S.; Liu, Y. Eye-tracking based adaptive user interface: Implicit human-computer interaction for preference indication. *J. Multimodal User Interfaces* **2012**, *5*, 77–84. [CrossRef]

30. Sulikowski, P. Evaluation of Varying Visual Intensity and Position of a Recommendation in a Recommending Interface Towards Reducing Habituation and Improving Sales. In *Advances in E-Business Engineering for Ubiquitous Computing, Proceedings of the 16th International Conference on E-Business Engineering, ICEBE 2019, Shanghai, China, 11–13 October 2019*; Lecture Notes on Data Engineering and Communications Technologies; Chao, K.M., Jiang, L., Hussain, O., Ma, S.P., Fei, X., Eds.; Springer: Cham, Switzerland, 2020; Volume 41, pp. 208–218.

31. Bigras, É.; Léger, P.M.; Sénécal, S. Recommendation Agent Adoption: How Recommendation Presentation Influences Employees' Perceptions, Behaviors, and Decision Quality. *Appl. Sci.* **2019**, *9*, 4244. [CrossRef]

32. Khusro, S.; Ali, Z.; Ullah, I. Recommender systems: Issues, challenges, and research opportunities. In *Information Science and Applications, Proceedings of the 7th International Conference on Information Science and Applications (ICISA), Ho Chi Minh, Vietnam, 15–18 February 2016*; Springer: Singapore, 2016; pp. 1179–1189.

33. Xu, S.; Jiang, H.; Lau, F. Personalized online document, image and video recommendation via commodity eye-tracking. In Proceedings of the 2008 ACM Conference on Recommender Systems, Lausanne Switzerland, 23–25 October 2008; pp. 83–90.

34. Conati, C.; Carenini, G.; Toker, D.; Lallé, S. Towards user-adaptive information visualization. In Proceedings of the Twenty-Ninth AAAI Conference on Artificial Intelligence, Austin, TX, USA, 25–30 January 2015; pp. 4100–4106.

35. Steichen, B.; Carenini, G.; Conati, C. Adaptive Information Visualization-Predicting user characteristics and task context from eye gaze. In Proceedings of the International Conference on User Modeling, UMAP Workshops, Montreal, QC, Canada, 16–20 July 2012; Volume 872.

36. Lerche, L. Using Implicit Feedback for Recommender Systems: Characteristics, Applications, and Challenges. Ph.D. Thesis, Technische Universität Dortmund, Dortmund, Germany, December 2016.

37. Zhao, Q.; Harper, F.M.; Adomavicius, G.; Konstan, J.A. Explicit or implicit feedback? engagement or satisfaction?: A field experiment on machine-learning-based recommender systems. In Proceedings of the 33rd Annual ACM Symposium on Applied Computing, Pau, France, 9–13 April 2018; ACM: New York, NY, USA, 2018; pp. 1331–1340.

38. Varga, E. Recommender systems. In *Practical Data Science with Python 3*; Apress: Berkeley, CA, USA, 2019; pp. 317–339.

39. Zhou, M.; Ding, Z.; Tang, J.; Yin, D. Micro behaviors: A new perspective in e-commerce recommender systems. In Proceedings of the Eleventh ACM International Conference on Web Search and Data Mining, Los Angeles, CA, USA, 5–9 February 2018; pp. 727–735.

40. Zdziebko, T.; Sulikowski, P. Monitoring Human Website Interactions for Online Stores. In *New Contributions in Information Systems and Technologies*; Advances in Intelligent Systems and Computing; Springer: Cham, Switzerland, 2015; Volume 354, pp. 375–384.

41. Tian, G.; Wang, J.; He, K.; Sun, C.; Tian, Y. Integrating implicit feedbacks for time-aware web service recommendations. *Inf. Syst. Front.* **2017**, *19*, 75–89. [CrossRef]

42. Liversedge, S.P.; Drieghe, D.; Li, X.; Yan, G.; Bai, X.; Hyönä, J. Universality in eye movements and reading: A trilingual investigation. *Cognition* **2016**, *147*, 1–20. [CrossRef]

43. Buscher, G.; Dengel, A.; Biedert, R.; Van Elst, L. Attentive documents: Eye tracking as implicit feedback for information retrieval and beyond. *ACM Trans. Interact. Intell. Syst.* **2012**, *1*, 1–30. [CrossRef]

44. Xu, S.; Jiang, H.; Lau, F.C. User-oriented document summarization through vision-based eye-tracking. In Proceedings of the 13th International Conference on Intelligent User Interfaces (IUI'09), Sanibel Island, FL, USA, 8–11 February 2009; ACM: New York, NY, USA, 2009; pp. 7–16.

45. Loyola, P.; Brunetti, E.; Martinez, G.; Velásquez, J.D.; Maldonado, P. Leveraging Neurodata to Support Web User Behavior Analysis. In *Wisdom Web of Things*; Zhong, N., Ma, J., Liu, J., Huang, R., Tao, X., Eds.; Web Information Systems Engineering and Internet Technologies Book Series; Springer: Cham, Switzerland, 2016; pp. 181–207.

46. Sheng, H.; Lockwood, N.S.; Dahal, S. Eyes Don't Lie: Understanding Users' First Impressions on Websites Using Eye Tracking. In *Human Interface and the Management of Information. Information and Interaction Design, Proceedings of the 15th International Conference, HCI International 2013, Las Vegas, NV, USA, 21–26 July 2013*; Lecture Notes in Computer Science; Yamamoto, S., Ed.; Springer: Berlin/Heidelberg, Germany, 2013; Volume 8016, p. 8016.

47. Sharafi, Z.; Shaffer, T.; Sharif, B.; Guéhéneuc, Y.-G. Eye-Tracking Metrics in Software Engineering. In Proceedings of the 2015 Asia-Pacific Software Engineering Conference, New Delhi, India, 1–4 December 2015; IEEE Press: Piscataway, NJ, USA, 2015; pp. 96–103.

48. Lee, J.; Ahn, J.-H.; Park, B. The effect of repetition in internet banner ads and the moderating role of animation. *Comput. Hum. Behav.* **2015**, *46*, 202–209. [CrossRef]

 *electronics*

Article

Generative Adversarial Network-Based Neural Audio Caption Model for Oral Evaluation

Liu Zhang, Chao Shu, Jin Guo, Hanyi Zhang, Cheng Xie * and Qing Liu

School of Software, Yunnan University, Kunming 650504, China; zhangliu@mail.ynu.edu.cn (L.Z.);
shuchao_shanty@126.com (C.S.); jinguo@mail.ynu.edu.cn (J.G.); miukkazhang@outlook.com (H.Z.);
liuqing@ynu.edu.cn (Q.L.)
* Correspondence: xiecheng@ynu.edu.cn

Received: 1 February 2020; Accepted: 2 March 2020; Published: 3 March 2020

Abstract: Oral evaluation is one of the most critical processes in children's language learning. Traditionally, the Scoring Rubric is widely used in oral evaluation for providing a ranking score by assessing word accuracy, phoneme accuracy, fluency, and accent position of a tester. In recent years, by the emerging demands of the market, oral evaluation requires not only providing a single score from pronunciation but also in-depth, meaning comments based on content, context, logic, and understanding. However, the Scoring Rubric requires massive human work (oral evaluation experts) to provide such deep meaning comments. It is considered uneconomical and inefficient in the current market. Therefore, this paper proposes an automated expert comment generation approach for oral evaluation. The approach first extracts the oral features from the children's audio as well as the text features from the corresponding expert comments. Then, a Gated Recurrent Unit (GRU) is applied to encode the oral features into the model. Afterwards, a Long Short-Term Memory (LSTM) model is applied to train the mappings between oral features and text features and generate expert comments for the new coming oral audio. Finally, a Generative Adversarial Network (GAN) is combined to improve the quality of the generated comments. It generates pseudo-comments to train the discriminator to recognize the human-like comments. The proposed approach is evaluated in a real-world audio dataset (children oral audio) collected by our collaborative company. The proposed approach is also integrated into a commercial application to generate expert comments for children's oral evaluation. The experimental results and the lessons learned from real-world applications show that the proposed approach is effective for providing meaningful comments for oral evaluation.

Keywords: oral evaluation; generative adversarial network; neural audio caption; gated recurrent unit; long short-term memory

---

1. Introduction

Oral evaluation is a language-testing process, which includes pronunciation accuracy, fluency, integrity, logical ability, understanding ability and so on. Among them, the evaluation of logical ability and understanding ability generally requires more personalized expert comments, rather than a single score. Oral evaluation plays an important role in the process of language learning. Now there are some products and standards available in the market for oral evaluation [1–3]. However, most of these approaches are based on the Scoring Rubric [4] that only focuses on the pronunciation characteristics but ignoring the semantic characteristics, such as context, content, logic, or understanding, of the oral expression. It leads to current oral evaluation that can only provide a single ranking score [2]. It cannot provide meaningful comments for the oral evaluation. In order to improve the quality of the oral evaluation, some companies thus hire massive experts to generate comments for the evaluation manually. However, it is expensive and inefficient that not all companies could bear it. Therefore, automated comment generation in oral evaluation becomes the emerging demand that markets are chasing. The emerging demand requires machines to imitate expert to generate expert comments for the oral expressions. It is a comprehensive problem that combines speech recognition [5], natural language generation [6], and deep learning [7]. This new interdisciplinary study is challenging. It not only requires the machine to recognize the audio features from oral speech but also requires the machine to understand the relationships between audio features and corresponding comments. There is still no mature product/method in the market that can automatically generate expert comments for oral evaluation.

With the rapid development of artificial intelligence technology [8–10], a new possible solution for automated oral evaluation emerges gradually. Our previous work had tried to apply the caption generation model to generate expert comment for the oral evaluation [11]. In this work, we optimize the previous model. In detail, a Neural Audio Caption Model (NACM) is proposed to generate expert comments from the oral audio. In NACM, based on Gate Recurrent Unit [12], an elaborate encoder-decoder structure is designed for the mapping learning between audio features and text features. Afterwards, a recurrent structure is designed by combing Generative Adversarial Network (GAN) [13] with NACM. The new model is called Generative Adversarial Network-based Neural Audio Caption Model (GNACM). Compared with the previous work, GNACM can produce more accurate and complete expert comment for the oral evaluation. Figure 1 shows the overall framework of the proposed approach.

As shown in Figure 1, the input of the model is the oral audio to be evaluated. Section 3.1 will introduce the detail of audio feature extraction. The output of the model is the generated comments according to the input oral audio. The mappings between oral audio and comments are trained in NACM. Section 3.2 will provide the structure of NACM. The generated comments are further trained through the Discriminator to improve the quality of the comments. Section 3.3 will explain the detail of GNACM. In summary, the work has the following contributions:

- We propose a model called NACM that can generate expert comment for the oral audio.
- Based on NACM, we propose an improved model called GNACM that can generate more accurate and complete expert comment for the oral audio.
- Beyond the Scoring Rubric approach, the work is the early try to generate expert comments for the oral evaluation.

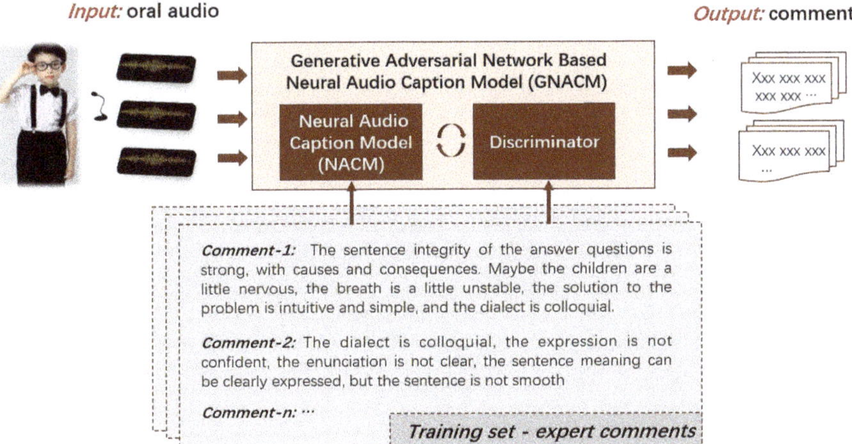

Figure 1. The overview of the proposed approach for the oral evaluation.

2. Related Work

In the related works, we first discuss the typical image caption generation model. The basic idea of image caption generation is also applied in the audio comments generation. Then, related technologies are discussed, including audio feature extraction and text generation approaches.

2.1. Caption Generation Model

Caption generation model is widely used in visual recognition for image description generating [14,15]. In the typical caption generation model, encoder-decoder architecture usually applies to generate captions for each feature vector. In the architecture, the encoder relies on deep neural networks to encode images, audio, or video to generate intermediate vectors. Then, the decoder accepts the intermediate vectors as input, perceives the intermediate vectors in turn, outputs the words one by one, and finally generates captions [16,17]. Vinyals O, et al. [18] proposed an encoder-decoder image caption generation method based on Convolutional Neural Network (CNN) [19] and Long Short-Term Memory (LSTM) [20] network. The approach extracts image features through a convolutional neural network. It then generates the target language through a long short-term memory network, whose objective function is the maximum likelihood estimation of the maximum objective description. Since the classic Neural Image Caption (NIC) model only accepts images as input at the beginning of the LSTM model. As the length of caption grows, LSTM will gradually lose the correspondence between caption and image features. Jia X, et al. [21] uses semantic information to guide the LSTM to generate descriptions at various moments. Here, descriptions indicate image caption. Various moments mean each moment of LSTM for words generation. You Q, et al. proposed a novel semantic attention model [22], which combines the mechanisms of top-down and bottom-up. The model uses responses from intermediate filters of the classification CNN to build a global visual description. In addition, the model runs a set of attribute detectors to obtain a list of visual attributes or concepts that are most likely to appear in the image. The advantage of this semantic attention model is the focus on these aspects and the use of global and local information to generate better caption. The sequence-to-sequence learning model for generating image captions has become popular, but systems for generating audio captions in the speech field are indeed rare. Therefore, this paper study the audio caption model to solve the problem of audio caption generation.

2.2. Audio Feature Extraction Model

The capture of audio feature information is closely related to the generated captions. Therefore, feature extraction is a crucial step in the caption generation task. With the development of deep learning, researchers have proposed a large number of acoustic model (AM) methods based on deep neural networks in speech recognition, which is generally divided into hybrid acoustic models and end-to-end acoustic models. Hinton G, et al. presented a pioneering work that applied deep neural networks in speech recognition tasks, and achieved a significant progress [23]. Alex Graves and Navdeep Jaitly described a system [24]. The system combined a deep bidirectional LSTM network structure and a connectionist temporal classification (CTC) [25] objective function. When the network is used in combination with the baseline system, compared with the training dataset used in the experiment (LDC corpus LDC93S6B and LDC94S13B in the Wall Street Journal corpus), the word error rate is reduced to 6.7%. Yangyang Shi, et al. proposed a method that replaces the traditional projection matrix with a higher-order projection layer [26]. Experimental results show that compared with the traditional LSTM-CTC end-to-end speech model, a higher-order LSTM-CTC model can bring a 3–10% decline in relative word error rate. It can be seen that the field of speech recognition is developing rapidly. Audio caption generation requires speech recognition-related technologies for speech processing, thereby completing the task of audio feature extraction.

2.3. Text Generation Model Based on Deep Learning

Text generation is an essential subtask of natural language processing. According to different inputs, automatic text generation can include text-to-text generation, meaning-to-text generation, data-to-text generation, and image-to-text generation. In the process of image-to-text generation at the generated text, a Recurrent Neural Network (RNN) or a recursive neural network is usually used to model the process of natural language sentence generation [27]. Socher, et al. [28] used recurrent neural networks to model sentences and used syntactic parse trees to highlight the model of actions (verbs). This method jointly optimized the image and text ends to characterize the relationship between objects and actions better. To unify the data of two different modalities under one framework, Chen and Zitnick [29] combined text information and image information into the same recurrent neural network and realized image-to-text and text-to-image bidirectional representation. To improve the quality of the generated text, Fedus W, et al. employed a Generative Adversarial Network (GAN). Compared with the maximum likelihood training model, this method can produce more realistic conditional and unconditional text samples to achieve good results [30]. The audio caption model studied in this paper is an essential branch of natural language text generation. The fidelity of the generated text determines the quality of the model.

3. The Approach

The design details of the method are described in this section. The preprocessing method for audios and comments is designed in Section 3.1. The neural audio caption model (NACM) is described in Section 3.2. A Generative Adversarial Network (GAN) is considered adding to the NACM structure. The GAN-based generative model (GNACM) is designed in Section 3.3.

3.1. Data Preprocessing

Data preprocessing is a critical step in model research. The input of this model is audio data, and the output is natural language. The specific methods for processing audio data and comment data will be described in the section.

3.1.1. Audio Feature Extraction

Audio feature extraction base on Mel Frequency Cepstral Coefficient (MFCC) features [31]. The first step is to divide the original audio into frames in time series, and then extract the MFCC features of each frame. MFCC simulates the processing characteristics of human ear to speech to a certain extent and is designed according to the knowledge of human ear auditory system. It is a general method for feature extraction in speech processing. MFCC feature extraction mainly comprises the following steps: Pre-emphasis, Framing, Windowing, FFT, Mel filter bank, computing DCT. After the audio above processing, each audio can be represented as a two-dimensional matrix ($N_{time} \times N_{mfcc}$), where N_{time} represents the number of frames in each audio, N_{mfcc} represents the feature dimension of MFCC. Next, according to the MFCC features of the first N frames and the next N frames of each frame, we calculate the deltas and delta-deltas for personalized features that preserve dynamic information effectively [32]. The approach through this step increases the connection between the previous and subsequent frames, thereby improving the representation of the feature. The length of different voice files causes the number of frames N_{time} to be different after feature extraction. To be suitable for batch calculation, it needs to uniform the number of frames in each audio. Thus, we set a hyperparameter N_{max} to indicate the maximum number of frames. Therefore, for audio with frames less than N_{max}, we pad zero to its feature matrix until the length reaches N_{max}. Finally, the MFCC feature sequence $\mathbf{A} = (\vec{a}_1, \vec{a}_2, ..., \vec{a}_n)$ is obtained after the source audio processing. Where $\vec{a}_x$ represents the MFCC feature vector of each frame of an audio.

3.1.2. Text Preprocessing

Text preprocessing converts natural language into vector form. The comments in the training set are first segmented in the text processing process, and the segmented result is loaded into the thesaurus. All the words appeared in the training set together to be a thesaurus. In the thesaurus, the first word is coded 0, and the second word is coded 1, and so on. The thesaurus holds a series of correspondences between words and codes. Therefore, the sentence corresponding to each audio can be expressed as $\vec{S} = (S_1, S_2, ..., S_n)$. Where S_x represents the xth word's code in a sentence. When the model performs natural language to vector conversion, the corresponding content can be retrieved by directly accessing the thesaurus.

3.2. Neural Audio Caption Model

The model is based on the neural audio caption method. The basic idea of the model is: input the audio MFCC feature sequence $\mathbf{A} = (\vec{a_1}, \vec{a_2}, ..., \vec{a_n})$ and encode it into a fixed-length $\vec{K}$, then decode the fixed-length vector and output the predicted evaluation $\mathbf{R} = (\vec{R_1}, \vec{R_2}, ..., \vec{R_n})$. The encoder of the model encodes MFCC features into learnable feature vectors. Then, these feature vectors are used to learn the correspondences with the training comments. Afterwards, the decoder of the model decodes the feature vector into comments. The complete structure of the neural audio caption method is shown in Figure 2. The model is divided into an encoder part and a decoder part.

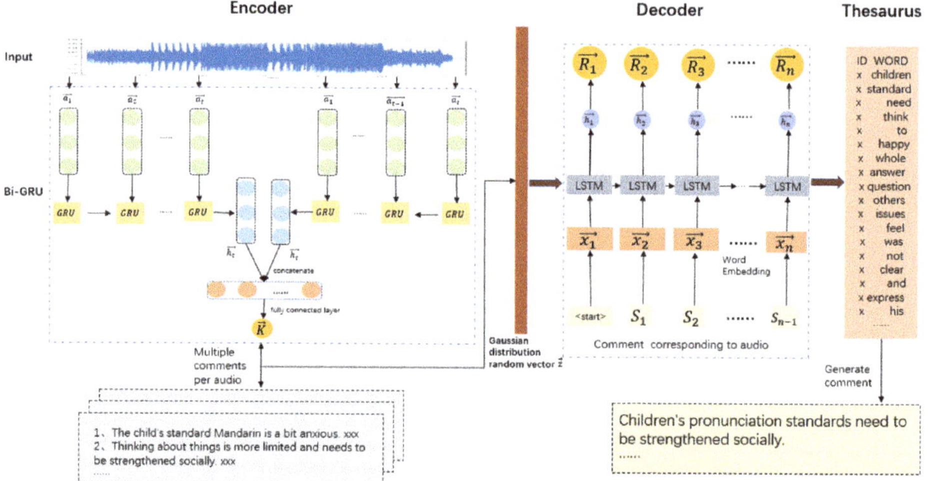

Figure 2. Neural Audio Caption Model. The encoder of the model encodes the MFCC feature sequence into a learning representation. Because one oral audio corresponds to multiple comments, a random vector with a Gaussian distribution is added after audio information encoding. The decoder model gets the input learnable audio features and corresponding comment vectors, and finally outputs comments.

3.2.1. Encoder

The feature encoder part in Figure 2 shows that the information encoding of the NACM model uses our Bi-GRU model. The Bi-GRU model is composed of GRU cells. Given the audio MFCC feature token sequence $\mathbf{A} = (\vec{a_1}, \vec{a_2}, ..., \vec{a_n})$, we use the encoder to encode the sequence information and generate a single representation $\vec{K}$. A GRU unit takes each token as input and outputs a hidden state computed by Equations (1)–(4). Then, a hidden state sequence is generated after the GRU network has computed from left to right along the input sequence. Meanwhile, a counterpart is calculated by another GRU unit computing in the reverse direction. By concatenating the last hidden state vectors and connecting the fully connected layer with activation, we finally get the encoded audio feature $\vec{K}$

with a dimension of 256. In this way, we obtain the holistic information of the sequence forward and backward. This process can be represented by Equation (5).

$$\vec{r}_t = \sigma(\mathbf{W}_r \cdot [\vec{h}_{t-1}, \vec{a}_t]) \tag{1}$$

$$\vec{z}_t = \sigma(\mathbf{W}_z \cdot [\vec{h}_{t-1}, \vec{a}_t]) \tag{2}$$

$$\vec{\tilde{h}}_t = tanh(\mathbf{W}_{\tilde{h}} \cdot [\vec{r}_t \circ \vec{h}_{t-1}, \vec{a}_t]) \tag{3}$$

$$\vec{h}_t = (1 - \vec{z}_t) \circ \vec{h}_{t-1} + \vec{z}_t \circ \vec{\tilde{h}}_t \tag{4}$$

The formula and output of GRU forward propagation are shown in Equations (1)–(4). Where $\vec{a}_t$ represents the current input vector at time t, $\vec{h}_{t-1}$ and $\vec{h}_t$ represent the hidden layer states at time $t-1$ and time t respectively. In Equations (1)–(4), $\mathbf{W}$ terms denote weight matrices and the weight matrices are defined in Equation (13), σ represents the sigmoid function and sigmoid is a non-linear function in neural network, $\vec{z}_t$ and $\vec{r}_t$ represent update gate and reset gate, $\vec{\tilde{h}}_t$ is the candidate hidden state at time t. $\circ$ indicates element-wise multiplication.

$$\vec{K} = tanh(\mathbf{W}_K \cdot \vec{h}_t + \vec{b}_K) \tag{5}$$

where $\mathbf{W}_K$ is weight matrix and $\vec{b}_K$ is bias vector. The bias vector is initialized to be 0. The weight matrices are defined in Equation (13).

3.2.2. Decoder

The decoder part designed in Figure 2 consists of LSTM cells. In practice, the same audio will be evaluated by different experts. In our training set, one audio corresponds to multiple evaluations. Thus, we concatenate a random vector $\vec{z}$ of a Gaussian distribution to the encoded audio feature $\vec{K}$ and use them as the initial hidden state to the LSTM network. Gaussian vector enables Decoder to generate multiple comments with the same input feature vector. During the training, Gaussian vector is initialized by standard normal distribution and we use the embedding of expected output from the training dataset as the input of LSTM network. The procedure can be formulated as Equations (6)–(11). Then we can get a hidden state sequence $\mathbf{h} = (\vec{h}_1, \vec{h}_2, ..., \vec{h}_n)$ after the LSTM unit has finished loop computing. Finally, we use a compositional operation to change the dimension to N_{voc}, and then getting the probability distribution for word in $\vec{R}_n$. The calculation formula is shown in Equation (12).

$$\vec{f}_t = \sigma(\mathbf{W}_f \cdot [\vec{h}_{t-1}, \vec{x}_t] + \vec{b}_f) \tag{6}$$

$$\vec{i}_t = \sigma(\mathbf{W}_i \cdot [\vec{h}_{t-1}, \vec{x}_t]] + \vec{b}_i) \tag{7}$$

$$\vec{\tilde{c}}_t = tanh(\mathbf{W}_c \cdot [\vec{h}_{t-1}, \vec{x}_t]] + \vec{b}_c) \tag{8}$$

$$\vec{c}_t = \vec{f}_t \circ \vec{c}_{t-1} + \vec{i}_t \circ \vec{\tilde{c}}_t \tag{9}$$

$$\vec{o}_t = \sigma(\mathbf{W}_o \cdot [\vec{h}_{t-1}, \vec{x}_t]] + \vec{b}_o) \tag{10}$$

$$\vec{h}_t = \vec{o}_t \circ tanh(\vec{c}_t) \tag{11}$$

$$\vec{R}_n = softmax(\mathbf{W}_R \cdot \vec{h}_t + \vec{b}_R) \tag{12}$$

The calculation formula and output of LSTM forward propagation are shown in Equations (6)–(11). Where σ is the sigmoid function, $\vec{h}_{t-1}$ is the hidden layer output at time $t-1$, $\vec{x}_t$ is the input at time t, $\vec{c}_{t-1}$ is the cell state at time $t-1$, $\vec{\tilde{c}}_t$ is the candidate value of cell state at time t, $\vec{f}_t$, $\vec{i}_t$ and $\vec{o}_t$ are the forget gate, input gate and output gate respectively. In Equation (12), $\mathbf{W}_R$ is weight matrix and $\vec{b}_R$ is bias vector. The bias is initialized to be 0 and the weight is initialized as Equation (13).

$$W \sim U[-\frac{1}{\sqrt{n}}, \frac{1}{\sqrt{n}}] \tag{13}$$

where U denotes the uniform distribution, n is the size of the previous layer.

According to the probability distribution for the output word in $\mathbf{R} = (\vec{R}_1, \vec{R}_2, ..., \vec{R}_n)$ and the coding vector $\vec{S} = (S_1, S_2, ..., S_n)$ of the actual sample, we can calculate the loss via L Equation (14) and update the parameters of this model.

$$L = \frac{1}{n} \sum_{i=1}^{n} -log(R_i[S_i]) \tag{14}$$

where vector $\vec{S} = (S_1, S_2, ..., S_n)$ represents real reviews from experts, each word is numbered. S_i denotes the code of the word whose index is i in the sentence and $R_i[S_i]$ is the probability of the word whose index is S_i in $\vec{R}_i$.

3.3. Generative Adversarial Network-Based Neural Audio Caption Model

The neural audio caption method based on data transformation can generate understandable and appropriate evaluations. To make the generated comment closer to the expert comment, a Generative Adversarial Network (GAN) is combined [33,34].

GAN-Based Neural Audio Caption Model is composed of two neural networks, a generative neural network and a discriminative neural network. It uses the NACM as its generator, which generates comments. Meanwhile, the discriminator evaluates them for authenticity. The goal of the generator network is to generate comments that are as similar to the samples in the training set as possible. The input of the discriminator network is the real sample or the output of the generator network. Its purpose is to distinguish the output of the generator network from training data as much as possible. Thus, GAN builds a sort of feedback loop where the generator is helping to train the discriminator, and the discriminator is helping to train the generator. They both get better together. With this structure, the generated comment is closer to the real evaluation. The structure of the model is shown in Figure 3.

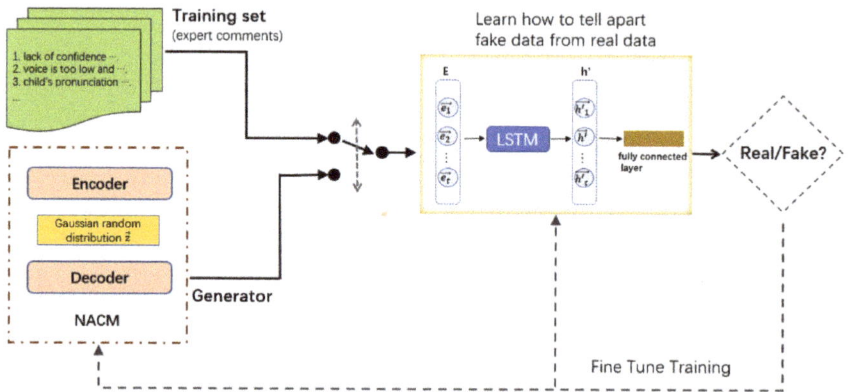

Figure 3. GAN-Based Neural Audio Caption Model. The model is divided into a generator and a discriminator. Among them, the generator follows the Neural Audio Caption Model. The discriminator uses an embedding layer, a deep LSTM layer and a fully connected layer.

3.3.1. Discriminator

The discriminator network of the model is composed of an embedding layer, a LSTM unit and a fully connected layer. The goal of training the discriminator is to maximize the probability of correctly classifying a given input as real or fake. The inputs of the discriminator are generated pseudo-expert comment and real comment. First, the fake/real comment is embed to be a matrix $\mathbf{E} = (\vec{e}_1, \vec{e}_2, .., \vec{e}_n)$ by a embedding layer, then a LSTM unit takes the embedding $\vec{e}_i$ of each word as input and outputs a hidden state $\vec{h}'_t$ computed by Equations (14)–(19). Last, we feed the last hidden state vector into a fully connected layer with sigmoid activation, outputting a scalar probability that the input comment is real.

$$\vec{f}_t = \sigma(\mathbf{W}_f \cdot [\vec{h}_{t-1}, \vec{e}_t] + \vec{b}_f) \tag{15}$$

$$\vec{i}_t = \sigma(\mathbf{W}_i \cdot [\vec{h}_{t-1}, \vec{e}_t] + \vec{b}_i) \tag{16}$$

$$\vec{\tilde{c}}_t = tanh(\mathbf{W}_c \cdot [\vec{h}_{t-1}, \vec{e}_t] + \vec{b}_c) \tag{17}$$

$$\vec{c}_t = \vec{f}_t \circ \vec{c}_{t-1} + \vec{i}_t \circ \vec{\tilde{c}}_t \tag{18}$$

$$\vec{o}_t = \sigma(\mathbf{W}_o \cdot [\vec{h}_{t-1}, \vec{e}_t] + \vec{b}_o) \tag{19}$$

$$\vec{h}'_t = \vec{o}_t \circ tanh(\vec{c}_t) \tag{20}$$

The loss function for the discriminator is used to judge the ability of the discriminator. The loss function is defined as:

$$L^D_{real/fake} = -(E(logD_{real/fake}(\vec{T})) + E(1 - log(D_{real/fake}(G(\mathbf{A}, \vec{z}))))) \tag{21}$$

where $G(\mathbf{A}, \vec{z})$ represents the comments generated by the generator. $D_{real/fake}(x)$ is the discriminator which outputs the scalar probability that x came from the truth sample rather than the generator, $\vec{T}$ represents the real sample.

3.3.2. Generator

The generator of the model inherits the structure from the Neural Audio Caption Model. The generator is divided into an encoder and a decoder. The encoder encodes the MFCC features into learnable feature vectors. Then, a Gaussian-distributed random vector $\vec{z}$ is connected to the encoded features. The decoder takes the encoded features as input and then generates comments.

The similarity between the real samples and the comments generated by the generator is the primary standard for measuring the generator compliance. Therefore, the comments generated by the generator must be similar to the real data to deceive the discriminator. The generator loss function is designed to evaluate this ability of the generator. The goal of training the generator is to maximize $D(G(\mathbf{A}, \vec{z}))$. The loss of the generator is defined as:

$$L^G_{real/fake} = E(-logD_{real/fake}(G(\mathbf{A}, \vec{z}))) \tag{22}$$

where $G(\mathbf{A}, \vec{z})$ represents the comments generated by the generator, and $D_{real/fake}(x)$ represents scalar probability that x came from the truth sample rather than the generator.

The results of supervision using only discriminators are uncertain. Therefore, we turn the learning problem into an optimization problem. Define a loss function to measure the distribution difference between the result and the actual sample to minimize the loss. The loss function is as follows:

$$L^G_{dis} = \propto \times (G(\mathbf{A}, \vec{z}) - \vec{T})^2 \tag{23}$$

where α is the balance factor and $\vec{T}$ represents the real sample.

Finally, we can calculate the loss via Equation (23) and update the parameters of generator network.

$$L^G = L^G_{real/fake} + L^G_{dis} \tag{24}$$

4. Case Study

This section describes the application of the Neural Audio Caption Model (NACM) and the Generative Adversarial Network-Based Neural Audio Caption Model (GNACM) to actual cases in detail. In Section 4.1, we describe the application scenario of the oral evaluation system. We explain the dataset required for model training and performance testing of the system in Sections 4.2 and 4.3. We introduce the application of the system in the entity enterprise, which fully shows the oral evaluation system's practical application and significance value in Section 4.4. The advantages and disadvantages of the model are thoroughly analyzed according to the experimental results in Section 4.5.

4.1. Scenario

At present, the education and training industry has maintained a high growth rate of nearly one trillion Chinese Yuan. According to statistics, the scale of China's training market in 2015 was about 882.1 billion Chinese Yuan. Language training institutions account for a large proportion in the training market, around 17.3%. The size of the education and training market in 2016 has exceeded one trillion Chinese Yuan, maintaining a growth rate of 13.1% per year. After investigation, the types of language training institutions are increasing in the market. The language training institutions mainly includes four categories: comprehensive curriculum education institutions, spoken language institutions, study abroad institutions, and minor language institutions. According to the survey of the language education market, we find that language training institution will be a long-standing industry in the future market.

The development prospects of artificial intelligence language training institutions are excellent [35–37]. Online language education directly hits the pain points of the industry. It solves many problems, such as time-consuming, labor-intensive, expensive, and one-to-one teaching. Although online language training systems overcome the disadvantages of manual education, it still has a lot of obstacles. For example, most current online language training systems can only score speech according to the principle of the Scoring Rubric or give a single evaluation according to the unified Scoring Rubric standards. Figure 4 shows the test process of the traditional language evaluation system.

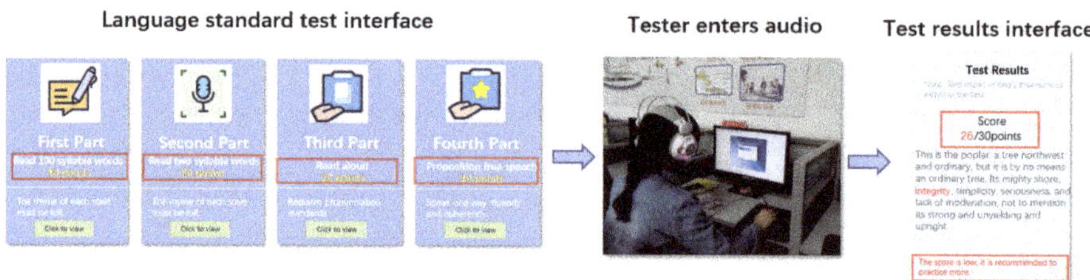

Figure 4. Test process of the traditional language evaluation system. The language standard test interface displays the four parts of the language test. The test contents marked by the red rectangle are made based on the gauge. The Test results interface displays the audio test results of "Read aloud".

With the continuous advancement of science and technology, artificial intelligence products have entered human life [38–40]. To solve the problems of the online education systems, we apply artificial intelligence technology to the online language evaluation systems. The development of an artificial intelligence online language evaluation system will solve the problems in the language education market. The Neural Audio Caption Model (NACM) and Generative Adversarial Network-Based

Neural Audio Caption Model (GNACM) in the paper can generate meaningful expert comments for the oral audio evaluation. The system based on NACM and the system based on GNACM meet the needs of the language training market. NACM and GNACM are applied in the oral evaluation system, which will realize the automation and intelligence of language evaluation.

4.2. Dataset

The first step in the development of the intelligent oral evaluation system is collecting relevant datasets, including audios and expert comments corresponding to audios. The data set affects the accuracy and professionalism of the comments generated by the final model. Therefore, the training set and testing set should include the objective phonemes of different people, different ages, and different environments. The data set was provided by children's language education institutions for this paper and is not publicly available. With the consent of the child's guardian, the oral audio of children 5–6 years old was manually collected. During the collection process, the children watch a cartoon video and say their thoughts. After the audio collection is complete, the relevant children's language experts are invited to comment accordingly. This dataset is called the children oral audio dataset.

4.3. Performance Testing

The NACM and GNACM proposed in this paper will be an essential model of the oral evaluation system. The oral evaluation system is used in enterprise software. Application software has particular evaluation indicators. Therefore, NACM and GNACM are tested using applied evaluation indicators. The evaluation indicators and evaluation results of the model are introduced in detail below in this paper.

4.3.1. Evaluation Metrics

The evaluation metrics of models include three aspects in this paper. The model evaluation metrics are the quality score of model generation comments, the average response time, and the scalability. The quality score of the model-generated reports will be manually reviewed and scored by language education experts. For the performance of the model, we assessment the model from the response time and scalability.

4.3.2. Evaluation Results

The experiments study two models NACM and GNACM. The dataset used is children oral audio dataset. The experimental process is shown in Figure 5. Children 5–6 years old watch cartoon video and answer questions. The child's oral audio is input into two models, and the models automatically generate comments. During the experiment, we assessment the model using the model evaluation metrics proposed in Section 4.3.1.

Relevant language experts evaluate the quality score of the comments generated by the models, as shown in Figure 6. The Score represents the score of the generated comment graded by human experts. The Expert number of the chart represents the expert number. The figure shows the comparison between the quality score of evaluation generated by the baseline model NACM and the quality score of evaluation generated by the GNACM model. The two systems output audio comments by inputting different audios. Multiple experts score the corresponding comments of the output multiple audios. The average score obtained is the system evaluation quality score corresponding to each expert. The process of collecting expert comments is that we first get the test results, then send the audio and generated comments to the anonymous experts, and then the experts send their score results to us, and finally sort out each expert score on the quality of the generated evaluation. Among them, we asked experts to rate the audio and comments given on a 100-point scale. As Figure 6 shown, we can find that the quality score of the comment generated by GNACM is better than NACM.

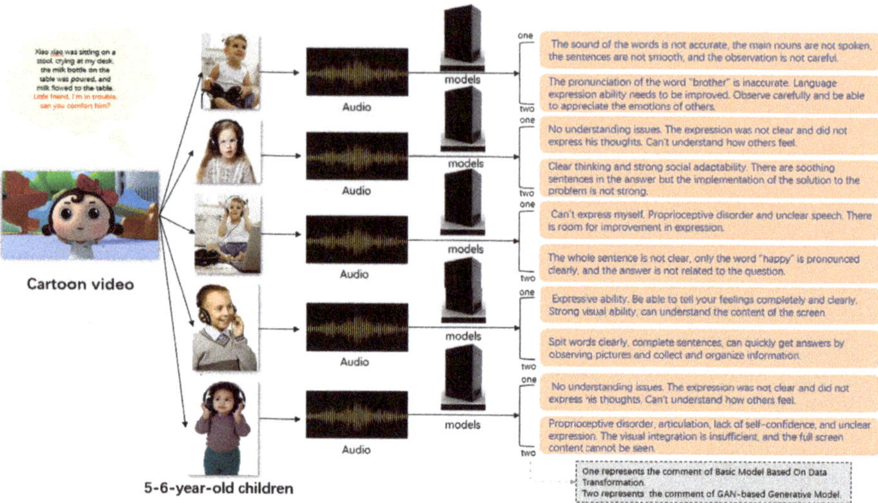

Figure 5. Experimental of an oral evaluation test for children. The participants are asked to watch a cartoon. In the cartoon, there are well designed oral questions. The participants answer the corresponding questions orally. Then, the system will automatically generate the expert comments for the participants.

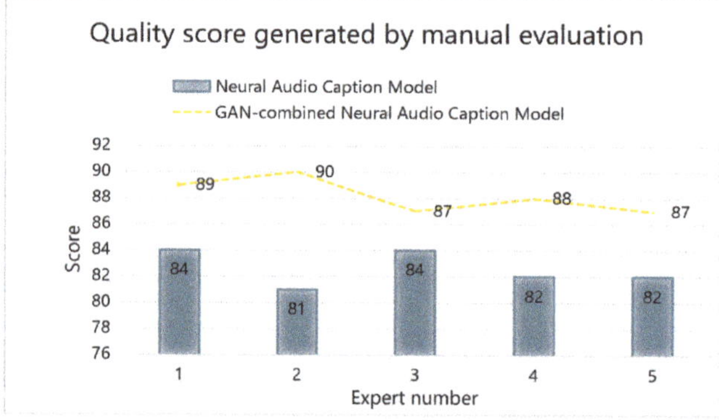

Figure 6. Quality score of evaluations generated by manual assessment models.

The average response time of the model intuitively reflects the performance in the application environment, as shown in Table 1. We tested all audio samples for NACM and GNACM and then calculated the average response time. The NACM model and the GNACM model run and tested on Windows10. The software environment is anaconda3, PyTorch V1.3.1, cuDNN V7.0, and CUDA V10.1. According to the data in Table 1, the model can meet the requirements for practical production.

Table 1. Average response time (ms) of the proposed models.

NACM	GNACM
129.75	125.50

The scalability of the model describes the system's ability to respond to load growth, as shown in Figure 7. The Response time of the chart represents the total time consumption. The Number of samples of the chart represents the number of test samples. The scalability of the model describes the system's ability to respond to load growth, as shown in Figure 7. The Response time of the chart represents the total time consumption. The Number of samples of the chart represents the number of test samples. For scalability evaluation, samples are called increasingly. The evaluation is used to test whether the models suitable for large scale deployment. As the number of test samples increases, the total response time is rising linearly. It means, in the deployment, response time for each audio sample can always be stable by applying acceptable computing resources. The scalability of the model can fully meet the needs of the online language education system.

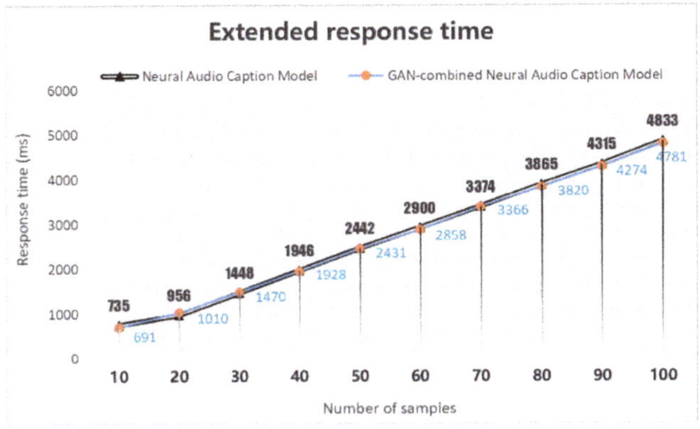

Figure 7. Scalability evaluation for NACM and GNACM.

4.4. Application

The artificial intelligence oral evaluation system is devoted to creating an online testing software that enhances human language proficiency. The software is convenient, fast, comprehensive, and inexpensive. With the development of computer technology, online education has become routine. Therefore, the intelligent oral evaluation system we develop has met the needs of human language education and reduced the cost of human and material resources. In this section, we will apply NACM and GNACM to the oral evaluation system on the market. The NACM is applied to the baseline system of the intelligence oral evaluation for children system. The example of GNACM is compared with NACM in application software.

4.4.1. Baseline System Based on NACM

Language education is an essential part of children's learning and life development. The design of the intelligence oral evaluation for children system will provide each child with a language-centric education platform with fluent language skills. The research and development of this artificial intelligence product is an important step to integrate intelligent technology in the field of children's education. The development of the intelligence oral evaluation system provides a solution to personalized services for children's home education and early childhood school education. Each child can use the software according to their own needs. Under the guidance of parents and teachers, they can test the development level of language skills. The system can also provide suggestions for children's language skills development. At the same time, parents and teachers can monitor the progress of children's language skills in real-time.

This paper proposes an NACM that automatically generates expert comments, which is first applied to the intelligence oral evaluation for children system. We apply the trained NACM to the "Speaking Practice" section of the intelligent oral evaluation for children system. The basic operating environment of the model is as follows. The software environment is anaconda3 (based on Python 3.7), PyTorch V1.3.1, cuDNN V7.0, and CUDA V10.1. The hardware environment is Intel (R) Core (TM) i7-9700 CPU @ 3.00 GHz (8 CPUs) 3.0 GHz, NVIDIA GeForce RTX 2070 SUPER 8 GB, 16 GB RAM. The system imitates a language expert to assess the user language ability and give personalized comments. Through the experimental research, the NACM-based generated comment is incomplete. In the next section, we introduce the use of the GNACM model to solve this problem.

4.4.2. GNACM for Children Oral Evaluation

The intelligent oral evaluation system is currently equipped with GNACM on the market. The server hardware environment is E5-2680v2 CPU @ 20 cores and 40 threads, the main frequency is 2.80 GHz, 128 G memory, four GeForce RTX ^{TM} 2080 Ti GPUs. To simplify the calling method, we deploy a web service on the server. Web service can quickly provide data transmission services to third parties due to the advantages of cross-platform and cross-language. The enterprise establishes a web service client and initiates a connection with our server, which transmits audio information. Then the web service program calls our deep learning module and returns the evaluation result to the client. We provide a web service interface to outside companies, and they can call the corresponding model and apply it to the corresponding function module of the system. Figure 8 shows a language education company that accesses our model by invoking the web APIs of the model. The enterprise uses the API we provided to call the model. The user can use the model with a mobile phone or PC and other devices, input oral audio, and transfer the data to the layers. The server and the model give comments, and the comments data is transmitted layer by layer and fed back to the user. Finally, the display interface displays comments. GNACM gives the audio comment for children oral evaluation is more accurate, similar to expert comment.

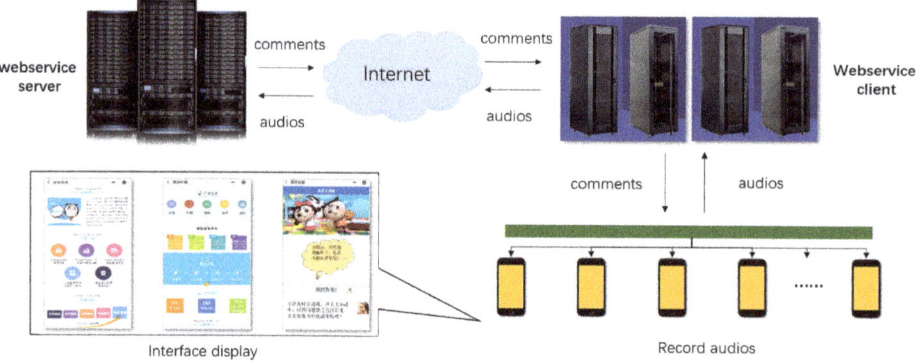

Figure 8. The way the third-party users to access the proposed model for an oral evaluation.

4.5. Lesson Learned

The study finds that people pay attention to the development of language education. There is a great demand for intelligent systems for language evaluation in the market. Therefore, this paper proposes two models of NACM and GNACM and applies them to the intelligent oral evaluation system. Through case analysis, we can find that the intelligent oral evaluation system has many advantages. (1) The development of the intelligent oral evaluation system will solve the problem of time-consuming manual participation of oral evaluation in the market. Compared with traditional online systems, the comments generated by the model are personalized and comprehensive. (2) The extraction and encoder of audio features are difficult. This paper uses the MFCC feature extraction to extract audio features and uses the Bi-GRU model to encoder audio features to solve these two problems. Moreover, in the audio preprocessing process, we use the method of setting the maximum frame and zero-padding to solve the problem of different lengths of audio feature sequences. (3) After NACM is completed, the GAN method is added to the NACM, which significantly improves the accuracy of the evaluation. The intelligent oral evaluation system is more applicable in the business model. (4) The performance test results of the model indicate that the proposed approach can generate meaningful expert comments for the oral audio evaluation. It is suitable for language learning and testing market. By adjusting the parameters and training set, the approach can be also applied in industrial applications. (5) Oral audio evaluation is widely used in various domains, including education, security, finance, and industry. The proposed approach can be further applied to worker status evaluation through analyzing oral questions and answers in many industrial environments.

The development of the intelligent oral evaluation system solves the pain points of intelligent evaluation of language education. However, deficiencies are also found during the research of the model. (1) Audio features are extracted through MFCC feature extraction. In the encoder, the accuracy of the audio feature encoding needs to be further improved. Encoder of the model can still be optimized in the future. At the time of comment generation, as the audio feature layer deepens, the audio features gradually weaken. It may lead the generated comment, to some extent, deviates from the audio. (2) The generated comments have a high similarity with the trained expert comments, resulting in inflexible generated comments. We want the model to generate a more personalized comment for different participants. GAN-based model is hard to generate varieties beyond the training set. Based on the deficiencies of the model, we will continue to study the model to improve the accuracy of the generated evaluation.

5. Conclusions

This paper proposes a generative model for the oral evaluation. Compared with the traditional method, the proposed model is more effective and efficient in oral evaluation. It can generate meaningful comments according to the oral audio without manual works. The proposed approach consists of two parts. The first part is Neural Audio Caption Model (NACM). It applies a Gated Recurrent Unit (GRU) to encode the audio features into the neural network. It also applies a Long Short-Term Memory (LSTM) model to discover the mappings between audio features and text features. The second part of the approach is the Generative Adversarial Network-Based Neural Audio Caption Model (GNACM). It uses the output of NACM as its input to improve the quality of generated comments. The proposed approach is evaluated in a real-world dataset. It also is applied in a commercial application. The evaluation results and the lessons learned from the application show that the proposed approach is effective and efficient in oral evaluation. In the future, we plan to apply the knowledge graph to process the content and context of oral audio. Therefore, the model will consider not only the audio analysis but also the semantic analysis for the oral evaluation.

Author Contributions: Conceptualization, C.X. and L.Z.; methodology, C.S., H.Z.; software, C.S., H.Z.; validation, C.X., J.G., and Q.L.; formal analysis, C.X.; data curation, L.Z.; writing—original draft preparation, L.Z.; writing—review and editing, L.Z. and J.G.; visualization, L.Z.; supervision, C.X. and Q.L.; project administration, C.X. and Q.L.; funding acquisition, C.X. All authors have read and agreed to the published version of the manuscript.

Funding: This word was funded by Scientific Research Fund of Yunnan Provincial Department of Education.

Acknowledgments: The authors would like to thank to all the reviewers who helped us in the review process of our work. Moreover, special thanks to International Conference on e-Business Engineering (ICEBE 2019) recommend the paper.

Conflicts of Interest: The authors declare no conflict of interest.

Abbreviations

The following abbreviations are used in this manuscript:

NACM	Neural Audio Caption Model
GNACM	Generative Adversarial Network-Based Neural Audio Caption Model
MDPI	Multidisciplinary Digital Publishing Institute
GRU	Gated Recurrent Unit
LSTM	Long Short-Term Memory network
NIC	Neural Image Caption
NAC	Neural Audio Caption

CNN	Convolutional Neural Network
AM	Acoustic Model
CTC	Connectionist Temporal Classification
RNN	Recurrent Neural Network
GAN	Generative Adversarial Network
MFCC	Mel Frequency Cepstral Coefficient
BP	Error Back Propagation
APP	Application
AI	Artificial Intelligence

References

1. Voice Evaluation. Available online: http://global.xfyun.cn/products/ise (accessed on 22 September 2019).
2. Smart Oral Evaluation-English. Available online: https://cloud.tencent.com/product/soe-e (accessed on 22 September 2019).
3. Computer Assisted Pronunciation Training. Available online: https://ai.youdao.com/product-assess.s (accessed on 22 September 2019).
4. Moskal, B.M.; Leydens, J.A. Scoring rubric development: Validity and reliability. *Pract. Assess. Res. Eval.* **2000**, *7*, 10.
5. Toshniwal, S.; Sainath, T.N.; Weiss, R.J.; Li, B.; Rao, K. Multilingual Speech Recognition with a Single End-to-End Model. *arXiv* **2018**, arXiv:1711.01694.
6. Gatt, A.; Krahmer, E. Survey of the state of the art in natural language generation: Core tasks, applications and evaluation. *Vestn. Oftalmol.* **2018**, *45*, 75–170. [CrossRef]
7. Lecun, Y.; Bengio, Y.; Hinton, G. Deep learning. *Nature* **2015**, *521*, 436–444. [CrossRef] [PubMed]
8. Kennedy, J.; Séverin, L.; Montassier, C.; Lavalade, P.; Irfan, B.; Papadopoulos, F. Child speech recognition in human-robot interaction: Evaluations and recommendations. In Proceedings of the 2017 ACM/IEEE International Conference on Human-Robot Interaction, Vienna, Austria, 6–9 March 2017; pp. 82–90.
9. Wang, J.; Kothalkar, P.V.; Kim, M.; Bandini, A.; Green, J.R. Automatic prediction of intelligible speaking rate for individuals with als from speech acoustic and articulatory samples. *Int. J. Speech Lang. Pathol.* **2018**, *20*, 669–679. [CrossRef] [PubMed]
10. Ma, Z.; Yu, H.; Chen, W.; Guo, J. Short utterance based speech language identification in intelligent vehicles with time-scale modifications and deep bottleneck features. *IEEE Trans. Veh. Technol.* **2018**, *68*, 121–128. [CrossRef]
11. Liu, Z.; Hanyi, Z.; Jin, G.; Detao, J.; Qing, L.; Cheng, X. Speech Evaluation based on Deep Learning Audio Caption. In Proceedings of the International Conference on e-Business Engineering, Bali, Indonesia, 21–23 December 2019; pp. 51–66.
12. Cho, K.; van Merriënboer, B.; Bahdanau, D.; Bengio, Y. *On the Properties of Neural Machine Translation: Encoder–Decoder Approaches*; Association for Computational Linguistics: Doha, Qatar, 2014; pp. 103–111.
13. Goodfellow, I.; Pouget-Abadie, J.; Mirza, M.; Bing, X.; Warde-Farley, D.; Sherjil, O.; Courville, A.; Bengio, Y. Generative adversarial nets. In Proceedings of the Advances in Neural Information Processing Systems, Montreal, QC, Canada, 8–13 December 2014; pp. 2672–2680.
14. Deshpande, A.; Aneja, J.; Wang, L.; Schwing, A.G.; Forsyth, D. Fast, diverse and accurate image captioning guided by part-of-speech. In Proceedings of the IEEE Conference on Computer Vision and Pattern Recognition, Long Beach, CA, USA, 16–20 June 2019.
15. Yang, X.; Tang, K.; Zhang, H.; Cai, J. Auto-encoding scene graphs for image captioning. In Proceedings of the IEEE Conference on Computer Vision and Pattern Recognition, Long Beach, CA, USA, 16–20 June 2019.
16. Sutskever, I.; Vinyals, O.; Le, Q.V. Sequence to sequence learning with neural networks. *Adv. Neural Inf. Process. Syst.* **2008**, *2*, 3104–3112.
17. Cho, K.; Van Merriënboer, B.; Gulcehre, C.; Bahdanau, D.; Bougares, F.; Schwenk, H.; Bengio, Y. Learning phrase representations using rnn encoder-decoder for statistical machine translation. *arXiv* **2014**, arXiv:1406.1078.
18. Vinyals, O.; Toshev, A.; Bengio, S.; Erhan, D. Show and tell: A neural image caption generator. In Proceedings of the IEEE Conference on Computer Vision and Pattern Recognition, Boston, MA, USA, 7–12 June 2015.

19. Phil, K. Convolutional Neural Network. In *MATLAB Deep Learning*; Apress: Berkeley, CA, USA, 2017.

20. Hochreiter, S.; Schmidhuber, J. Long short-term memory. *Neural Comput.* **1997**, *9*, 1735–1780. [CrossRef]

21. Jia, X.; Gavves, E.; Fernando, B.; Tuytelaars, T. Guiding the Long-Short Term Memory Model for Image Caption Generation. In Proceedings of the IEEE International Conference on Computer Vision, Santiago, Chile, 7–13 December 2015.

22. You, Q.; Jin, H.; Wang, Z.; Fang, C.; Luo, J. Image captioning with semantic attention. In Proceedings of the IEEE Conference on Computer Vision and Pattern Recognition, Las Vegas, NV, USA, 27–30 June 2016.

23. Hinton, G.; Deng, L.; Yu, D.; Dahl, G.E.; Mohamed, A.R.; Jaitly, N. Deep neural networks for acoustic modeling in speech recognition: The shared views of four research groups. *IEEE Signal Proc. Mag.* **2012**, *29*, 82–97. [CrossRef]

24. Graves, A.; Navdeep, J. Towards end-to-end speech recognition with recurrent neural networks. In Proceedings of the International Conference on Machine Learning, Beijing, China, 21–26 June 2014; pp. 1764–1772.

25. Graves, A.; Santiago, F.; Gomez, F. Connectionist temporal classification: Labelling unsegmented sequence data with recurrent neural networks. In Proceedings of the 23rd International Conference on Machine Learning, Pittsburgh, PA, USA, 25–29 June 2006.

26. Shi, Y.; Hwang, M.Y.; Lei, X. End-to-end speech recognition using a high rank lstm-ctc based model. In Proceedings of the IEEE International Conference on Acoustics, Speech and Signal Processing, Brighton, UK, 12–17 May 2019; pp. 7080–7084.

27. Karpathy, A.; Fei-Fei, L. Deep visual-semantic alignments for generating image descriptions. *IEEE Trans. Pattern Anal. Mach. Intell.* **2014**, *39*, 664–676. [CrossRef] [PubMed]

28. Socher, R.; Karpathy, A.; Le, Q.V.; Manning, C.D.; Ng, A.Y. Grounded compositional semantics for finding and describing images with sentences. *Trans. Assoc. Comput. Linguist.* **2014**, *2*, 207–218. [CrossRef]

29. Chen, X.; Zitnick, C.L. Learning a recurrent visual representation for image caption generation. *arXiv* **2014**, arXiv:1411.5654.

30. Fedus, W.; Goodfellow, I.; Dai, A.M. Maskgan: Better text generation via filling in the_. *arXiv* **2018**, arXiv:1801.07736.

31. Upadhya, S.; Cheeran, A.N.; Nirmal, J.H. Discriminating Parkinson diseased and healthy people using modified MFCC filter bank approach. *Int. J. Speech Technol.* **2019**, *224*, 1021–1029. [CrossRef]

32. Mingyi, C.; Xuanji, H.; Jing, Y.; Han, Z. 3-d convolutional recurrent neural networks with attention model for speech emotion recognition. *IEEE Signal Process. Lett.* **2018**, *25*, 1440–1444.

33. Liu, F.; Zheng, J.; Zheng, L.; Chen, C. Combining attention-based bidirectional gated recurrent neural network and two-dimensional convolutional neural network for document-level sentiment classification. *Neurocomputiong* **2019**, *371*, 39–54. [CrossRef]

34. Yan, S.; Xie, Y.; Wu, F.; Smith, J.S.; Lu, W.; Zhang, B. Image captioning based on a hierarchical attention mechanism and policy gradient optimization. *arXiv* **2018**, arXiv:1811.05253.

35. Dalim, C.; Samihah, C.; Sunar, M.S.; Dey, A.; Billinghurst, M. Using augmented reality with speech input for non-native children's language learning. *Int. J. Hum. Comput. Stud.* **2019**, *134*, 44–64. [CrossRef]

36. Schepens, J.; van Hout, R.; Jaeger, T. Florian. Big data suggest strong constraints of linguistic similarity on adult language learning. *Cognition* **2019**, *194*, 104056. [CrossRef]

37. Cho, K.; van Merrienboer, B.; Bahadanau, D.; Bengio, Y. On the properties of neural machine translation: Encoder-decoder aaaroaches. *arXiv* **2014**, arXiv:1409.1259.

38. Chen, C.; Gao, L.; Xie, X.W.; Wang, Z. Enjoy the most beautiful scene now: A memetic algorithm to solve two-fold time-dependent arc orienteering problem. *Front. Comput. Sci.* **2020**, *14*, 364–377. [CrossRef]

39. Manikandan, R.; Patan, R.; Gandomi, A.H.; Sivanesan, P.; Kalyanaraman, H. Hash polynomial two factor decision tree using IoT for smart health care scheduling. *Expert Syst. Appl.* **2019**, *141*, 112924. [CrossRef]

40. Pan, J.S.; Xi, T.; Jiang, R. Emotional Effects of Smart Aromatherapeutic Home Devices. In Proceedings of the International Conference on Applied Human Factors and Ergonomics, Washington, DC, USA, 24–28 July 2019; pp. 498–503.

MDPI
St. Alban-Anlage 66
4052 Basel
Switzerland
Tel. +41 61 683 77 34
Fax +41 61 302 89 18
www.mdpi.com

Electronics Editorial Office
E-mail: electronics@mdpi.com
www.mdpi.com/journal/electronics